AF322876

DEVELOPMENTAL PSYCHOBIOLOGY

DEVELOPMENTAL PSYCHOBIOLOGY

New Methods and Changing Concepts

Edited by

Harry N. Shair

Gordon A. Barr

Myron A. Hofer

New York   Oxford
OXFORD UNIVERSITY PRESS
1991

Oxford University Press

Oxford   New York   Toronto
Delhi   Bombay   Calcutta   Madras   Karachi
Petaling Jaya   Singapore   Hong Kong   Tokyo
Nairobi   Dar es Salaam   Cape Town
Melbourne   Auckland

and associated companies in
Berlin   Ibadan

Published by Oxford University Press, Inc.,
200 Madison Avenue, New York, New York 10016

Oxford is a registered trademark of Oxford University Press.

Library of Congress Cataloging-in-Publication Data
Developmental psychobiology : new methods and changing concepts /
[edited by] Harry N. Shair, Gordon A. Barr, Myron A. Hofer.
p.   cm.   Includes bibliographical references and index.
ISBN 0-19-505649-3
1. Developmental psychobiology—Methodology.   I. Shair, Harry N.
II. Barr, Gordon A.   III. Hofer, Myron A., 1931–
[DNLM: 1. Animals, Newborn—physiology.   2. Animals, Newborn—
psychology.   3. Behavior, Animal.   4. Psychology, Experimental—
methods. QL 763 D4894]
QP360.D493   1991   90-7956
152—dc20
DNLM/DLC

9   8   7   6   5   4   3   2   1
Printed in the United States of America
on acid-free paper

PREFACE

A number of years ago, one of the editors (M.A.H.) was attending an international psychiatric conference at which three Nobel laureates had been invited to give their personal views on the essence of creativity in science. The first speaker, Linus Pauling, recounted how, a few years previously, he had given up working on a particularly puzzling problem in organic chemistry and immersed himself in a different question. Six months later, when he was idly waiting in a restaurant for a friend, the solution to the forgotten problem began to occur to him. In a few minutes he was able to write on the back of the menu all the essential points for his theory of the covalent bond. The second speaker, Albert Szent-Gyorgy, arose, took off his coat (because it had become quite warm in the auditorium), rolled up his sleeves, and launched into an impassioned exposition of how a scientist must fall in love with his work. He described how he had abandoned himself to the research that led him to discover the structural features of muscle proteins which enable muscles to contract rapidly. Finally, it was Edgar Adrian's turn. He remarked that it was all very well putting thoughts out of your mind, or becoming enraptured with them, but so far as he was concerned it was the advent of new methods that had led to scientific creativity. Without the means of detecting extremely low levels of electrical current flow in nerve tissue, he would never had made any of his discoveries. To a young scientist, passion for one's work and the unconscious mind may seem more appealing sources of creativity, but Adrian's perspective may be the wisest of the three, and certainly the route with the most universal applicability.

This book proceeds along these lines of thought to describe a number of the new methods that have emerged in the study of developmental processes over the past 10 to 15 years and to relate how the application of these methods has begun to lead to new concepts and ways of thinking about a number of issues in the field of developmental psychobiology. We are still far from a comprehensive theory of development, but the methods described in this book have already changed our concepts of the nature of development and promise much more.

We hope this book will not be like other edited volumes that document recent progress in a field. Instead, we hope these chapters will *lead* to progress, for the methods described are all relatively new and have thus far had only limited application. It was most difficult for the authors to describe how these methods have altered their concepts of development. In part this was because once one has seen things in a new way it is not easy or rewarding to recapture former, less enlightened modes of thinking. But it seemed useful to us to illustrate how new methods have resulted in new ideas in our field, and we thought if each author described how illuminating the experience had been, it would encourage readers to find their own ways to employ these methods.

Thus, our hope is that readers will quickly see ways in which these methods can be employed to approach the numerous intriguing questions that abound in the study of developing organisms. For, ultimately, it is the *questions* that are important. Vincent Dethier (1962), in his charming book *To Know a Fly*, observed: "Too many research workers have no questions at all to ask, but this does not deter them from doing experiments. They become enamored of a new instrument, acquire it, then ask only 'What can I do with this beauty.' " The contributing authors to this volume were selected not only for the new methods they developed but also for how they applied these methods to important questions. Others have made equally important contributions, and any single book can give only a representative sampling. We hope that the readers will profit from these examples and that they will be encouraged to modify these methods and to invent new ones themselves. One of the greatest joys of using new methods is the unexpected, unimaginable results that can occur; this is what attracts many of us into science. However, such information is not always easy to integrate into our present understanding, and the challenge of devising and using new methods is to use the results to change our ways of thinking and to construct new ideas and concepts. This process is the central theme of the chapters of this book, and it is a creative process in which we all can participate.

The research methods described here use live infant animals of several species. We do so in the belief that the resulting information provided by these methods will provide important information and long-term benefits for the medical and psychological health of both human and animal species. The importance of animal studies for the progress of medicine in general and psychology in particular have been well documented (Leader & Stark, 1987; Miller, 1985), and although developmental psychobiology is a younger science, health benefits from research in this field can be documented. The authors of the chapters in this book, because they are leaders in our field, have contributed in large measure to the progress of the science of developmental psychobiology.

For both humane and scientific reasons, researchers in all disciplines have the responsibility to provide optimal conditions for their experimental subjects. Thus, each scientist adheres to guidelines provided by the relevant scientific organizations that govern their work and by regulations of funding agencies and federal law. These include the guidelines provided by the International Society for Developmental Psychobiology, which are reproduced here:

1. Animals are used to enable researchers to answer questions and solve problems that cannot realistically be addressed without their use.
2. Animals are treated humanely during all phases of research. When an experimental procedure requires stressful stimulation, it will be applied judiciously to meet the objective of the experiment while minimizing distress.
3. Healthy animals are the best subjects for studies on live animals; they give the most reliable data, thereby reducing the overall number of animals needed to provide meaningful and reliable experimental findings.
4. The primary investigator responsible for day-to-day conduct of the research is responsible for appropriate care and use of each animal. Personnel working with animals should be supervised by well-trained and experienced scientists who are familiar with government regulations and aided by a veterinary staff, trained animal-care technicians, and appropriate institutional facilities.

It is the responsibility of each investigator to ensure the use of the appropriate numbers of subjects, with the appropriate care and maintenance of live animals. This is most important for individuals beginning their scientific careers. In particular, it is the responsibility of each investigator planning to implement the methods described in this book to obtain appropriate training prior to using any of these techniques, both to ensure good science and humane animal care. We hope and trust that each reader of this volume will take these principles to heart.

ACKNOWLEDGMENTS

We wish to acknowledge the fact that the field of developmental psychobiology is a shared enterprise. Progress comes from the efforts of many scientists, each working singly or in small groups, but in a cooperative spirit to advance our understanding of development. To all scientists in the field of developmental psychobiology, each of whom has contributed to the progress described in this volume, we are grateful.

In addition to our institutions and grants listed elsewhere, we would like to thank Phyllis Blackman for her administrative and secretarial assistance.

REFERENCES

Dethier, V. (1962). *To Know a Fly* (p. 21). San Francisco: Holden-Day.
Leader, R. W., & Stark, D. (1987). The importance of animals in biomedical research. *Perspectives in Biology and Medicine, 30,* 470–485.
Miller, N. E. (1985). The value of behavioral research on animals. *American Psychologist, 40,* 423–440.

CONTRIBUTORS

Gordon A. Barr
Department of Psychology
Biopsychology Doctoral Program
Hunter College, City University of
    New York
Department of Developmental
    Psychobiology
New York State Psychiatric Institute
    & Columbia College of
    Physicians and Surgeons
Columbia University
New York, NY 10032

Stephen C. Brake
Denver Therapeutic Institute
671 Grant
Denver, CO 80203

Thomas J. Carew
Departments of Psychology and
    Biology
Yale University
New Haven, CT 06520

Joseph T. Coyle
Departments of Psychiatry and
    Behavioral Sciences,
    Neurosciences, & Pediatrics
The Johns Hopkins University
    School of Medicine
Baltimore, MD 21205

Jaime Diaz
Department of Psychology
University of Washington
Seattle, WA 98195

Bennett G. Galef, Jr.
Department of Psychology
McMaster University
Hamilton, Ontario L8S 4K1
Canada

William T. Greenough
Departments of Psychology and Cell
    and Structural Biology
Beckman Institute
University of Illinois at Urbana-
    Champaign
Urbana, IL 61801

Nicholas R. S. Hall
Department of Medical
    Microbiology and Immunology
University of South Florida
Tampa, FL 33613

Myron A. Hofer
Department of Developmental
    Psychobiology
New York State Psychiatric Institute
    & Columbia College of
    Physicians and Surgeons
Columbia University
New York, NY 10032

Richard L. Hyson
Department of Otolaryngology
University of Washington
Seattle, WA 98195

Michael A. Leon
Department of Psychobiology
University of California at Irvine
Irvine, CA 92717

Emilie A. Marcus
Departments of Psychology and
   Biology
Yale University
New Haven, CT 06520

Timothy H. Moran
Department of Psychiatry and
   Behavioral Sciences
The Johns Hopkins University
   School of Medicine
Baltimore, MD 21205

Michael M. Myers
Department of Developmental
   Psychobiology
New York State Psychiatric Institute
   & Columbia College of
   Physicians and Surgeons
Columbia University
New York, NY 10032

Thomas G. Nolen
Department of Biology
University of Miami
Coral Gables, FL 33114

Maureen P. O'Grady
Department of Psychiatry and
   Behavioral Medicine
University of South Florida
Tampa, FL 33613

C. B. Phifer
Louisiana Scholar's College
Northwestern State University
Natchitoches, LA 71457

Catherine H. Rankin
Department of Psychology
University of British Columbia
Vancouver
British Columbia V6T 1Y7
Canada

Scott R. Robinson
Center for Developmental
   Psychobiology
Department of Psychology
SUNY Binghamton
Binghamton, NY 13901

Edwin W. Rubel
Department of Otolaryngology
University of Washington
Seattle, WA 98195

Jerry W. Rudy
Department of Psychology
University of Colorado
Boulder, CO 80309

Evelyn Satinoff
Departments of Psychology,
   Physiology and Biophysics
Program in Neural and Behavioral
   Biology
University of Illinois
Champaign, IL 61820

Harry N. Shair
Department of Developmental
   Psychobiology
New York State Psychiatric Institute
   & Columbia College of
   Physicians and Surgeons
Columbia University
New York, NY 10032

Anita M. Sirevaag
Departments of Psychology and Cell
   and Structural Biology
Beckman Institute
University of Illinois at Urbana-
   Champaign
Urbana, IL 61801

William P. Smotherman
Center for Developmental
    Psychobiology
Department of Psychology
SUNY Binghamton
Binghamton, NY 13901

Norman E. Spear
Center for Developmental
    Psychobiology
Department of Psychology
SUNY Binghamton
Binghamton, NY 13901

Donald J. Stehouwer
Departments of Psychology and
    Neuroscience
Center of Neurobiological Sciences
University of Florida
Gainesville, FL 32611

Mark A. Stopfer
Departments of Psychology &
    Biology
Yale University
New Haven, CT 06520

Regina M. Sullivan
Developmental Psychobiology
    Laboratory
Department of Psychology
University of Oklahoma
Norman, OK 73019

Christina L. Williams
Department of Psychology
The Johns Hopkins University
Baltimore, MD 21218

Donald A. Wilson
Developmental Psychobiology
    Laboratory
Department of Psychology
University of Oklahoma
Norman, OK 73019

Contents

II. PHYSIOLOGICAL SYSTEMS

I

BEHAVIORAL SYSTEMS

A.

REGULATORY PROCESSES

The four chapters in this section describe different methods that have revealed unexpected behavioral processes hidden within the observable interactions of the mother-infant relationship in the laboratory rat. As the authors make clear, the methods were simply devised in order to get a closer look at one or another behavior, but their application led them to new ways of thinking about development.

The most broadly applicable method, described by Myers in Chapter 1, sets out a strategy for identifying and analyzing developmental processes by which early experiences affect long-term outcome. He shows how the tight genetic control provided by homozygous strains can be used to identify likely environmental effects, and how systematic behavioral observations early in development can be used to shed new light on biological differences in adulthood. These insights are embodied in a novel strategy that combines features of standard genetic, behavioral, and physiological methods in a developmental framework focused on the early mother-infant interaction.

Hofer in Chapter 2, provides an introduction to microsurgery that has obvious applicability to physiological studies, but goes on to show how microsurgery can be used to explore, at a new level of analysis, how infants organize their behavior in relation to their mother. Selective lesioning of sensory systems revealed unexpected stages in information processing by two different systems, each of which controlled portions of a goal-directed behavioral sequence—nipple attachment and nursing.

In Chapter 3, Shair describes methods for applying the naturalistic approach, often used in the study of behavior, to the study of physiological systems in infant rats freely interacting with their dams. The application of these methods led to the discovery of unexpected relationships between sleep/wake state regulation and nursing behavior and between behavior and blood pressure regulation.

Finally, Brake, in Chapter 4, describes methods for recording and measuring a behavior that is ordinarily invisible, although crucial to the infant's survival—sucking. His methods led to changes in our concept of sucking, which had been thought to be a simple reflex or fixed-action pattern, and led to changes in our concept of milk transfer, which had been thought to be controlled primarily by the dam. Other methods for the study of ingestive behavior are described by Phifer in Chapter 11.

In applying these methods, novel processes were discovered, which surprised the authors, and, in some cases, were counterintuitive. These processes were found to guide or control the behavioral interactions and the biological systems of dam and pup within certain limits. In this sense they may be referred to as regulators of the interaction and of the underlying neural systems of both participants. Since qualitative and quantitative differences in the mother-infant interaction are known to have long-range effects on development, these processes in the long run function as regulators of development. They constitute a level of influence on development that is probably of equal importance to learning, but much less extensively studied.

1

Identifying Relationships Between Early Life Experiences and Adult Traits

MICHAEL M. MYERS

The goal of this chapter is to describe research methods that should be of particular interest to investigators studying developmental processes. The emphasis is on research designs rather than special measurement techniques. These designs constitute a strategy for identifying specific elements of normal early environmental experiences that are likely to be involved in modifying characteristics of physiological and behavioral systems in adult animals.

A rich variety of animal studies conducted by investigators in the field of developmental psychobiology have demonstrated that manipulations of the early environment create long-lasting effects, a phenomenon that every parent "knows" to be true. However, at the end of experiments in which manipulations have produced significant effects, these researchers all too often sound like parents when they are forced to conclude, "It must have been something we did." The important inference drawn from these studies is that early life events *can* change adult characteristics. Yet there are few examples in which variability within the normal range of early experiences has actually been shown to be linked to individual differences in specific adult traits. The methodologies I describe are intended to address this issue and spawn a more comprehensive understanding of how naturally occurring events of early life mediate long-lasting effects.

Although individual elements of these strategies are not new, using them in the proposed sequence provides a systematic way to begin investigations pertinent to a broad range of questions. The approach involves, first, identifying adult characteristics of research animals that are likely to have been influenced by early life experiences. If we limit what constitutes early life experiences to the preweaning environment, a powerful approach for determining which adult traits have been shaped by early experiences is to study litter effects in homozygous strains of rats or mice. If littermates are more similar to each other than they are to nonlittermates, I argue that this can be taken as evidence for an early environmental effect, at least with regard to experiences confined to the context of the mother-litter environment. Next, salient features and events in the early environment must be characterized. Because the focus of my approach is effects of early life experiences that are a normal part of the preweaning environment, this means recording behavioral activities in the maternal nest. The strategy, however, could be applied to other types of questions. For example, What are the long-term effects of differences in early postweaning social interactions? After quantifying the routine events that comprise the early environment, the next step is to look for statistical relationships between the frequencies of occurrence of early naturally occurring events and quantitative variations in the adult trait of interest. A variety of multivariate techniques such as multiple linear regression and factor analysis are useful at this stage. If interesting relationships emerge, further studies can be designed to discover the succession of events that will ultimately form an explanation for the coupling of early life events with the adult trait of interest.

I have used many of these proposed designs for studying early environmental contributions to adult blood pressure and will draw from this work to illustrate how these strategies can be implemented. However, because I wanted to determine if the methods I am proposing would, in fact, work for variables other than blood pressure, I have reexamined data from my experiments to determine if variations in specific kinds of naturally occurring early experiences might account for individual difference among animals in adult weight. My examples draw on data from studies using spontaneously hypertensive (SHR) and Wistar Kyoto (WKY) rats and various crosses of these inbred strains, but this fact is only incidental to the main points of my discussion.

ADULT TRAITS AND EARLY EXPERIENCES

How can you tell if an adult trait may have been affected by early experiences? It should be obvious that if one is to study how naturally occurring experiences affect expression of adult traits, the first item on the agenda is to identify traits that are likely altered by things happening early in life. What kinds of traits would these be? Certain behavioral characteristics would undoubtedly be interesting—for example, indexes of personality, fearfulness, aggressiveness, or cognitive abilities. Or perhaps we should look in the physiologic domain for individual differences that might be related to early experiences—for example, measures of stress responsiveness, disease susceptibility, or cardiovascular reactivity. Any of these would be possible to study, and the choice would depend only upon your area of interest.

Is there a way to determine whether adult traits are likely to have been influenced by experiences of early life that is not dependent upon which specific traits are being studied? Though no single strategy will work in all cases, there is one method that will often work. We can ask whether members of families are statistically more similar to each other than they are to members of other families. This approach is even more powerful if we ask this question using homozygous strains of research animals, most commonly rats or mice. In these multiple offspring species, the question of familial similarity translates into a statistical issue: Are there significant "litter effects" for the trait of interest? The fundamental assumption of these analyses is that if all animals in the study are genetically identical, then similarities among littermates exist because littermates have shared similar experiences; on the other hand, differences among animals from different litters must arise from differences in environmental experiences. With the proper experimental designs, we can delineate the important environmental experiences to the context of the naturally occurring activities of mothers and their litters.

It is interesting to note that there is no exact equivalent of this approach using human subjects. Studies of monozygotic twins reared either together or in foster families are in some ways similar but less powerful because, unlike homozygous rats from different litters, human twins are genetically identical only to each other—they are not the same as twins from different families.

The fact that significant litter effects are common has been viewed more as a problem or potential confound than as an important clue (Chapman & Stern, 1979). If members of a given litter are different from members of another litter, even though each litter was given the same treatment, how can you be sure that a significant treatment effect is not just an effect of "random" differences between litters? The answer, of course, is to account for this source of variance in the statistical model used to evaluate treatment effects. Unfortunately, this means that, in the face of significant litter effects, large numbers of litters will be needed to have sufficient statistical power to detect treatment effects. This is because the numbers of litters rather than individuals will determine the error degrees of freedom. In other words, significant litter effects increase the amount of work we must do to study effects of treatments. For the purposes of studying links between early experiences and development, however, differences among litters with regard to variables of interest are critically important. By studying these variables, we can be more certain that early experiences have a measurable effect and that the path between these experiences and the adult trait is one that can be traced. If there is no significant litter effect with regard to some particular adult measurement, this does not mean that the environment has, or could have, no effect on this variable. Yet it does mean we are likely to have a much more difficult time establishing the nature of coupling between the adult trait and early life causes for individual difference in this trait.

Measurements of adult body weight provide a good illustration of what a significant litter effect looks like. Figure 1.1 shows adult body weights, measured at 20 weeks of age, of 50 genetically identical (SHR) male rats, 5 animals from each of 10 litters. Examination of this figure reveals that the 5 rats from some of these litters display quite a broad range of weights. This is particularly true for litters 3, 6, and 8. However, for most of the litters the range of weights for animals within the litter is narrow, and littermates are more similar to each other than they are to animals from other litters. For example, compare litters 1 and 10. Analysis of variance of these data, testing for an effect of litter, indicates that 43% of the total variation in adult body weight is accounted for by the presence of litters within the data and that it is unlikely that this effect is due to random sampling error ($p = .004$).

The reason this clustering of litter body weights is so important to our goal becomes apparent when we consider why it is that littermates are similar to each other and—even more to the point—why litters are different. Three hypotheses come to mind. First, perhaps littermates have more body-weight-influencing genes in common with each other than they do with animals from other litters. For example, if the mother and father of litter number 1 had genes that contributed to the animals' being light while the mother and father of litter 10 had genes that predisposed the animals to becoming heavy, then it would not be surprising to see the offspring of these parents distributed as they were. But since the parents of these animals all came from a strain that has been inbred for well over 50 generations and is therefore considered to be genetically homogeneous (Rapp, 1983), this explanation is not tenable.

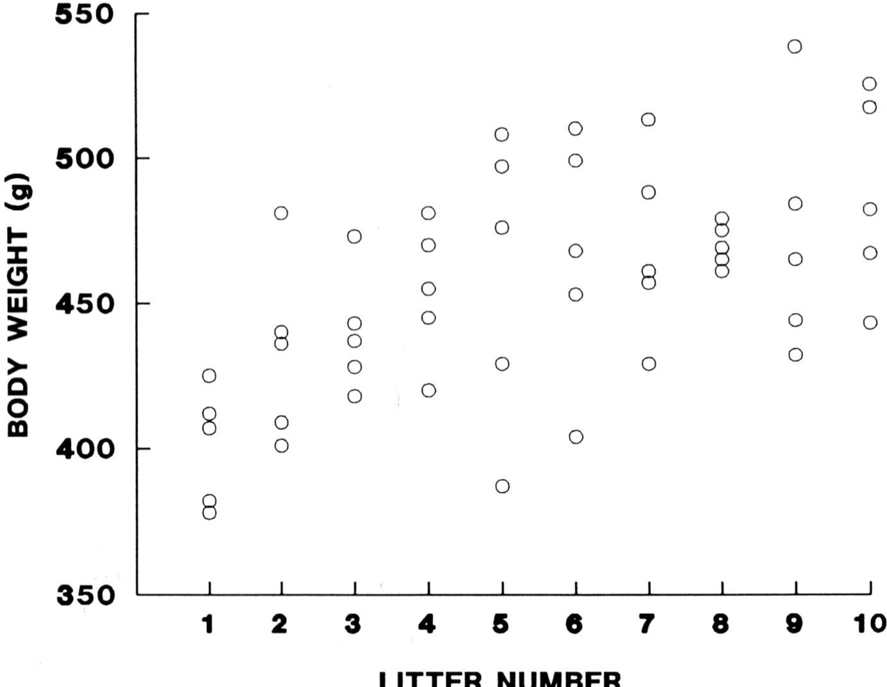

Figure 1.1. Adult body weights of 50 male SHR rats at 20 weeks of age. The data are grouped by litter with 5 males in each litter.

The second possibility is that animals within litters are similar to each other because they share experiences in utero or early during postnatal life that influence development of physiological and/or behavioral mechanisms involved in determining adult body weight. If we were convinced that this possibility accounts for the effect of litter membership on adult body weight, then we could begin to search for the specific experiences of early life that might account for this effect. However, a third possibility for explaining littermate similarities is that life experiences shared by these siblings much later in life could be responsible for this effect. For example, if littermates were group housed after weaning, then perhaps the effect could be explained by the fact that cages placed on the bottom of racks were cooler than ones on the top and these animals were lighter because they expended more energy staying warm. Since our goal is to show that differences in experiences prior to weaning account for litter effects, it is necessary to either make certain that every animal has the same postweaning experiences or, since this is virtually impossible, try to randomize the experiences of later life.

One strategy for dealing with this issue is to use assignments to housing conditions as a way of manipulating the degree to which littermates have shared postweaning experiences. In the experiment from which the data of Figure 1.1 were drawn, a design that combined randomized and littermate postweaning housing was employed. In this study, three of the animals from each litter were housed together at weaning while the other two animals in the litter were each housed with two nonlitter-

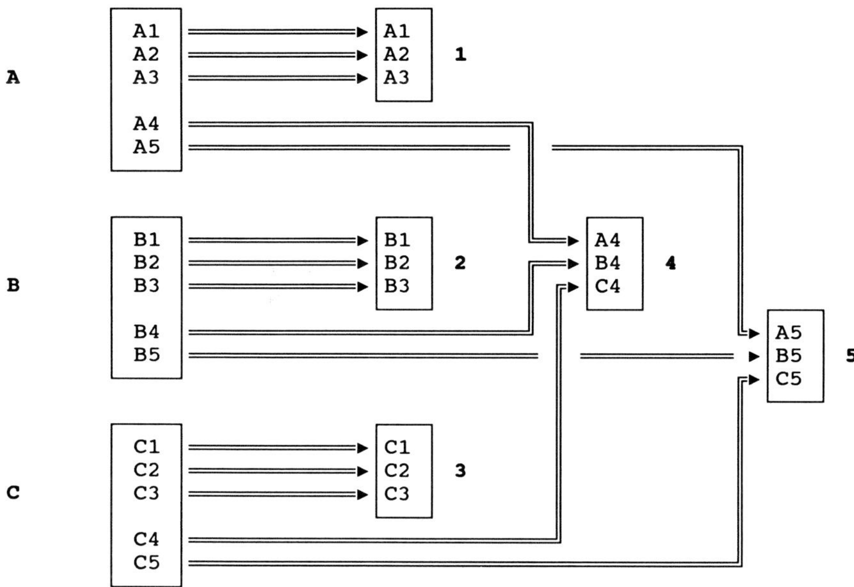

Figure 1.2. A scheme for postweaning housing of littermates that allows for isolation of preweaning effects.

mates. Figure 1.2 illustrates this strategy by showing how the animals from the first three litters in this study were housed.

From this figure it can be seen that rats A1, A2, and A3 were littermates that shared postweaning as well as preweaning environments, which is also true of rats B1, B2, and B3 and C1, C2, and C3. In contrast, the other pair of rats from each litter—that is, A4 and A5, B4 and B5, and C4 and C5—were littermates whose common cage experiences ended at weaning because they were housed with nonlittermates. When an analysis for litter effects was run using data from those animals that shared only preweaning environments (e.g., cages 4 and 5), the effect of litter on adult body weight was still significant. Indeed, for these animals 66% of the variance in adult weight was attributed to litter membership ($p = .003$). Thus, if anything, litter effects were stronger (66% vs. 43% of total variance) when animals from the same litter were exposed to different postweaning environments. This result strongly suggests that variations among individuals in adult body weight might well be linked to events during the preweaning period of development.

Preweaning Litter Effects and Shared Experiences of the Pre- or Postnatal Period

How do we determine if preweaning litter affects are attributable to shared experiences in the prenatal and postnatal periods? Thus far I have described an approach for determining whether adult traits are likely to have been influenced by environmental events prior to weaning. Of course, the preweaning period includes both the prenatal and lactational periods. In order to determine whether litter effects are related strongly to one or the other periods of development, it is necessary to use cross-fostering techniques at birth. The most powerful strategy for making this distinction is to pool animals from different mothers on the day of birth and then randomly foster these animals to new postnatal mothers. In this way, the experiences that are shared because of litter membership will end at the time of birth. In turn, new litters will be formed, and these fostered littermates will not have shared in utero environments. An analysis of variance for the effects of pre- and postnatal litter membership would allow the dissection to be made.

Although I have not conducted an experiment in exactly this way, I have run a small pilot study in which a conceptually similar question was asked. Does the strain of the prenatal mother affect adult blood pressure or body weight? To answer this question, I cross-fostered genetically identical animals that had been exposed to different prenatal environments (i.e., mothers of different inbred strains) to postnatal mothers that were all of the same strain. The experiment was performed with two types of F_1 hybrid pups: those produced by mating SHR males with WKY females and those produced by mating WKY males with SHR females. At birth, both types of F_1s were cross-fostered to WKY dams. In this small experiment, we found there was no effect of strain of the prenatal mother type on the F_1's adult blood pressure or body weight. This suggests that litter effects on F_1 body weights, which are highly significant, are probably associated with variability in the postweaning environment and not due to in utero effects. However, because there may be no important differences between strains in prenatal environments but significant differences in the in utero environments among mothers, this experiment does not fully address the question originally raised. Experiments in which randomized cross-fostering of homozygous pups at birth, followed by randomized postweaning housing, is the best strategy for isolating the source of adult litter effects to the early postnatal period.

MEASURING EARLY EXPERIENCES

Results from experiments designed as those just described will increase confidence in the conclusion that body weight, blood pressure, or other variables of interest are affected by, as yet unspecified, events or experiences occurring during the preweaning period. From these experiments we can proceed with trying to find out whether this effect is associated with some particular type of experience. At this stage, we could follow intuition and measure variables we suspect might explain our litter effects. For example, preweaning effects on adult body weight might reasonably be assumed to be related to early feeding experiences. Indeed, we have known for many years that adult weight and fat disposition is related to early weight gain (Johnson, Stern, Greenwood, Zucker & Hirsch, 1973). Thus, we might expect that measuring the amount of time pups spend nursing would provide a good index behavior for the proximal cause of the long-term effect. However, beware of the "it's intuitively obvious to the most casual observer" philosophy of science. As Hofer (1983) has pointed out, sources of regulation of developing systems are often embedded in events and social interactions that are not intuitively obvious. For example, perhaps individual differences in adult body weight of males are related to differences in adult levels of testosterone. Since Moore (1984) has found that sexual development of male rats is influenced by the amount of genital licking they receive from their mothers, perhaps differences in adult body weight, blood pressure, and so forth, are a function of individual differences in maternal licking. The fact that such less than obvious possibilities exist suggests that we should consider another approach for identifying early life determinants of long-term effects.

The approach I refer to is essentially the "shotgun" method. In practice, this method involves measuring as many behaviors as you can think of that might be related to the observed differences among litters and then determining whether any of these variables are related to the effect of interest. There is no easy way of doing this. You must simply watch mothers and their litters and record in great detail what you see. One technique is to videotape mother-infant interactions and then carefully score the tapes at a later date. Although this method is certainly rigorous and offers the advantage of adding items to your checklist after the experiment is completed, it is very time consuming. My colleagues and I have used a second approach that involves recording behavior at specified time intervals on a predetermined behavior checklist. In our studies we recorded 18 categories of behavior that could be easily observed (Myers, Brunelli, Squire, Shindeldecker, & Hofer, 1989). Included were various types of nursing postures; grooming; eating; drinking in the nest, out of the nest; and so forth. Using this method, maternity cages were observed twice each day during the first 20 days after birth. During each of the twice daily observation sessions, ten 5–10-second "snapshot" observations of each cage were made, and the activities that predominated during this brief period were recorded on the checklist. In this way, 400 observations of each maternity were made during the first 20 days of life, and a fairly detailed record of the most common early experiences of the pups being studied was gathered.

A few comments about this technique are worth making. There are no rules concerning how often one should sample. As it turns out, 400 observations worked well in the study just described. In fact, in a subsequent study we cut the number of observations in half and obtained results comparable to those in our first study (Myers, Brunelli, Shair, Squire, & Hofer, 1989). However, it is best to err on the side of too many observations than too few. For example, taking records once each minute for 2–3 hours would probably have provided a better characterization of differences among mothers. This would have allowed for estimates of time between nursing bouts and average time away to be made—variables we did not record in our studies. Because there are substantial circadian differences in behavior, it is also important, especially during the early stages of pursuing this strategy, to make these measurements at least twice each day, one time when the lights are on and one time with the lights off.

Individual Differences in Maternal Behavior

The first step in analyzing the observation data collected is to look for significant differences among litters with regard to the categories of events that were recorded. This can be done using repeated measures analysis of variance, with age and mother (actually, maternity cage) as the primary sources of variance. Why is this important? Recall that we have observed differences among litters in some adult trait and are trying to relate these differences to variations in early life experiences. If our observations indicate there are no significant differences among litters for the behavioral interactions recorded, there is no basis for pursuing possible connections. However, this will not likely happen. Even within homozygous strains, most items on our checklist exhibited significant litter effects (Myers, Brunelli, Squire, Shindeldecker, & Hofer, 1989). Note that with 18 tests for significance we need to protect against possible random differences by adjusting the probability level we accept as being significant. A conservative strategy is the Bon Foroni adjustment, which simply involves dividing the traditional alpha by the number of tests performed; that is, $0.05/18 = 0.0025$.

If there are significant litter effects, this means that superimposed upon patterns of behavior common to all mothers and their litters are interactions of individual mothers and their litters that differ quantitatively from each other. This result supports the idea that pups from different litters are exposed to variations in naturally occurring early life experiences, and we now have a list of those specific interactions that characterize at least some of these differences.

INVESTIGATING LINKAGE BETWEEN EARLY EXPERIENCES AND ADULT TRAITS

The next step in the strategy is to determine whether any individual category of experience taken from the observed behavioral interactions, or some combination of experiences, is related to the adult trait of interest. At this point we run into a problem. If we have found 10 different categories of behavioral interactions that differ among litters, we will want to examine each of these categories, and perhaps some combinations, as correlates of the adult trait. In principle this is easy: We simply run correlation or multiple linear regression (MLR) analyses. Although the mechanics of running MLR analyses have in the past been a formidable obstacle, these analyses are now available on many powerful statistical packages that can be run on desk-top computers. This ready availability, however, does not solve our problem. We may have too many variables and want to perform too many tests for the number of litters in the study and thus run the risk of finding chance effects. Of course, the preferred approach to this dilemma is either to test only a limited number of specific hypotheses or to increase the number of subjects tested such that we have 6–8 litters per variable, say 60–80 litters. If you take the position that no relationship will be considered established until it is replicated, a less conservative strategy may be acceptable, but some rules of thumb may be helpful for deciding just how loose these preliminary analyses can be. One rule relies on computation of what is called the "shrunken" R^2 (Cohen & Cohen, 1983) or the adjusted R^2 in SYSTAT (SYSTAT Manual, 1987). This statistic is designed to produce an estimate of R^2 when a multiple regression analysis is repeated using a new sample drawn from the same population. The formula for the adjusted R^2 is as follows:

$$ADJ\, R^2 = 1 - (1 - R^2)\, \frac{n-1}{n-k-1}$$

where R^2 is the value obtained on the first analysis, n is the number of subjects, and k is the number of predictor variables. The strategy for this rule is to determine whether, given a projected multiple R, n, and k, a replication would likely be significant. For example, if you anticipate a strong effect, say a multiple R of 0.6, using 4 variables and 20 subjects, the adjusted R^2 will be 0.19 and the expected multiple $R = 0.44$, which is very nearly a p of 0.05. However, if the number of predictor variables is increased to 6, the adjusted R becomes 0.25 and is clearly not significant. By iterative substitution of various values we find that with strong effects a ratio of 5–6 subjects/predictor variable is about the minimum ratio that should be accepted. A second, more formal, approach is to conduct power analyses for R^2 (Cohen & Cohen, 1983). Using a desired power of 0.80, a significance criterion of 0.05, and 4 predictor variables, power analyses indicate that an n of 26 will be required. Although a ratio of n/predictor variables is not strictly linearly related to power, iterative substitutions into power equations indicate that 5–6 subjects per predictor variable is the minimum ratio that will yield reproducible results.

Other factors will help with this problem. First, it is likely that several of the variables measured during the preweaning period will prove to be redundant. For example, in our work (Myers, Brunelli, Squire, Shindeldecker, & Hofer, 1989), mother-in-nest and the predominant nursing posture (blanket-nurse) were highly correlated, which suggested it was not necessary to pursue relationships between both of these variables and adult blood pressure. Although we again run a risk of falsely accepting a significant relationship from running multiple correlations, if some correlations are particularly strong (i.e., > 0.90) and they clearly have face validity—for example, mothers in the blanket nursing posture are almost always in the nest—then it is probably safe to exclude one of these variables from further analysis and thereby increase legitimacy of subsequent regression modeling.

A similar approach would be to conduct factor, or principal component, analyses on the observation variables as a step toward data reduction. Again, you may not be able to satisfy rigorous criteria for these analyses, but if there were dimensions that were strong (e.g., eigenvalues > 3.0) and that made a great deal of sense, then factor scores could be substituted for individual variables, and these would become the potential correlates of the adult trait that interests you. These are reasonable ideas, but since there may have been no a priori hypotheses in this strategy, I would emphasize the one rule that cannot be neglected: After you have determined that a relationship is significant, you must replicate the finding. Having said that, I will proceed to show the results of some preliminary analyses I conducted while preparing this chapter. This continues the discussion of what accounts for litter differences in adult body weight.

These analyses are based on a small set of data from 44 F_1 (SHR × WKY) rats, two males and two females from each of 11 litters. Although postweaning housing and preweaning litter assignment were not randomized in this study, data from my other studies clearly suggested that adult body weight was being influenced by events between birth and weaning, and I therefore proceeded with these analyses. Recall that matings between SHRs and WKYs are not segregate matings, and F_1 pups are genetically identical to each other, with each chromosome being a hybrid of the two parental genomes. Thus, although the F_1 population is not a homozygous strain, it is well suited to the strategy I have outlined. In this study there were significant litter effects with regard to adult body weight, and this was generally true for males and females, although the litter effect for females was only marginally significant (females, $p = .07$; males, $p = .03$). Furthermore, analyses of mother-infant behavioral interactions, which were recorded 200 times during the preweaning period, indicated there were significant differences among the mothers with regard to many of the variables.

For this example I decided to test two variables as correlates of adult body weight. I chose these two variables because I believed they were probably index behaviors for two different types of potential effects. One variable is what we and others have called the *arched nursing posture*. Although relatively infrequent in occurrence, the arched nursing posture is nearly always associated with the pups' "stretch response," which is a good behavioral marker for milk let-down (e.g., Lincoln, Hill, Wakerly, 1973; Lau & Henning, 1985). Thus, looking at this variable was an indirect test of the hypothesis that litter differences in adult body weight are related to feeding in infancy. The second variable I chose to examine was pup-licking. Unfortunately, we had not restricted our definition of pup-licking to licking of the anogenital region, but since much of the pup-licking we recorded was of this type, I believed this analysis would provide a preliminary test of another hypothesis based on

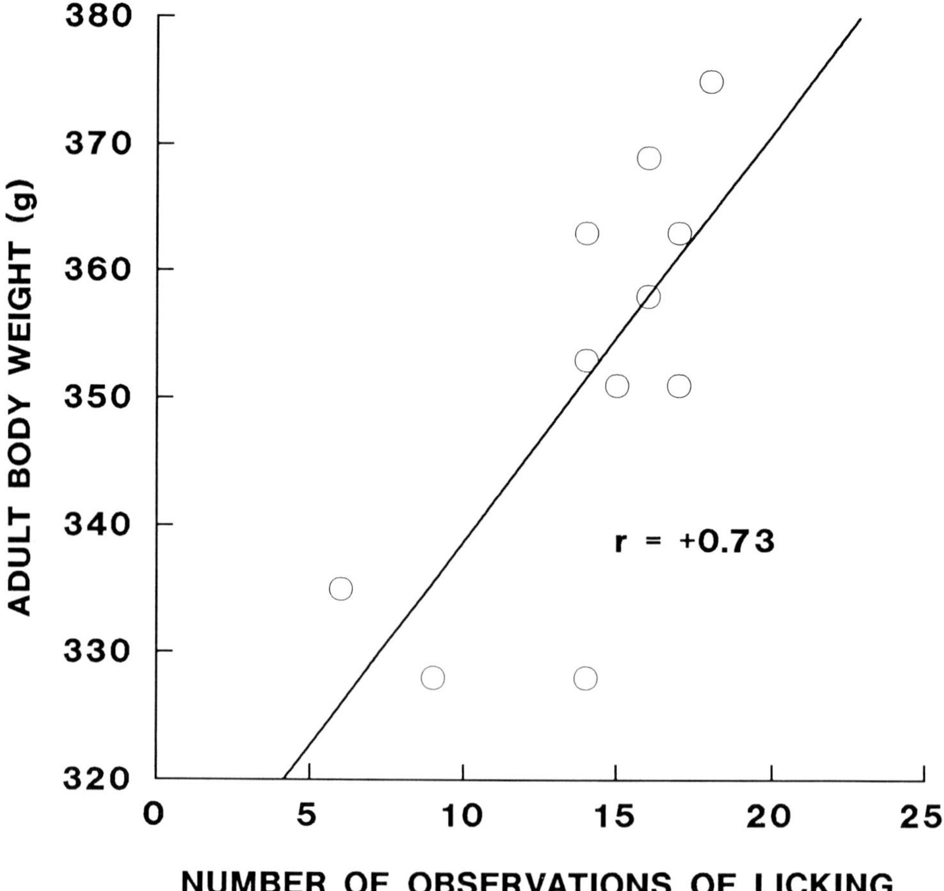

Figure 1.3. A scatter plot showing the relationship between the number of times mothers were observed licking their pups and the mean body weights of the male offspring.

the findings of Moore (1984), which show that sexual development of male rats is influenced by the amounts of anogenital licking they receive from the their mothers during infancy. Since adult body weight is clearly sex-linked, I reasoned that perhaps adult body weight might be correlated with amounts of maternal licking. This hypothesis had a second component. If licking was correlated with adult body weight, we might expect that this would hold true only for males.

The results from these analyses were not what I had expected. First, there was no significant relationship between the amount of arched nursing we observed and adult body weight of either males or females. There was not even a tendency for these correlations to be significant. This lack of a relationship, of course, does not mean that early feeding patterns do not influence subsequent feeding patterns and in turn adult body weight. Perhaps only during certain periods of lactation does this effect hold true. Other analyses should be run in which the amounts of arched nursing are broken down into blocks of days and the analyses then looked at again. Or perhaps the arched nursing posture is not such a good index of food intake. This could be determined by correlating infant weight gain with the frequency of arched nursing.

With regard to licking, the results were quite interesting. There was no significant correlation between licking and adult body weight of the females ($r = +.24$), but for the males the correlation was surprisingly strong ($r = +.71$, $p = .011$). Figure 1.3 shows the scatter plot of the litter-mean adult body weights of the males from these 11 F_1 litters versus the number of times their mothers were observed licking any member of the litter. I am not aware of any theory that links a specific type of mother-infant interaction to sexual development and, in turn, to the determination of adult body weight; but our strategy provides data that suggest this might be an interesting chain of events to study.

CAUSALITY AND EARLY EXPERIENCE

The last topic is what to do after a correlation between a specific early experience and an adult trait has been discovered. Naturally, I saved the most difficult issue for last. What we have done is develop a strategy that can lead to the identification of specific kinds of experiences that are correlated with adult traits. Two important questions remain. Can you determine whether the relationship is a causal one? How do you pursue the nature of the connection between these events?

After finding a significant relationship between some specific type of interaction between mothers and their young and some specific adult trait of the offspring, I believe there are three basic ways to proceed. The first entails experimental manipulation of the early experience of interest and then following the outcome of this manipulation. In practice, this step is not as simple as it sounds. Manipulating one aspect of maternal behavior without influencing many others is quite difficult. Nonetheless, in some cases this approach can prove to be fruitful. For example, by constructing a harness that restricted the mother's ability to lick her pups, Moore was able to demonstrate the impact of maternal licking upon sexual development (Moore, 1984). A similar strategy could be used to pursue the potential relationship we saw between licking and adult body weight. Another experimental procedure that would likely be useful for these purposes is artificial rearing (Hall, 1975; Diaz, chap. 16, this volume.) Using this technique, we could conceivably control many of the variables that were not of interest while varying the amounts of other types of experiences. Of course, we would not be surprised if the expected effects were not found because the artificially reared rat might not respond in the same way as a normal animal. Because of the limitations of the specific experimental manipulation approach, we need other strategies so that we can converge on a causal hypothesis.

A second way to proceed with this problem is in essence a behavioral counterpart of a type of analysis my colleagues and I have used to test for genetic cosegregation of behavioral traits and blood pressure (Hendley, Atwater, Myers, & Whitehorn, 1983; Yongue & Myers, 1988). In this paradigm, the hypothesis that there is a common genetic basis for two traits that are known to be different in two inbred strains is tested by reshuffling the genomes of the two inbred strains via segregate matings and then measuring the correlation between the two traits. If the two traits are not correlated in segregate populations—that is, do not cosegregate—the hypothesis of common genetic basis can be rejected. I first considered this approach for environmental linkage studies following a suggestion from Gerry Turkewitz, but as yet have not tried it. This strategy would involve doing something, not necessarily anything very specific, to the mother-infant environment that would disturb the normal course of events. By modifying room temperature, lighting, noise, or amounts of daily handling, we might be able to induce some mothers to lick their pups more and some less. In other words, we would try to find a treatment that did not necessarily alter the mean of behavior, but rather, would change variance. Although these manipulations would likely change many aspects of the mother-infant interactions, the idea is that a relationship can be causal only if the specific preweaning event (e.g., licking) maintains a stable correlation with the adult trait of interest in the face of these disturbances. If it does not, the causal hypothesis for the observed relationship can be ruled out.

The last strategy is the "go for it" approach. The basic idea here is to assume that the relationship observed is likely to be causal in nature and immediately begin to try to identify the mechanisms that might underlie the observation. This is the approach I have taken in my recent work. We observed that the frequency of arched nursing is positively correlated with offspring adult blood pressures (Myers, Brunelli, Shair, Squire, & Hofer, 1989; Brunelli, Squire, Shindeldecker, & Hofer, 1989). This suggested that something interesting might be going on during feeding that influenced development of the cardiovascular system. Indeed, we found that during feeding blood pressures of young rat pups increase dramatically (Shair, Brake, Hofer, & Myers, 1986), that these increases in pressure are much greater in animals predisposed to develop hypertension (Myers & Scalzo, 1988), and that blood pressure responses to feeding can be inhibited by administering sympathetic blockers (Scalzo & Myers, in press). Thus, even though we have not proved that the relationship between frequency of arched nursing and adult blood pressure is causal in nature, assuming that it might be has led us to new insights about the importance of feeding behavior and cardiovascular function in young animals. Our goal now is to pursue these observations using the other follow-up strategies I have described in order to learn more about how feeding alters development of the cardiovascular system.

CONCLUSION

Why are the strategies I have presented important? At one time I believed the greatest impediment to advances in understanding development and the role played by the environment in shaping developing systems was the lack of general principles that account for the dramatic changes organisms undergo as they mature. The evolutionary biologists had the guiding theory of natural selection; the physiologists had homeostasis; and the molecular biologists had the DNA, RNA, protein triad. What developmental scientists had was the observation that organisms change in very predictable ways as they age. This appeared to be a long way from a promising theoretical basis for understanding development. What was as disturbing as the lack of understanding of how all members of a species change so predictably during development was the simultaneous lack of understanding of how individual differences that are so obvious emerge at the same time.

Confusion was fueled by the concepts of genetic and environmental determinism and the endless debate over which of these was at the heart of a general theory of development. We have, for the most part, escaped from this epicontroversy and accept the obvious: Genes and environment interact. This realization, however, has done little to promote a comprehensive theory of development. Patrick Bateson (1987) summed it up well: "To purpose, as sometimes has been done, that in development everything interacts with everything else is hardly an encouraging piece of advice to anybody who seeks to understand what is going on."

Perhaps a general theory is not the essential ingredient missing in our search for a better understanding of developmental processes. To a great extent, I believe the general principles necessary to explain development have been known for some time. We know that development is a recursive process with functions and characteristics of each progressive stage being dependent upon the current configuration of hierarchically organized systems. These systems emerge within certain species-specific constraints, which are shaped by prior experiences and by ongoing events to an extent allowed by the current organization. New connections and relationships are stabilized through repetitive experience; others are eliminated according to rules that are, in essence, derivatives of the idea of natural selection.

So if general principles are not in short supply, what is missing? I believe the answer to this question is actually quite simple and has been already summarized by, again Bateson, when he suggested that what is needed is "to translate a general and valid point about ontogeny into particular examples of how the developmental processes actually work." (Bateson, 1987) In other words, I believe we need to attend to the difficult business of demonstrating how specific naturally occurring experiences shape the development of specific traits. What we need are a few good examples. By focusing upon these specific illustrations, general rules and mechanisms that govern progressive maturation will emerge. Within my own area of interest I acknowledge that I am a long way from fulfilling this goal. Yet the strategies I have described have been aimed exactly at this point and have been helpful. I believe they may provide a generally useful approach to others interested in trying to gain a better understanding of how particular stimuli or experiences guide development.

REFERENCES

Bateson, P. B. (1987). Biological approaches to the study of behavioral development. *International Journal of Behavioral Development, 10*, 1–22.

Chapman, R. H., & Stern, J. M. (1979). Failure of severe maternal stress or ACTH during pregnancy to affect emotionality of male rat offspring. Implications of litter effects for prenatal studies. *Developmental Psychobiology, 12*, 255–267.

Cohen, J., & Cohen, P. (1983). *Applied multiple regression correlation analysis for the behavioral sciences*. Hillsdale, NJ: Erlbaum.

Hall, W. G. (1975). Weaning and growth of artificially reared rats. *Science, 190*, 1313–1316.

Hendley, E. D., Atwater, D. C., Myers, M. M., & Whitehorn, D. (1983). Dissociation of genetic hyperactivity and hypertension in SHR rats. *Hypertension, 5*, 211–217.

Hofer, M. A. (1983). The mother-infant interaction as a regulator of infant physiology and behavior. In L. A. Rosenblum & H. Moltz (Eds.), *Symbiosis in parent/offspring interactions*. New York: Plenum.

Johnson, P. R., Stern, J. S., Greenwood, M. R. C., Zucker, L. M., & Hirsch, J. (1973). Effect of early nutrition on adipose cellularity and pancreatic insulin release in the Zucker rat. *Journal of Nutrition, 103*, 738–743.

Lau, C. & Henning, S. J. (1985). Investigation of the nature of the "stretch response" in sucking rats. *Physiology and Behavior, 34*, 649–651.

Lincoln, D. W., Hill, A., & Wakerly, J. B. (1973). The milk-ejection reflex of the rat: An intermittent function not abolished by surgical levels of anesthesia. *Journal of Endocrinology, 57*, 459–476.

Moore, C. (1984). Maternal contributions to the development of masculine sexual behavior in laboratory rats. *Developmental Psychobiology, 17*, 347–356.

Myers, M. M., Brunelli, S. A., Shair, H. N., Squire, J. M., & Hofer, M. A. (1989). Relationships between maternal behavior of SHR and WKY dams and adult blood pressures of cross-fostered F_1 pups. *Developmental Psychobiology, 22*, 55–67.

Myers, M. M., Brunelli, S. A., Squire, J. M., Shindeldecker, R. D., & Hofer, M. A. (1989). Maternal behavior of SHR rats and its relationship to offspring blood pressures. *Developmental Psychobiology, 22*, 29–53.

Myers, M. M., & Scalzo, F. M. (1988). Blood pressure and heart rate responses of SHR and WKY rat pups during feeding. *Physiology and Behavior, 44*, 75–83.

Rapp, J. P. (1983). Genetics of experimental and human hypertension. In J. Genenst, O. Kuchel, P. Hamet, & M. Cantin (Eds.), *Hypertension: Physiopathology and treatment* (pp. 582–598). New York: McGraw-Hill.

Scalzo F. M., & Myers, M. M. Pharmacologic blockade of blood pressure responses to milk in infant SHR and WKY rats. *Physiology and Behavior*, in press.

Shair, H. N., Brake, S. C., Hofer, M. A., & Myers, M. M. (1986). Blood pressure responses to milk ejection in the young rat. *Physiology and Behavior, 37*, 171–176.

SYSTAT Manual. (1987). Evanston, IL: Systat, Inc.

Yongue, B. G., & Myers, M. M. (1988). Cosegregation analysis of salt appetite and blood pressure in genetically hypertensive and normotensive rats. *Clinical and Experimental Hypertension*. Part A, *Theory and Practice, A10*, 323–343.

2

Microsurgery:
Window on Hidden Developmental Processes

MYRON A. HOFER

Older concepts of development viewed the infant from the perspective of the adult and emphasized the gradual emergence of the flexible and complex behavior characteristic of mature forms. But the past two decades of research in the development of behavior and of its underlying biological systems has made us aware that development is not continuous and that, instead, early stages in mammalian life are organized to adapt the young to a series of successive inherited environments. We have come to realize that infants have unusual and highly adaptive mechanisms fitted to the particular requirements of each environment. These physiological and behavioral systems are often organized differently at each early stage, and much of development is now thought to consist of successive transformations along a series of these transient phases of organization.

Microsurgery has played an important role in the discovery of these novel early organizational systems because of the small size of infants of laboratory species such as the rat. The capability of placing catheters in blood vessels and into the gastrointestinal tract, implanting recording electrodes into muscle and on the cortical surface of the brain, of making selective lesions of autonomic and peripheral nerves, and of removing endocrine glands even in animals only an inch long has allowed us to begin to see more precisely how development takes place. It has become clear that the mammalian infant's physiology and behavior are organized to adapt it to the environment provided by the nursing mother. This social relationship is thus the very system that needs to be studied in order to understand this stage of development. The problem is how to take the mother-infant interaction apart experimentally, piece by piece, to see how it works at the biological as well as the behavioral level. At the time we began our studies in the late 1960s, microsurgery for humans was just beginning and had not yet been applied to small-animal experimental work. The minute size of the infant rat discouraged researchers from attempting any of the standard surgical approaches that had been used for decades in physiological and behavioral studies with dogs, cats, monkeys, and, to a limited extent, adult rats. The cost and ethical complications of working with infant monkeys, dogs, and cats were daunting. In addition, the prevailing view of infants as amorphous and unformed held out little promise that developing microsurgical techniques for infant rats would be worth the effort. And there was no precedent for applying them to the analysis of early social relationships.

But there was another set of assumptions that discouraged a direct analytic approach to biological processes within the mother-infant interaction when we began our work. It was widely believed that to the extent that the mother-infant interaction had developmental importance, this was through the psychological aspects of the relationship, not the physiological. Maternal separation and deprivation, for example, were thought to exert their effects through psychological processes having to do with attachment and its emotional components. Biological processes clearly existed in infants, such as growth and metabolism, ingestion, temperature regulation, and sleep, but these were considered to be "genetically programmed," little influenced by the mother-infant interaction and largely irrelevant to psychological development. The prejudice was that the human mother-infant relationship was solely the product of subtle cognitive and affective processes, whereas lower animals did not really have a relationship, but were caught up together in an interplay of reflexes along a genetically programmed schedule like mechanical dolls.

This picture was changed for us when we found that biological systems in the infant rat as well as behaviors were affected by separation of pups from their dams. We needed microsurgery in order to study a range of biological systems in these small animals (e.g., blood pressure by arterial cannula, sucking pattern by electromyogram), and we particularly needed microsurgery when it came to asking more precisely how the absence of the mother resulted in such changes (e.g., vagotomy, adrenalectomy). The realization that separation may exert its effects by different processes than those involved in psychological attachment and the discovery of "multiple regulators" within the mother-infant interaction directly resulted from the experiments made possible by microsurgery. (For reviews, see Hofer, 1983, 1987.) For example, we found that the infant rat's autonomic cardiovascular control systems are delicately regulated by the dam through the milk she supplies. Nutrients in milk act on gastrointestinal interoceptors over neural afferent and efferent pathways that were discovered through microsurgical approaches to vagus and splanchnic sympathetic chains, baroreceptor nerves, spinal cord, endocrine glands, and the gastrointestinal tract. Cardiac rate and arterial vasoconstriction were sensitively tuned over a 30% range by this form of maternal regulation (see Hofer, 1984, for review and Hofer, 1985, for newer methods).

Thus, the first social relationship between the mammalian mother and infant is a complex, highly organized system that we are just beginning to understand in some detail. In addition to physiological systems, the synchronous and enmeshed behavioral interactions are made possible by a number of hidden processes that depend on information exchange over defined sensory pathways. Microsurgery offers a variety of approaches that can reveal these behavioral processes. One approach, which I describe below, involves the use of microsurgery to ask how infants initiate and maintain nipple attachment at the outset of each nursing bout. Selective lesioning of peripheral sensory systems in the snout area produced a series of surprising and specific modifications in the pup's behavior at different ages, which have given us a new picture of the infant's perceptual and motivational processes during nursing bouts in the first few postnatal weeks.

Our experience in working with the method of microsurgery has been far from a dry technical exercise. Instead, it has opened a window on behavioral and physiological processes that otherwise would have been hidden from view and generated new ideas that in turn have suggested new ways to use the method.

GENERAL METHODS

Microsurgery is a method with general applicability to a wide variety of scientific questions in early development, as suggested in the introduction to this paper. For example, these techniques allow the implantation of electrodes for recording physiological functions (e.g., electromyogram, electrocorticogram) and of cannulas for administering and withdrawing nutrient and biologically active substances as well as for recording intraluminal pressure (e.g., intraarterial blood pressure). Furthermore, it allows the sectioning of nerves supplying key sensory fields, reflex pathways, or motor output and the removal or transplantation of sources of endocrine regulatory substances. Infants can later be returned in their surgically altered states to the home cage litter situation for subsequent observations, recordings, and sampling. General guides to surgical procedures on adult laboratory rats and a source of further references may be found in Farris and Griffith (1963) and Kraus (1980), to microsurgical technique in Acland (1990) and Olszewski (1984), to anesthesia in Gilman, Goodman, Rall, and Murad (1985), and to animal care considerations in the NIH Guide.

Central to microsurgery is, of course, a good binocular stereoscopic microscope, allowing three-dimensional viewing with a range of different magnifications, a long arm stand, and a long enough focal length to permit the use of hands and instruments in the space between the lens and the subject (at least 8 cm and optimally 15–20 cm). Highly desirable features are extensive depth of focus, high resolving power, a zoom magnification capability, a co-observer tube for an assistant, and foot control of focus and zoom. A device for foot control of focus attaching to standard microscopes is made by Goodfellow Technology. Good simple instruments are the Bausch and Lomb "stereozoom" line now made by Leica ($1,000 up), and the more desirable are operating microscopes with motorized stands made for human microsurgery such as those produced by Aus Jena, the former East German division of Zeiss (which sell for $10,000–$15,000—considerably less than comparable Swiss or West German equipment). The Swiss division of Leitz, Wild produces a wide range of high-quality instruments from $3,000 up, including instruments with dual viewing stations for learning procedures.

Next to the optics in importance is a strong light source without the heat that will dry out delicate tissues. Fiber optics have made a critical difference here. There are two types with different properties. The gooseneck type allows two or more light sources to be positioned to cast light from any angle and distance simply by bending the flexible segmented metal sheaths of the fiber optic tubes. This allows angled light, which enhances the three-dimensional details in the operative field, while the second light source can prevent dark areas of shadow. More expensive is coaxial illumination, delivered directly down the path of view of the microscope—a great advantage when operating at the bottom of an unavoidably deep dissection. Since light originates from a fiber optic circle, there are no shadows, and the operator's hands are less likely to come between the light source and the field. The disadvantage is that light from this source is necessarily "flat" and lacks the feature-detecting quality of angled light sources. Both types are available from Cole Parmer.

Microsurgical instruments are now available from a number of firms (Circon, Fine Science Tools, Roboz). Before investing in microscissors at $150 to $200 each or indeed any expensive instruments, ask a salesperson to allow you to try out demonstrators of several sorts in the kind of work you are doing. Microforceps and probes of several different sizes with various curved and angled tips, microtip needle holders, and microscalpels are the basic instruments needed.

Because of the delicacy of tissues in young animals, scissors are much more useful than microscalpels, and in newborn and fetal rats the jellylike consistency of the tissues makes scalpel blades almost useless. A great deal can be done with "blunt dissection," a gentle division of tissue performed between microforceps and a blunt, angled probe. This approach allows one to free the organs of interest from the many layers of delicate supporting tissues and the numerous collections of fat that are found in young animals. Bleeding is actually minimized, since the fine veins and arteries are stretched before severing, promoting spasm and closing off at the site of interruption. The drawback of blunt dissection is that the traction exerted on delicate structures may threaten immediate postoperative return of function that is essential. Thus, a careful watch must be kept, and scissors used in the vicinity of vital nerves. Arteries, veins, and nerves must not be grasped with forceps but gently manipulated with a blunt, angled probe. Since infant tissues exude moisture that obscures and blurs one's view of tissue detail under the microscope, repeated drying of the operative field is necessary. Q-tips are all right for low-power work, but the much smaller dental compressed-cotton absorbent points (Johnson & Johnson; Schein) are extremely useful for fine work under high power because they absorb fluids and blood without expansion and loss of shape. One of these points can be grasped between the tips of microforceps and used as a stabilizing probe to hold tissue against the action of another forceps or probe while at the same time absorbing any fluids released by the blunt dissection. A fortunate property of young rats is their remarkable hemostatic capabilities, allowing incision of skin, peritoneum, and so forth, without pausing to tie off "bleeders." Occasionally, minor arteries and veins need to be cauterized prior to cutting. This can be done with the fine point of a miniature soldering iron or (more expensive) an electrocautery unit (Harvard Apparatus). The last line of defense against bleeding is to pack Gelfoam (Upjohn) tightly around the site.

It is not necessary or even advisable to heat sterilize microsurgical instruments; heat destroys finely machined edges. Disposable microscalpel blades are now available (Rudolph Beaver, Inc., from Harvard Instruments). We have found that microsurgical instruments are best cleaned individually, first with water to remove blood and then with 100% alcohol to sterilize, applying these gently with gauze pads. Placing instruments together in metal trays for heat or alcohol soaking will allow them to strike against each other, destroying delicate, smooth cutting surfaces and fine points.

Surgical needles and sutures for microsurgery come as single sterile units in various sizes, suture types, and needle shapes (Ethicon, Inc.). They vary considerably in price. For most work I have found that a curved, flat needle with 6-0 braided silk is the easiest to use. Skin and peritoneum can be closed with a continuous suture since it does not easily slip, and a single square knot reliably holds. Silk cannot be left in place for longer than a week before tissue reaction begins to be a problem, but most acute experiments on rapidly developing animals do not last as long as this. Silk is a great deal easier to work with than the non-tissue-reactive nylons and other synthetic fibers that do not easily lie flat as knots are laid down, tend to spring suddenly out of reach during manipulation, and require three or four half knots to avoid working loose. Monafilament nylon, however, comes in much finer fibers with smaller needles (e.g., 10-0) for very fine tissues and for chronic internal use. Acland (1990) has written a superb manual that describes current microsurgical suturing techniques using adult rats as subjects. It contains many practical suggestions and useful product information.

Cyanoacrylate ester adhesive (e.g., Duro Super Glue) is gaining widespread use as a substitute for sutures (Stickrod, Soyke, & Weeks, 1984). The need for daily construction of disposable microapplicators, the stiffness of bonded areas, and the need for tissue dryness limits its usefulness, but it is particularly helpful when tissues are too delicate for sutures or where long incisions must be joined rapidly.

The key to conducting surgery easily is adequate exposure of the operative site. Proper positioning of the animal and stable sources of retraction are important elements in assuring this exposure. Except with the most expensive model microscope fitted with a co-observer tube, it is not possible to have an assistant provide retraction. Variable-position mechanical retractors can be made in the following way. A short right angle bend or hook is made at the end of a short (3 cm) length of 20–30-gauge stainless steel wire. The straight end is joined to a length of elastic cord of the same length, which exerts gentle tension, and in turn is joined to a beaded chain or cord that fits into tapered slots of movable cleats positioned along the edge of a small operating table (e.g., Harvard Instruments, No. 50-1239). The table itself is provided with thermostatically regulated heat control. Infant rats can be firmly positioned using gauze pads and "micropore" tape.

Anesthesia for infant rats has a number of special features not found in the literature on adult animal surgery (e.g., Kraus, 1980). The most striking of these is the ease of use of hypothermia made possible by the infant's high relative surface area, resistance to hypoxia, and unusual characteristics of thermoregulatory control. Neonates can be operated on while circulation and breathing are temporarily arrested, providing freedom from interference by bleeding and respiratory movements. Surgery can be conducted in the thorax without concern for miniaturizing a mechanical respiratory system. Rewarming is quick, and return to apparently normal functioning takes less than an hour. This method is generally thought to be limited to the first postnatal week, but we have found hypothermia to be a useful adjunct to inhalation anesthesia in older infants up to 2 1/2 weeks old when freedom from bleeding and respiratory movements is required. An excellent recent article is available on this technique (Phifer & Terry, 1986). The easiest anesthetic to use for terminal physiologic experiments is urethane given by intraperitoneal injection. This agent has a very broad dose range in rats before respiratory and/or cardiovascular depression occurs. Quite the opposite is true for barbiturates, the standard injectable anesthetic for adults, which in infants requires very careful titration of dosage at each age, and even then, individual differences between pups may result in unforeseen anesthetic deaths.

Urethane has another use in providing "passive" social companions for postoperative pups during recording procedures: It induces a prolonged unresponsive state lasting up to 24 hours from a single injection.

For most operative work, inhalation anesthesia is the easiest and safest. A nose cone can be fashioned from an appropriately sized plastic tube with a gauze insert and a small port cut in the wall into which anesthetic can be dropped from a syringe. Methoxyflurane (Metofane, Pitman-Moore) is probably the best and easiest agent to use, although induction and recovery are not as rapid as with ethyl ether, and terminal respiratory depression is more common than with ethyl ether. In the minute quantities needed for infant rodent surgery, the flammability of ethyl ether is not a great liability so long as it is kept in tightly closed containers and applied from a closed syringe into a semiclosed nose cone delivery system. Remember that ether vapors are heavier than air! When rapid recovery of a normal behavioral state is needed, ethyl ether anesthesia remains the agent of choice.

All inhalation anesthetics stimulate secretions, and a microsuctioning device should be available for periodic use since infants may simply stop breathing when these obstruct the airway. Soft flexible tips and stiffer shafts of various sizes can be formed by shrink-fitting (first dipping the silastic in ethyl ether to expand it) a section of silastic tubing (Dow Corning) over a 1–2 cm polyethylene tubing (Clay Adams) fitted onto the 23-gauge needle of a 5–10 ml hypodermic syringe. Suction is created by rapidly withdrawing the syringe plunger with the flexible tubing in the infant's posterior pharynx. If too much anesthesia is inadvertently administered, resuscitation of the infant is nearly always possible using a flexible rubber or plastic tube, about 15 cm long, fitting over the animal's snout at one end, the experimenter gently providing alternating positive and negative breath pressure from the other end of the tube while watching the infant's thoracic movements to gauge depth of lung expansion.

If pups are to be returned to the dam postoperatively, allow them sufficient time to recover within the home cage nest area so that temperature and skin odors are restored toward normal before returning the dam. A thermoregulated heating pad under the cage floor is advisable. Removing the dam from the home cage during surgery is preferred in general, so that littermates do not gain too great a nutritional advantage on the operated pups and the dam is less likely to single out the operated pups and treat them differently following reunion. Pup weight gain over 24 hours is a good "bottom line" indicator of how much or little the mother-infant interaction is disturbed by the surgery.

APPLICATION: SENSORY INTEGRATION IN NURSING BEHAVIOR

When researchers first set out to study the early development of rats in the 1960s, they spent a great deal of time trying to feed rat pups away from their mothers using the artificial nipple and bottle method that had proved so effective for other infant mammals, including human babies. But rat pups did not suck on artificial nipples. This was attributed at the time to an inability to produce an artificial nipple that closely enough mimicked the compressibility, texture, and elasticity of the rat dam's teat.

In 1967, Rouger, Tobach, and Schneirla published work suggesting that rat pups depended on olfaction for successful nursing. Using microsurgical techniques for olfactory bulbectomy, they found that anosmic newborn rats lost weight progressively and died after return to their mothers. In exploring this phenomenon, Pauline Singh and I (Singh, Tucker, & Hofer, 1976) found that chemically lesioning the olfactory epithelium with zinc sulphate nasal irrigation had the same effects on survival as bulbectomy and that these treated pups would not attach and suck on their dam's teats even when teats were placed in their mouths. In 1976, Martin Teicher in Elliott Blass's laboratory (Teicher & Blass, 1976) and our group (Hofer, Shair, & Singh, 1976) independently published strong evidence for a maternal ventral pheromone that elicited nipple attachment and sucking in infant rats.

What had been missing from the artificial nipples was an odor that few imagined existed. Only after microsurgery had revealed the unexpected lethal effects of anosmia on suckling rats did this possibility seem worth further consideration. And once discovered, it seemed to explain fully nipple attachment in young rats. Hungry anosmic pups were observed to pass their lips and nose directly over a protruding teat without mouthing or attaching to it, as did normal pups if the nipple had been washed with solvents that removed the ventral pheromone. Even when an anesthetized dam's teat was placed deep in an anosmic pup's mouth, the pup would attach and suckle. Furthermore, shaving and cooling the dam's ventrum had little effect on nipple attachment (Blass, Teicher, Cramer, Bruno, & Hall, 1977). Clearly, tactile cues were not sufficient, but we had not asked whether they were necessary. Thus, the data seemed to encourage us to conclude that tactile cues played a minor role in nipple location and attachment, and it was only the application of microsurgery to this sensory system that led us to realize that pups needed more than their olfactory systems to initiate nursing.

Methods

In order to understand the role of a given sensory system in the mediation of behavior, one must often first develop an experimental setting and procedure that allows the behavior to be studied repeatedly, using methods of observation that make possible detailed recording of behavioral sequences (see chap. 6 by Spear & Rudy and chap. 7 by Galef, this volume). These methods are then used to analyze the specific behavioral changes that follow nerve section. Such observations are often necessary to interpret the gross deficits that may occur when lesioned pups are returned to the litter situation (e.g., weight loss, pups out of the nest, poor survival).

Once a sensory field is identified as likely to be important, an anatomy text for the rat, such as Greene (1963) or Hebel and Stromberg (1986), will provide information on which nerves are likely to be involved and their approximate course to the sensory end organ. Since surgical approaches to peripheral nerves have generally not been previously established for rats, investigators must invent their own. I advise beginning with dissection of a dead pup, which provides a surgical field free from bleeding, tissue fluids, and respiratory movements. Placing the pup in the refrigerator for a day or two after a lethal overdose of anesthetic prevents troublesome bleeding during exploratory dissection. Once you have identified the nerve, its branch points, and its course and relations to other important structures in a wide dissection, one or more operative approaches will suggest themselves. Be sure you have found and identified all other nerves in the vicinity so that you can be sure to avoid cutting them in your later surgical approach. Choose a site for sectioning that is not too deep and that will allow the nerve to be fully exposed by retracting other structures such as blood vessels and other nerves. Ideally, it should be possible to remove a short section of nerve or displace the cut ends some distance from each other to avoid rapid regeneration. If studies longer than 5–7 days are planned, cut ends can also be separated by a thin plastic sheet of Parafilm (Fisher Scientific).

After the choice of site is made on the basis of a wide dissection, a much smaller operative incision and approach are made on another animal, modifications are made, and finally the approach is attempted under anesthesia. Bleeding should not be a problem with careful retraction of major vessels, but in some applications hypothermic anesthesia will be helpful. In order to visualize very small nerve fibers, a dry operative field is essential (see above "General Methods"). Nerves should never be gripped with forceps, except those about to be sectioned. All other nerves should be gently retracted, without stretching, using a blunt right angle or hooked probe.

Verification of selective interruption of a sensory nerve can be done in three ways. First, when the animal has recovered from anesthesia, or after testing, the area of sensory anesthesia can be mapped by pinching the skin with microforceps to assess deep pain sensitivity and fine touch sensitivity by using a fine stiff hair or wire. By pressing the tip of a 3–4-cm length of the wire against the skin until the wire bends, a uniform and reproducible stimulus is presented to a tiny area of skin. A flinch response to this stimulus will usually be elicited from a pup that is quietly resting in the experimenter's hand, except in denervated skin areas. By repeated pinching and probing, the area of anesthesia can be mapped out and recorded. Second, autopsy can be used to verify nerve section, although after several days scar tissue can make nerves and other landmarks very hard to find. Third, autoradiographic studies of central neural structures using metabolic markers such as cytochrome oxidase will demonstrate retrograde denervation effects in appropriate sensory relay fields.

For comparison studies, control operations must be done on littermates in which all steps are taken to expose the nerve but without actual sectioning. (This should *not* be a sham operation with only skin incision and anesthesia, because the dissection necessary to expose the nerve may unintentionally damage other nerves and blood supply to distant organs, exerting effects that could mistakenly be attributed to the selective nerve lesion.) A third, unoperated companion group will allow assessment of such inadvertent effects of surgery on later behaviors.

Finally, behavioral tests must be devised to assess the precise nature of the changes induced by loss of a given sensory or motor function. Details of these methods as applied to infraorbital nerve section can be found in Hofer, Fisher, and Shair (1981), including descriptions of the behavioral testing procedures used to obtain the results described below.

Alternative Methods

Instead of surgical nerve section, peripheral sensory fields can be temporarily inactivated with subcutaneous injection of procaine or one of its analogs. This is easy but has numerous limitations. First, it is nonspecific; the injected solution diffuses widely and will block all nerve conduction in the vicinity of the sensory field, including other sensory areas and even motor functions. In the case of the vibrissal pad, this means that facial motor nerve conduction would necessarily be blocked along with infraorbital afferents, since the two overlap in distribution. The motor deficit then could be responsible for any behavioral changes subsequently observed rather than the sensory deficit, or the two could interact. In addition to their nonspecificity, local anesthetics are absorbed and can have effects at distant sites such as the CNS, so that control injections in other sites are required, and results in these animals may be difficult to interpret. Finally, the effects of local anesthetics are transient. This has one advantage in that individual animals can be tested after as well as before the sensory interruption. But it precludes any study of long-term effects, including an assessment of whether the sensory process is crucial to survival, and could not have revealed the age-specific adaptations found in the studies described below.

Interpretation of Results of Deafferentation

As with lesions of the CNS, nerve section does not simply produce an animal in which all systems are unchanged except the affected one. Loss of a source of afferent stimulation can have general and/or specific effects on the subsequent functioning of other systems that go beyond those attributable to the discrete loss of sensory information produced by a particular lesion. For example, adult rats extensively deprived of sensory input from the interior mouth by multiple cranial nerve sections show a profound reduction in their motivation to eat, and they may starve despite an abundance of nutrient and despite their continuing ability to find and ingest solid food and water (Jacquin & Zeigler, 1983). Unexpected interactions between systems are likely to be found only if they are looked for.

Thus, careful behavioral analysis must be done on deafferented infants, both acutely and after a period of time has elapsed in order to assess compensatory and adaptational effects. As illustrated in our studies below, the stage in development at which surgery is done markedly affects outcome, and the results can reveal the operation of unexpected developmental processes. Reorganizational effects can be detected by careful behavioral observation, then analyzed functionally by selective lesioning of suspected compensatory systems and structurally by neuroanatomical-chemical studies of underlying central neural substrates.

Results of Snout Denervation Lead to Conceptual Changes

The surprising findings summarized in the introduction to this section did not establish an exclusive role for olfaction in the initiation of nursing. Observation of normal pups' behavior and the early (prenatal) maturation of sensory hairs of the vibrissal pad suggested that tactile function in the snout was also likely to be involved. The sensory nerve supplying this area is the infraorbital branch of the trigeminal nerve and consists of a broad band of fibers emerging from the floor of the orbit, branching out to the external nares, vibrissal pad, and upper lip. These areas were found to become anesthetic after section of the infraorbital nerves, while the interior of the mouth, lower lip, and chin were not affected.

When nerve-sectioned 1-week-old pups were returned to their dams, they all lost weight progressively, and none survived a week. Pups studied soon after recovery from surgery were observed closely on an anesthetized dam and were found unable to attach to her teats following denervation surgery. Initially they reacted normally to the dam, with intense behavioral arousal, orienting to her ventrum and vigorously scrambling and probing into her fur. Neither of these responses had been present in anosmic pups. The deficit of infraorbital-lesioned pups was in the phase of the attachment sequence more proximal to the teat: They failed to open their mouths and to lick when probing the teat itself, and they failed to grasp it and suck. One-week-old pups, in addition, showed wide, uncontrolled head-sweeping movements and occasionally fell from the ventrum in poorly coordinated frantic scrambling.

These results established that 1-week-old pups require tactile as well as olfactory information for suckling. Both these modalities are apparently necessary for nipple attachment. Without them the pup is unable to use the sensory information from the tongue and intraoral area to establish a seal around the teat or to initiate sucking. Olfactory and tactile senses are used at different stages of the behavioral sequence by which the infant attaches to the mother. Olfaction mediates behavioral arousal and provides distal cues for orientation to the nipple-bearing region of the maternal ventrum, and more proximal odor cues elicit probing with the nose into the ventral fur. At this point, tactile stimulation from contact of the snout with the ventrum and the teat takes over, mediating an inhibition of locomotor "search" behavior: rapid, highly focused probing, licking, and finally grasping of the teat. This stage of proximal cue detection cannot be bypassed in the 1-week-old, even if the experimenter places the teat deep in the pup's mouth so that it fully stimulates intact tongue, lower lip, and interior oral receptive fields. Snout-denervated pups at this age simply let the teat fall from their mouths. We can conclude either that the pup must pass through the mouthing, licking, and grasping phase for the behavior sequence to continue or that the upper lips and nares must normally convey continuous information that is crucial for recognition that what is in the mouth is indeed a teat. In the absence of this information after infraorbital nerve section, the pup treats the teat as if it were a stick of wood or a blade of grass, although continuing to respond to the dam's ventrum as a highly exciting object.

Thus, selective sensory lesions have given us new insight into the different processes by which the 1-week-old pup uses sensory information in order to perform a single, smooth behavioral sequence. We are currently investigating the final stages of the sequence by selective lesions of nerves innervating the tongue, oral cavity, and lower lip.

We could next ask questions about how such a sensory information processing system develops. Are pups born with these same sensory dependencies? Can older pups learn to compensate for loss of one or another modality? As described, when infraorbital nerve section was done 7 days after birth, no pups were able to obtain enough milk from their dams to survive for a week. We expected newborns to succumb even more quickly. However, 7 of 10 neonates given the same operation within 12 h of birth survived, gaining weight steadily from the first day of the 1-week observation period, although at a much reduced rate than their control-operated littermates. Apparently, the crucial role of the infraorbital nerve is acquired during the first few days after birth. Once acquired, loss of this sensory pathway is fatal until later-maturing systems allow compensatory adaptations (see below). It is our current hypothesis that infraorbital-lesioned neonates utilize the intact sensory field of chin and lower lip to carry out the sensory information processing role normally acquired by the snout, upper lip, and vibrissae.

Results with 12-day-olds showed evidence of a different form of adaptation. Infraorbital nerve-lesioned pups at this age all showed severe weight loss in the first two days, but then recovered and showed a rate of weight gain that paralleled their control-operated littermates. We found no evidence of nerve regeneration. How then did they accomplish this? We found that recovered pups showed greatly increased rates of licking when first placed on the ventrum. This behavior replaced probing with the snout. When the tongue contacted the teat, they rapidly focused on it, took it into their mouths, and began sucking. This adaptation was different in timing, form, and mechanism from that seen in nerve-lesioned neonates.

Finally, results of infraorbital nerve section were studied in pups that obtained their nutrient by free feeding rather than nursing. Two groups of 17-day-old pups were compared—a group that had been weaned to solid food and a water bottle and another group that were housed in a cage that allowed mothers, but not pups, access to solid food and water. Both groups were observed over the next 7 days, one group required to obtain all nourishment by nursing and the other by free feeding. Only the nursing group showed weight loss following nerve section. This demonstrated that the sensory processes involved in the early stages of free feeding on solid food are different from those involved in nursing. Clearly, the developmental transition to free feeding involves a major shift in sensory information processing, the neural basis for which can now be explored.

In these studies, microsurgery was used to reveal unexpected roles for sensory innervation of a specific skin area in the control and integration of a complex behavioral search program, roles that changed in at least four steps between birth and weaning. Behavioral study of adaptations of the pups to loss of sensory information from the altered area at different ages can help us explore the development of motivation and of plasticity in the control of behavior during early life. Finally, this use of microsurgery provides a new way to discover processes within the observable interaction between parent and infant that control and regulate the relationship.

CONCLUSIONS AND PROSPECTS

The evolution of the operating stereoscopic microscope has literally opened our eyes to a vast array of possibilities for experimental research on early developmental processes in physiology and behavior. For the experimental surgeon, the microscope turns an infant rat into an adult dog. Studies that before were possible only in large infants like sheep and monkeys can now be done on more convenient and rapidly developing species with multiple births such as the rat. The results of these studies are changing our concept of development by revealing unexpected processes in the behavior and physiological organization of immature forms. As illustrated in our studies of nipple attachment and feeding behavior, the role of specific neural input in this behavior differs sharply at four different ages of nursing rats. Many unusual forms of physiological regulation have also been found to take place within the early social relationship of pup and dam, each unique to a given stage. Early development now appears to consist of a series of highly specialized adaptations within the environment provided by a social relationship that is gradually evolving along a relatively predictable path. Studying the development of an organism over time has become more like a comparative study of different species. The challenge for the future is to understand the mechanisms of the transformations that take place along this changing trajectory.

Microsurgery has not yet been widely applied to the study of early developmental processes, and the prospects for its role in this area are impressive. Its potential for the future includes: (1) implanting measuring devices for functions that would otherwise be unobservable, often in the freely behaving organism; (2) conducting analytic physiological studies that interrupt, reroute, and transplant control pathways in order to reveal the mechanisms by which experiences alter the young organism; and (3) discovering unexpected behavioral processes that guide development, hidden within the observed behavioral interactions of the infant with its parents and early social companions; and (4) applying these basic approaches to the study of the early development of genetic animal models of human disease that are emerging at an accelerated pace as a result of genetic engineering.

The optics of the microscope have now improved to the point that the limiting factor is the size of the surgical instruments available. This applies not only to the operating instruments such as scissors, probes, needles, and sutures but even more importantly to the cannula tubes, electrodes, flow meter sensors, and other devices of behavioral physiology. For example, the analysis of motor control by electromyogram cannot be done in rodent infants until flexible, coated wires are produced that are much smaller in diameter than those now available. Small diameters of cannulas and wires can be achieved, but inevitably other necessary qualities such as flexibility, tensile strength, and resistance to destruction by tissue fluids are lost. Finally, the surgeon's hand control under high magnification is a limiting factor. Available electronic micromanipulators are capable of only a few simple movements and do not provide tactile feedback. What is needed is a device that transforms normal finger movements to microforceps with reduced amplitude and force but with retention of feedback to the surgeon's fingers. Such a device may be forthcoming from research in robotics. The electronics industry has been very successful in miniaturization, and a similar future is clearly in store for experimental surgery in developmental psychobiology. But in the meanwhile a multitude of important questions are waiting to be asked about early developmental processes for which the microsurgical methods at hand are well suited.

ACKNOWLEDGMENTS

The author is supported by a Research Scientist award and a research project (MERIT) award from the National Institute of Mental Health.

REFERENCES

Acland, R. D. (1990). *Microsurgery practice manual* (2nd ed.). St. Louis: Mosby.

Blass, B. M., Teicher, M. H., Cramer, C. P., Bruno, J. P., & Hall, W. G. (1977). Olfactory, thermal, and tactile controls of suckling in preauditory and previsual rats. *Journal of Comparative and Physiological Psychology, 91*, 1248–1260.

Farris, E. J., & Griffith, J. O., Jr. (1963). *The rat in laboratory investigation* (2nd ed.). New York: Hafner Publishing Co.

Gilman, A. G., Goodman, L. S., Rall, T. W., & Murad, F. (1985). *The Pharmacological Basis of Therapeutics* (7th ed.). New York: Macmillan.

Greene, E. C. (1963). *Anatomy of the rat*. New York: Hafner Publishing Co.

Hebel, R. E., & Stromberg, M. W. (1986). *Anatomy and embryology of the laboratory rat*. Worthsee: BioMed Verlag.

Hofer, M. A. (1983). The mother-infant interaction as a regulator of infant physiology and behavior. In L. A. Rosenblum & H. Moltz (Eds.). *Symbiosis in parent-offspring interactions* (pp. 61–75). New York: Plenum.

Hofer, M. A. (1984). Early stages in the organization of cardiovascular control. *Proceedings of the Society of Experimental Biological Medicine, 175,* 147–157.

Hofer, M. A. (1985). Nutrient control of cardiac rate in infant rats: Role of arterial baroreceptors. *American Journal of Physiology, 249* (*Regulatory Integrative Comparative Physiology, 18*), R443–R448.

Hofer, M. A. (1987). Shaping forces within early social relationships. In N. Krasnegor, E. Blass, M. Hofer, & W. Smotherman (Eds.), *Perinatal development: A psychobiological perspective.* New York: Academic Press.

Hofer, M. A., Fisher, A. & Shair, H. (1981). Effects of infraorbital nerve section on survival, growth, and suckling behaviors of developing rats. *Journal of Comparative and Physiological Psychology, 95,* 123–133.

Hofer, M. A., Shair, H., & Singh, P. (1976). Evidence that maternal ventral skin substances promote suckling in infant rats. *Physiology and Behavior,* 131–136.

Jacquin, M. F., & Zeigler, H. P. (1983). Trigeminal orosensation and ingestive behavior in the rat. *Behavioral Neuroscience, 97*(1), 62–97.

Kraus, A. L. (1980). Research methodology. In H. J. Baker, J. R. Lindsey, & S. H. Weisbroth (Eds.). *The laboratory rat*: Vol. II. *Research applications.* New York: Academic Press.

Olszewski, W. L. (1984). *CRC handbook of microsurgery*: Vol. I and II. Boca Raton, FL: CRC Press.

Phifer, C. B., & Terry, L. M. (1986). Use of hypothermia for general anesthesia in preweanling rodents. *Physiology and Behavior, 38,* 887–890.

Rouger, Y., Tobach, E., & Schneirla, T. C. (1967). Development of olfactory function in the rat pup. *American Zoologist, 7,* 792–793.

Singh, P. J., Tucker, A. M., & Hofer, M. A. (1976). Effects of nasal ZnSO4 irrigation and olfactory bulbectomy on rat pups. *Physiology and Behavior, 17*(3), 373–382.

Stickrod, G., Soyke, J., & Weeks, J. R. (1984). Cyanoacrylate ester adhesive: A versatile tool in experimental surgery. *Physiology and Behavior, 32*(4), 695–696.

Teicher, M. H., & Blass, E. M. (1976). Suckling in newborn rats: Eliminated by nipple lavage, reinstated by pup saliva. *Science, 193,* 422–425.

NOTE: As this book was going to press (February 1991), the first volume of *Manual of Microsurgery on the Laboratory Rat*, edited by von Dongen, J. J., Remie, R., & van Wunnik, G. H. J., was announced by Elsevier Science Publishers, New York City. This book contains drawings and descriptive text for a variety of cannulation procedures in adult rats, as well as general principles.

3

New Information About Early Feeding and Motivation: Techniques for Recording Sucking in Infant Rats

STEPHEN C. BRAKE

Feeding interactions between mother and infant consist of suckling and nursing behaviors (Hall, Hudson, & Brake, 1988). Suckling is the behavior of the infant that contributes to the procurement of milk from the mother's teat. Nursing is the physiological and behavioral activity of the mother that promotes milk transfer. Both suckling and nursing are essential to the survival of altricial mammals. They are, first and foremost, the means by which nutrient is provided to dependent young. They also encourage physical contact between mother and offspring, which allows the transfer of life-sustaining heat to the young (Leon, Adels, & Coopersmith, 1985; Leon, Croskerry, & Smith, 1978). Physical contact may also play a vital role in stimulating the infants' growth (Schanberg, Evoniuk, & Kuhn, 1984). Finally, feeding interactions provide a context for important early learning (Amsel & Stanton, 1980; Brake, Shair, & Hofer, 1988; Galef & Henderson, 1972; Gubernick & Alberts, 1984; Kehoe & Blass, 1986; Pedersen, Williams, & Blass, 1982; Sullivan, Hofer, & Brake, 1986).

Clearly, the constellation of behaviors that comprise early feeding is just as deserving of study as adult feeding behaviors. However, these behaviors are also somewhat more difficult to study. Over the past few years investigators have employed a variety of techniques that were designed to allow inferences to be drawn about neonatal feeding behavior in rats. Few, however, focused directly on the young rats' primary feeding responses. For example, many studies employed "before-and-after" techniques. These techniques addressed such questions as, How much did the rat pup weigh before suckling from the dam's teat and how much did it weigh after suckling? The answer to this question was intended to reveal, indirectly, how much milk the pup had consumed while attached to the teat, and, by inference, something about the mechanics of sucking and milk withdrawal. In such studies, however, little attention was paid to what the pup actually did while attached to the teat. Other studies employed "latency" techniques, which addressed such questions as, Does the pup attach to a teat more quickly if deprived of milk than if not? One could infer, from the answer to this question, something about the pup's motivation to feed. Again, however, no attention was given to how the pup behaved once it was actually attached to the teat and engaged in the work of withdrawing milk.

Here I describe methods we have developed for studying directly the rat pups' sucking behavior. The chapter illustrates how the techniques we employed in our investigations allowed us to rely less on inference and therefore revealed new information about what pups do during the time they are attached to the dam's teats. I also show how this information subsequently influenced our thinking about theoretical constructs, such as "motivation."

WHAT ARE THE BEHAVIORS INVOLVED IN THE TRANSFER OF MILK?

The transfer of milk from dam to pups involves a series of reciprocal behaviors. Behavior of the dam results in responses from the pup, which then affect the behavior of the dam, and so on. The result is an intricately balanced symbiotic system (Brake et al., 1988; Alberts & Gubernick, 1983). However, in thinking of ways in which to capture and record the interactions, it is easier initially to consider the component parts separately (see Table 3.1).

Table 3.1 Sequence of nursing and suckling behaviors

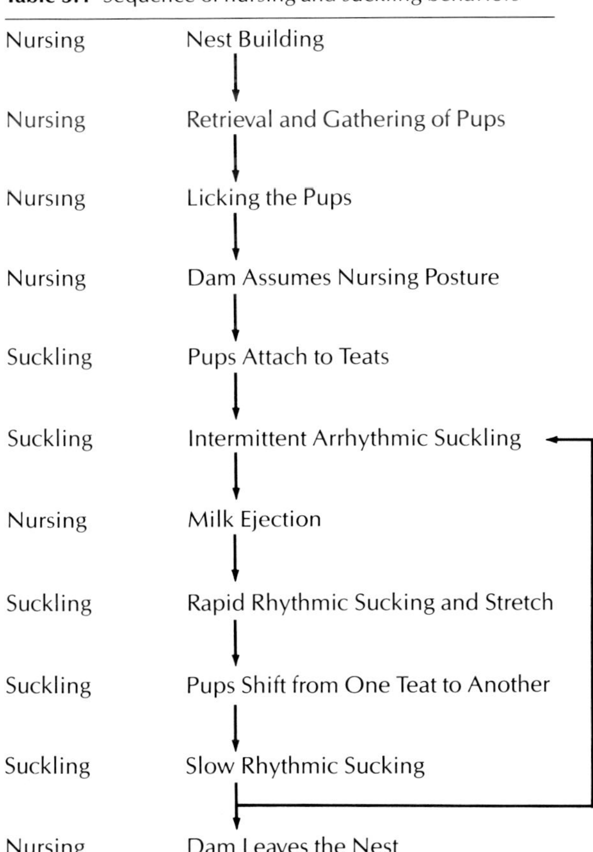

Nursing Behaviors

Nursing (maternal) behaviors that contribute to the transfer of milk include nest building and retrieving, licking, assuming the nursing posture (crouching), and releasing milk. Retrieving or gathering the pups together in the nest precedes the transfer of milk. The dam rarely will assume the nursing posture without first having assembled most of her pups in the nest. This is not to say a pup or two out of a litter of 10–15 might not miss being retrieved or will not stray after a bout has begun, but generally most will be gathered before a bout begins. During the first few days of a litter's life, licking seems to play a crucial part in the feeding interaction. When the dam licks her ventrum she deposits on her nipples a chemical contained within her saliva. The odor of this chemical acts as a beacon that guides the pups' probing of the ventrum toward a nipple and allows them to attach (Singh & Hofer, 1978; Teicher & Blass, 1976). By assuming the nursing position, which usually consists of crouching over the pups, the dam displays her entire ventrum to the searching pups and allows all or most of them to settle on a nipple. Sometimes what began as a crouch may end with the dam almost lazily reclined on her flank, but even in these cases most nipples are displayed and available.

Shortly after assuming the nursing position, and after the pups have attached to the teats, the dam lapses into slow-wave sleep (Voloschin & Tramezzani, 1979). Within a few, minutes (anywhere from 10 to 20 min), the dam releases milk to the pups (Drewett, 1983; Lincoln, 1983; Shair, Brake, & Hofer, 1984). Milk contained in the mammary glands is let down into the terminal galactophores. This is accomplished through the action of contractile tissue surrounding the glands. The active contractions last from 5 to 15 s. The milk does not, however, "squirt" into the pups' mouths on its own, nor will it seep from an unsuckled teat. The milk must be withdrawn by the pups through active sucking (which is a "suckling" behavior; see next section). It appears that milk is made available at each teat simultaneously. The volume of milk made available at each teat, however, is not precisely known. On average, each letdown results in about 0.1–0.3 ml of milk, as assessed by weighing suckling pups immediately before and after milk letdowns. There may be considerable variance in the amount of milk released from one teat to another or from one letdown to another. For example, dams deprived of suckling pups for a few hours often release greater amounts of milk upon reunion with pups than dams that have not been deprived of pups. However, the amount of variance from situation to situation, or teat to teat, has never been carefully documented. The second milk letdown may occur anywhere from 2 to 20 min following the first, and subsequent letdowns are generally spaced from 5 to 15 min apart. Nursing bouts are more frequent and last longer when the pups are younger than when they are older, but, generally, a bout may last anywhere from 15 to 60 min, so that 2 to 8 milk letdowns may occur during a bout.

Suckling Behaviors

Suckling behaviors (behaviors of the pup) include locating the nipple, attaching to it, maintaining a "seal" about the nipple once attached, sucking in between milk letdowns, making postural adjustments and "anticipatory" movements immediately preceding milk letdowns, sucking and making postural adjustments during milk letdowns, and shifting from one nipple to another following milk letdowns.

Pups locate the dam's nipples by using a combination of olfactory and tactile cues. Initially, pups are attracted to an odor contained in parturient dam's saliva (and other pups' saliva) that is normally deposited on the nipple (Teicher & Blass, 1977). This olfactory signal guides them to within a few millimeters of the teat. Then they rely on proprioceptive tactile cues to pinpoint the teat's location. These cues are provided by sweeping their snouts across the area of the teat (Hofer, Fisher, & Shair, 1981). Attaching to the teat requires that the pups open their mouths and securely fasten about the nipple. Forelimb movements, or "treadling," usually accompany attachment and may aid the pup in maintaing the proper orientation toward the nipple while grasping it with the mouth (Drewett, Statham, & Wakerley, 1974). As the pup grasps the teat and begins to attach, it exerts 5 to 10 s of vigorous negative pressure (displaying a pattern of sucking we have called *rapid rhythmic sucking*, or RRS; Brake, Tavana, & Myers, 1986; Pelchat & Brake, 1987; see below). Maintaining a seal about the nipple once attached seems to require the presence of the pup's own saliva. The fluid seems to act as a "gasket" and helps ensure a steadfast grip about the teat (Epstein, 1986).

Once attached, the pup does not suck constantly, unless quite hungry or suckling-deprived (Brake, Wolfson, & Hofer, 1979; Brake, Sager, Sullivan, & Hofer, 1982). Instead, the pup sucks periodically between milk letdowns. We have identified two forms of this sucking, which we call *slow rhythmic sucking*, or SRS, and *arhythmic sucking*, or AS (Brake et al., 1979; 1986; Brake & Hofer, 1980). These forms of sucking may serve different functions (see below).

A few seconds prior to milk letdown, the pup begins to treadle, as it did when first attaching to the teat. It also begins to engage is RRS, just as it did when first grasping the teat with its mouth. The eliciting stimulus of this behavior is not the arrival of milk to the pup's mouth. If the tip of the dam's teat is ligated to prevent milk flow, the pup will still treadle. Interestingly, if the base of the teat is ligated, the pup will not treadle. This suggests that engorgement of the teat with milk signals the pup to awaken (the pup is usually asleep while attached to the teat; Shair et al., 1984), treadle, and begin sucking. This may ensure that the pup is able to obtain a maximum amount of milk during the brief milk delivery.

During the milk delivery itself, the pup assumes a position called the *stretch*, which is a rigid extension of all four limbs and an arching of the back (Blass, Hall, & Teicher, 1979; Drewett & Trew, 1978). The pup also engages in RRS for the duration of the milk letdown (Brake et al., 1986; 1988; see also Wakerley & Drewett, 1975). The amount of milk consumed by pups during milk letdown varies from 0.1 to 0.2 ml, as assessed by before-and-after weighings. Deprived pups often consume more milk than sated pups, but not always. Furthermore, teats may vary in the amounts of milk they make available (see above and Phifer, Chap. 11, this volume).

At the completion of the letdown, pups often leave the nipple to which they have been attached, locate another nipple, and attach to it. This is called *nipple shifting* and is more characteristic of older pups than younger ones. When it occurs, all pups generally shift at the same time (Cramer, Blass, & Hall, 1980). Once reattached, the pup usually engages in SRS, sometimes for a few seconds, sometimes for a few minutes, depending on its privational status. Older pups may also tend to nipple shift between milk letdowns, as if searching for a milk-laden teat.

TECHNIQUES THAT HAVE BEEN USED TO ASSESS SUCKLING

Many of the behaviors the pups display while suckling were hidden from us until measures were devised that detected them. Thus, our thinking of how the pup acquires milk and the degree to which the pup may control milk acquisition has changed as our methodologies have changed. For example, we did not know that hungry pups engaged in more forceful sucking during milk delivery than sated pups until we had a technique for measuring RRS. In fact, we did not know of the existence of different patterns of sucking, each with its own function, until we developed techniques for measuring the behavior directly. Instead, it was generally thought that a pup's sucking was similar to a fixed-action pattern that varied little from one circumstance to the next. We have discovered, quite to the contrary, that sucking varies significantly from situation to situation and even from pup to pup. We now know that these variations in sucking are largely responsible for a pup's relative success or failure in obtaining milk.

In this section, I review various ways investigators have assessed suckling behaviors and how our thinking about such issues as motivation and goal-directed responding have been influenced by the assessment techniques that have been used.

Nipple Location and Attachment: Assessing the Value of Milk to the Pup

The pups' ability to locate a teat is usually assessed with a latency measure. Pups are placed with a dam (often anesthetized), and a timer is started. Once a pup has attached to a teat, the timer is stopped. Long attachment latencies may reflect difficulty in locating the teat. For example, if a teat has been thoroughly washed, latencies may be quite long (Singh & Hofer, 1978; Teicher & Blass, 1976). Similarly, if certain of the pup's afferent nerves have been sectioned, long latencies are observed. This is how we came to know that both olfactory and tactile cues are essential in allowing the pup to find and attach to a teat (see Hofer, chap. 2, this volume).

Long latencies, however, may also reflect the pup's motivation or inclination to locate and attach to a teat, not simply its ability to do so. For example, after about 10–15 days of age, pups that have been deprived of the dam display shorter latencies for locating and attaching to a teat than do nondeprived pups (Blass et al., 1979). Furthermore, if the nature of the deprivation period is experimentally altered, we can understand even more about what "motivation" might mean in a suckling context. For example, if during a "deprivation" period pups are allowed to attach to a teat and suck but are prevented from receiving milk, their latencies to attach to a teat during a subsequent timed test are longer than those of pups that were deprived simply by separation from the dam (Brake et al., 1982). Thus, a small change in the methodologies employed in latency tests revealed that the pup's motivation to suckle may be just as strong as its motivation to ingest.

Latency tests of a different kind provide additional information about the perceived value of milk to a suckling pup. When pups are required to traverse alleyways in order to attach and suckle a dam, one can demonstrate that the motivation to feed becomes more important with age. The age at which this becomes noticeable to an investigator, however, depends upon the methodology the investigator employs. For example, older pups run down straight alleyways more quickly if "reward" consists of receiving milk while suckling rather than simply the opportunity to attach to a teat, but younger pups do not. This shift in the value of milk seems to occur around 10 or 11 days of age (Amsel, Letz, & Burdette, 1977). However, changing the methodology slightly may provide a somewhat different conclusion about the age of the shift. Thus, if pups younger than 15 days are allowed a choice between "wet" and "dry" suckling at the end of the alleyway (now a Y-maze), they show no preference between the choices (Kenny, Stoloff, Bruno, & Blass, 1979). When left to draw conclusions about the value of milk from these different sets of experiments, it is difficult to decide at what age milk becomes "rewarding."

Still another methodology suggests that pups younger than 15 days do indeed perceive milk as rewarding. In this procedure, a novel odor was paired with the delivery of milk to suckling pups in a classical conditioning paradigm. The novel odor was the CS, and milk was the US. The goal was to determine if the CS would gain salience as a "secondary reinforcer" through associative learning. This was tested by subsequently giving the pup a choice between the previously novel CS target odor and another, more familiar, odor and observing how much time the pup spent in the presence of both odors. Pups 11–13 days of age clearly associated milk received while suckling with the novel odor and subsequently showed a learned preference for that odor (Brake, 1981; Pelchat & Brake, 1982). These results would seem to indicate that milk is an important reinforcer to suckling pups prior to 15 days of age.

Unfortunately, the results of all of these experiments do not establish how pups react to milk delivery while they are attached to a teat, and so we are left without critical information about the rewarding value of milk to a suckling pup. Furthermore, the experiments do not reveal how pups actually obtain milk by sucking. Can they willfully consume more milk if hungry or otherwise deprived? It seemed to us that assessments of the pup's nipple attachment behavior provided an incomplete picture of the forces at work in the suckling situation. We believed that until we knew how the pup behaved after it attached to the teat, we would not fully know what motivated feeding.

Sucking: Further Assessments of Motivation and the Value of Milk

Techniques that my colleagues and I devised to measure the pup's sucking responses are described in the next few paragraphs. Until our work, sucking responses were generally viewed as rather immutable, innate motor patterns. We found that the information provided by new techniques changed our thinking about the degree to which a suckling infant could be motivated by milk and the degree to which it exercised control over how much milk it obtained.

The Recording Techniques

EMG. We have employed two techniques for the study of sucking in rat pups (Brake et al., 1986; see also Loeb & Gans, 1986). The first is a measure of the electromyographic activity of the pup's jaw muscle, recorded with the use of thin silver-wire electrodes and a polygraph (Brake et al., 1979; Brake & Hofer, 1980). These electrodes, implanted in the digastric muscle, provide a reliable indicator of when the

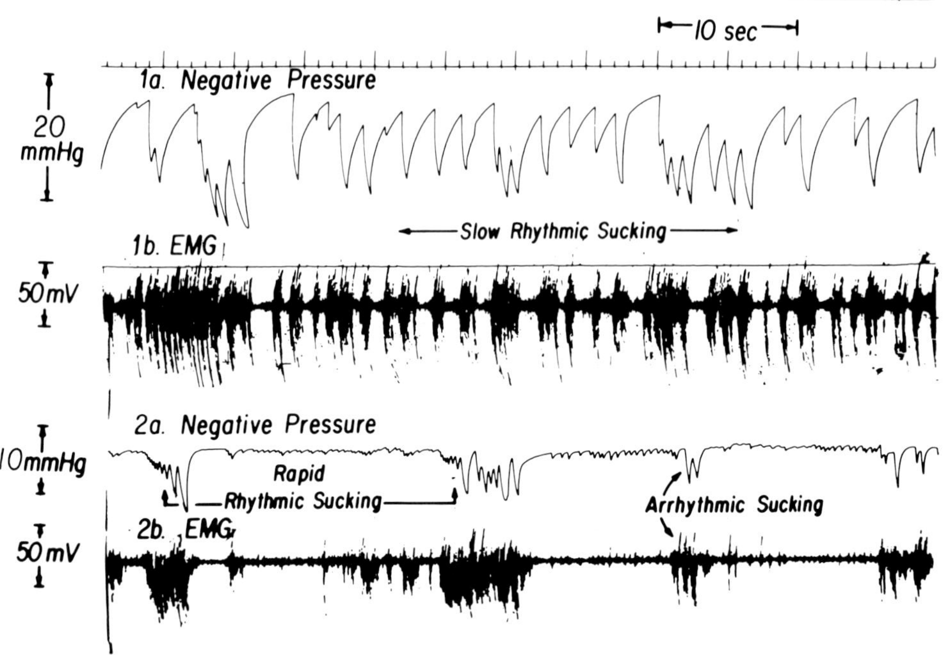

Figure 3.1. Polygraph recordings of sucking. Lines 1a and 2a are recordings of intraoral negative pressure; lines 1b and 2b are electromyographs. Lines 1a and 1b were recorded simultaneously, as were lines 2a and 2b. (From "Sapid Savvy in Sucklings: The Effect of Quinine Hydrochloride on Intraoral Negative Pressure and Intake by 11–13-Day-Old Rat Pups" by M. L. Pelchat and S. C. Brake, 1987, *Developmental Psychobiology, 20*, pp. 261–275. Copyright by John Wiley & Sons. Reprinted by permission of the publisher.)

pup is sucking, but do not provide information about the strength or amplitude of sucking (Figure 3.1).

Prior to the start of a test, pups are anesthetized with ether for the installation of the EMG electrodes. A small incision is made along the midline of the bottom of the pup's jaw, exposing the left and right digastric muscles. Two thin, Teflon-coated silver wires (Medwire, Ag-5T) are twisted together and inserted into one of the muscles. The electrode leads are externalized through a small puncture wound to one side of the incision, and the incision is sutured. The electrodes are held fast with a drop of collodion placed at exit puncture. The procedure requires about 10 min, and pups recover within 30–60 min.

Sucking is usually assessed during a test in which pups are allowed to attach to the teats of an anesthetized dam. Urethane (2 g/10 ml water; 1 ml/100 g) is the anesthetic we have used for long-duration tests. Anesthesia blocks naturally occurring milk letdowns and allows us to program milk delivery by administering oxytocin to the dam (i.v. injections of Pitocin, 0.0015 U) or by infusing milk through a posterior tongue cannula (see below).

Electromyographs are obtained with the use of a Grass Model 7 polygraph. The electrode leads are channeled to a wide-band AC preamplifier and driver amplifier. Optimum sensitivity and time-constant values on the amplifiers are determined through pilot observations and left unchanged during tests. The signals are calibrated so that the maximum range is 20 μ for the EMG signal. Paper speed is set at 5 mm/s.

Intraoral Negative Pressure. The second recording technique involves the use of a cannula (PE-10 polyethylene tubing, Clay Adams) placed at the level of the intermolar eminence on the posterior portion of the pup's tongue, according to the method first described by Hall and Rosenblatt (1978). By connecting the free end of the cannula to a pressure transducer and polygraph, one can detect changes in the intraoral negative pressure created by the pup on the nipple (Brake et al., 1986; Pelchat & Brake, 1987). This technique provides information about when the pup is sucking, as well as how strongly the pup sucks (see Figure 3.1).

The tongue cannula is prepared much as described by Hall and Rosenblatt (1978) and Phifer (see Phifer, chap. 11, this volume). Recordings of sucking are obtained by connecting the free end of the tongue cannula to a Gould-Statham pressure transducer (we used model P23Dc, but newer models are available from Spectramed), which, in turn, is connected to a low-level DC preamplifier and driver amplifier. The signals are calibrated so that the maximum range is 20–30 mm hg. Paper speed is set at 5 mm/s.

Scoring the Polygraph Recordings. EMG records are scored by two trained raters. Each rater independently identifies episodes of electromyographic activity based on criteria for amplitude and duration of pen deflections. A score reflecting total amount of EMG activity is then calculated. Finally, EMG episodes are categorized by type (see below; Brake et al., 1979).

Negative pressure recordings are analyzed similarly. Trained raters evaluate the polygraph tracings and identify episodes of intraoral negative pressure changes based on amplitude and duration of pen deflections. A score reflecting total sucking time is then calculated, and the episodes are categorized by type. In addition, an estimate of total sucking effort is obtained. This measure takes into consideration both the amplitude of the pen tracing and its duration. It is derived by calculating the area under the curves created by the pen deflections. This is done with an image-analyzing device (for example, the Sigma-Scan Measurement System from Jandel Scientific). An electronic stylus is used to trace the hard copy of the polygraph recordings, and a microprocessor converts the area within the tracing to numerical scores. We call the resulting measure *integrated relative pressure* (IRP) because the units are expressed in millimeters squared, not millimeters of hectograms. To give the reader an idea of the intraoral negative pressures that correspond to these numbers, a pup that displayed 10 sucks over 10 s, each with an amplitude measured at 10 mm hg, would receive a score of approximately 2,500 mm^2 (see Brake et al., 1986; Pelchat & Brake, 1987).

Negative pressure data may also be analyzed by computer. We have written and installed several computer programs that allow on-line collection of data from the polygraph using a DEC PDP-11/23 computer system. The data are digitized and stored on disk for later analysis. A separate set of analysis programs serves to partition the data collected from each animal into bins of time (of arbitrary duration) and then derive several summary parameters for each bin. These parameters include total sucking time, average amplitude (negative pressure), average duration each suck, and total number of sucks.

Patterns of Sucking

We are able to identify two basic types of sucking patterns from the electromyograph: arhythmic sucking (AS) and rhythmic sucking (RS; see Figure 3.1). As we have

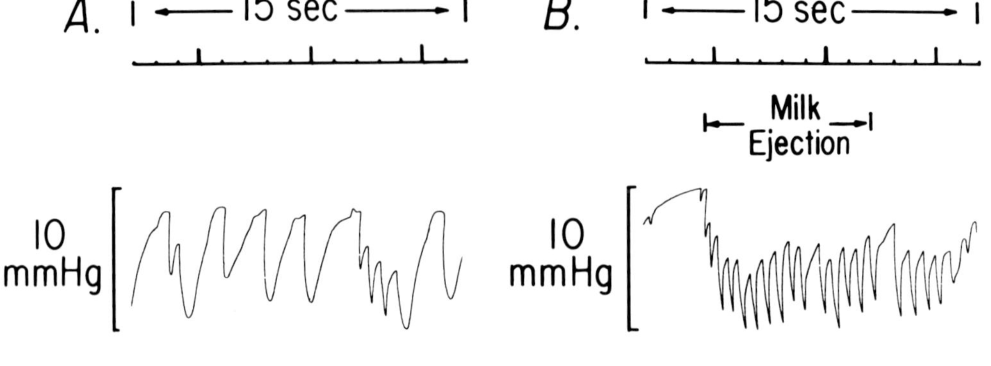

Figure 3.2. Examples of rapid rhythmic sucking (RRS) and slow rhythmic sucking (SRS) from intraoral negative pressure recordings. (From "A Method for Recording and Analyzing Intraoral Negative Pressure in Suckling Rat Pups" by S. C. Brake, S. Tavana, and M. M. Myers, 1986, *Physiology and Behavior, 36*, pp. 575–578. Copyright by Pergamon Press. Reprinted by permission of the publisher.)

described elsewhere (Brake et al., 1986), AS can be described as brief bursts of EMG activity (2–5 s) that occur infrequently (a burst every 10–20 s). RS consists of a series of regularly alternating decreases and increases in EMG activity lasting at least 5 s and often much longer. The correlation coefficient between raters for identifying episodes of AS is .76; for RS it is .92.

We are able to distinguish at least three types of sucking patterns with the use of the negative pressure recordings. As mentioned previously, one of these patterns is AS, similar to that identified by the EMG measure. AS sucking consists of 1–2 sucks (negative pressure peaks) of relatively long duration (1–2 s each) within a span of 2–5 s (see Figure 3.1). The other two patterns are both rhythmic sucking patterns (see Figure 3.2). The first of these, as discussed above, we call *slow rhythmic sucking* (SRS). This pattern is very similar to the RS pattern identified by the EMG measure. It consists of a series of relatively long-duration sucks (each suck lasting at least 1 s) with no more than 2 s elapsing between each two sucks in the series. Each series lasts a minimum of 5 s, but might last much longer (up to several minutes). The second rhythmic pattern we call *rapid rhythmic sucking* (RRS). RRS consists of a series of short-duration sucks (each suck less than 1 s in duration) with no more than 1 s elapsing between individual sucks. These episodes can be as short as 3 s, but can last as long as 15 s.

In most cases, the correlation between patterns identified by the EMG technique and the negative pressure technique is quite good. For example, in pups from which both measures are taken simultaneously, the overall correlation coefficient between the two measures is .82 (Brake et al., 1986). However, the EMG measure usually produces a tracing characteristic of AS when RRS occurs, and so does not distinguish AS from RRS. The negative pressure technique also provides information about the force of sucking, which the EMG technique does not. This is illustrated in Figure 3.3. These segments of sucking are of equal length and similar patterning but vary greatly in amplitude. The maximum negative pressure exerted in the "strong" example is

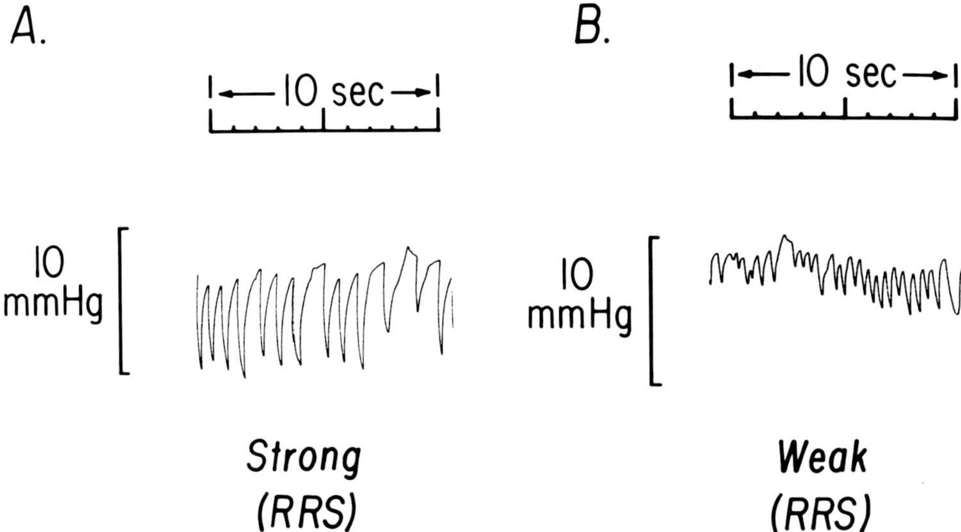

Figure 3.3. Examples of RRS that vary in amplitude. (From "A Method for Recording and Analyzing Intraoral Negative Pressure in Suckling Rat Pups" by S. C. Brake, S. Tavana, and M. M. Myers, 1986, *Physiology and Behavior, 36,* pp. 575–578. Copyright by Pergamon Press. Reprinted by permission of the publisher.)

9.5 mm hg while in the "weak" example it is 6.5 mm hg. IRP analysis reveals that differences in the average pressures exerted is even greater. The IRP for 10 s of the "strong" example was 2,480; for 10 s of the "weak" example, 990.

The disadvantage of the negative pressure technique is that it introduces a foreign presence (the cannula) into the pup's mouth while it is sucking. We have not been able, however, to detect significant differences in sucking or other suckling behaviors between pups equipped with the tongue cannula or the EMG electrodes.

Functions of the Different Sucking Patterns: New Information About Regulation of Intake

The pup engages in 10–20 s or RRS as it attaches to the teat. The function of this sucking may be to grasp the teat firmly during the jostling among littermates and dam that occurs during times of nipple attachment. One should consider, too, that pups often detach from one teat and attach to another immediately following a milk ejection. Thus, the RRS displayed when the pup attaches may also function to withdraw any residual milk in the ducts of the teat.

Once attached, it may be as long as 5–20 min before a milk ejection occurs, and even longer if the nursing bout has just begun (Shair et al., 1984). During these intervals, pups may display a number of different sucking patterns. Pups that have not been deprived of milk (or not deprived of the opportunity to attach to a teat) display a burst of AS every 20–60 s. This is a very languid form of sucking, often exhibited while the pups are asleep (Shair et al., 1984) and may be analogous to a muscle "twitch." Its function, if any, is not entirely clear. One possibility is that AS may be one way pups communicate to the sleeping dam that they are still present and attached (dams will not let down milk if pups are not suckling at her ventrum).

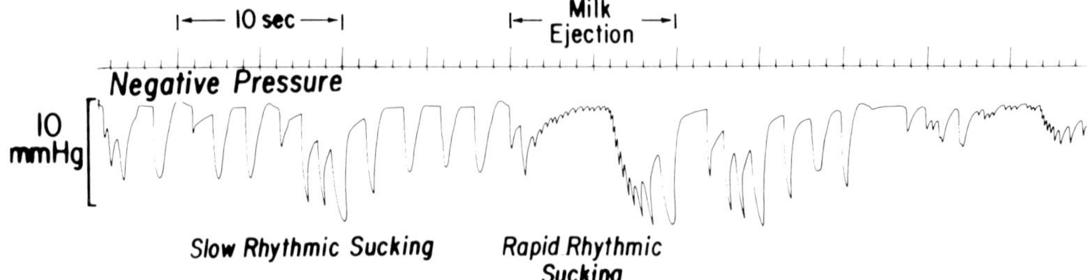

Figure 3.4. Sucking before, during, and after a milk ejection. (From "A Method for Recording and Analyzing Intraoral Negative Pressure in Suckling Rat Pups" by S. C. Brake, S. Tavana, and M. M. Myers, 1986, *Physiology and Behavior, 36,* pp. 575–578. Copyright by Pergamon Press. Reprinted by permission of the publisher.)

Hungry pups, on the other hand, are likely to engage in lengthy periods of SRS. This is a very active and effortful form of sucking and is not usually seen while the pup is asleep (Brake et al., 1988; Shair et al., 1984). These bouts of SRS can be sustained, without interruption, for several minutes. SRS may facilitate the withdrawal of residual milk from the ducts of the teat. SRS may also be the pups' way of communicating their privational status to the dam, and, perhaps, shorten the interval between milk ejections (although this has not been demonstrated).

Pups begin to engage in RRS at the onset of milk letdown. This sucking continues throughout the duration of the delivery (and the accompanying "stretch" response; see Figure 3.4). The function of this sucking is to withdraw milk from the teat. A strong correlation exists between sucking amplitude and consumption of milk during milk ejection (Brake et al., 1988).

The changes in sucking pattern and force between sated and hungry pups reflect differences in privational status (or motivation) that are often not reflected by other measures. For example, 11–13-day-old pups spend equivalent amounts of time attached to a teat whether food-deprived or sated, even though they display extremely different sucking patterns between milk ejections (Brake et al., 1979, 1982, 1988). Without the use of the polygraph recording techniques, this difference in the pups' feeding behaviors would not have been detected. One might have concluded, without the tracings of sucking, that pups of this age simply are not affected by deprivation. Similarly, we have found that sated pups suck less forcefully than do hungry pups during milk deliveries (Brake et al., 1988). Again, this is particularly interesting since we have observed no differences between the two groups of pups in the amount of time they spend attached to the teat. Although it is likely that pups sometimes adjust intake by leaving or staying attached to a nipple (Cramer & Blass, 1983), our results support the notion that they control intake more often by varying RRS. This conclusion could not have been reached without examining the sucking patterns of the pups.

Individual Differences in Sucking: New Information About Feeding in Response to Privational Challenge

Polygraph recordings of sucking reveal that pups display striking individual differences in feeding from the teat. One variable that reliably increases individual differences in the amount and amplitude of sucking displayed is privational status of the pups. For example, the variance in sucking amplitude during milk delivery is as much as three times greater in deprived pups than in sated pups (Brake et al., 1988). This is due to the fact that while most deprived pups engage in vigorous RRS, some do not. To be sure, the number of deprived pups displaying robust RRS is much greater than the number of sated pups displaying robust RRS (as demonstrated by the shift in mean amplitude of sucking between the two groups). And yet a surprising number of deprived pups do not display strong RRS. A few deprived pups actually display weaker RRS during milk delivery than do sated pups. This may mean that while many pups can compensate for a deprivational challenge quite well, some cannot. These findings stand as a reminder that when the equilibrium of a system is disturbed (such as when pups are deprived of food) individuals will inevitably respond in different ways to the challenge. Still, we were impressed with the magnitude of individual variation displayed by deprived pups (see below).

The fact that individual pups may differ greatly in their sucking suggests that a pup's rate of growth may depend largely on its competence as a feeder. Pilot studies seem to support this hypothesis. We have correlated measures of sucking taken at 2 days of age with subsequent rate of growth and found that many of these correlations are positive and significant. For example, the correlation between mean amplitude of RRS during milk delivery and grams gained per day is $r = .52$, $p < .05$, and the correlation between mean sucking amplitude (for the entire test) and grams gained per day is $r = .71$, $p < .01$. These data indicate that a pup's rate of growth can be predicted at an early age by analyzing one aspect of the pup's suckling behavior. This is a possibility we would not have entertained without the subtle power provided by the intraoral negative pressure recordings.

CONCLUSIONS: HOW NEW TECHNIQUES PROVIDED NEW INFORMATION ABOUT EARLY FEEDING

The use of the polygraphic recording techniques has changed the way we think about infantile feeding and motivation. Had we relied on other measures, such as latency to attach to the nipple, time spent attached to a nipple, amount of milk consumed, and so forth, we might not have realized how privational status affects pups in such profoundly different ways. We might have assumed that a pup's sucking responses are fixed and invariable. Or we might have assumed that ingestion is regulated only by the mother's supply of milk or by the amount of time pups spent attached to a teat. Or we might have assumed that pups vary little in their responses to privational challenges.

Instead, we now know that the strength of a pup's sucking plays a large role in regulating ingestion. We also know that pups differ greatly in their response to privational challenge and that a pup's rate of growth might be predicted by individual differences in the strength of sucking.

A final anecdote may serve to summarize these points. In the course of our studies, we, like many other investigators in the area, were bewildered time and time again by the odd and annoying fact that hungry pups and sated pups often consumed about the same amounts of milk, on average, from lactating dams, as assessed by before-and-after weighings of the pups. Perhaps this should not have been surprising since hungry and sated pups often spend equivalent amounts of time attached to teats, and teats may vary in the amount of milk they deliver. Still, we were unwilling to accept that milk was of so little importance to hungry and growing pups, or that intake is so strictly limited by the dam. We wondered whether our failure to demonstrate increased consumption in hungry pups was due to use of poor methodologies: Determining intake by repeated weighings is fraught with difficulties and interrupts the natural feeding sequence. Furthermore, we could not know if pups were attached to "good" teats or "bad" teats. All true, perhaps, but it was not until we were able to examine the record of a pup's sucking profile that we were forced to the realization that some pups just do not suck very well in response to the challenge of deprivation. Thus, those hungry pups did not consume much more than sated pups. On the strength of this and similar findings, we concluded that a pup's motivation to feed does indeed affect the amount of milk it consumes. This simple fact had been obscured by our inability to examine closely the extremely different ways that individual pups responded to motivational challenge. The development of new techniques for recording sucking provided the means for such careful examination.

ACKNOWLEDGMENTS

Part of the work described in this chapter was conducted while the author was with the Department of Psychiatry at Montefiore Medical Center, and part was conducted at the Department of Developmental Psychobiology at the New York State Psychiatric Institute, College of Physicians and Surgeons, Columbia University. Portions of this work were funded by Grant MH32429 from the NIMH. The author wishes to thank Marcia Pelchat, Harry Shair, Michael Myers, Patricia Kenney, Regina Sullivan, and Myron Hofer for their assistance and support.

REFERENCES

Alberts, J. R., & Gubernick, D. J. (1983). Reciprocity and resource exchange: A symbiotic model of parent-offspring relations. In H. Moltz and L. A. Rosenblum (Eds.), *Symbiosis in parent-offspring interactions*. New York: Plenum.

Amsel, A., Letz, R., & Burdette, D. R. (1977). Appetitive learning and extinction in 11-day-old rat pups: Effects of various reinforcement conditions. *Journal of Comparative and Physiological Psychology, 91*, 1156–1167.

Amsel, A., & Stanton, M. (1980). Ontogeny and phylogeny of paradoxical reward effects. In J. S. Rosenblatt, R. A. Hinde, C. Beer, & M. Busnel (Eds.), *Advances in the study of behavior*. New York: Academic Press.

Blass, E. M., Hall, W. G., & Teicher, M. H. (1979). The ontogeny of suckling and ingestive behaviors. *Progress in Psychobiology and Physiological Psychology, 8*, 243–299.

Brake, S. C. (1981). Suckling infant rats learn a preference for a novel olfactory stimulus paired with milk delivery. *Science, 211*, 506–508.

Brake, S. C., & Hofer, M. A. (1980). Maternal deprivation and prolonged suckling in the absence of milk alter the frequency and intensity of sucking responses in neonatal rat pups. *Physiology and Behavior, 24*, 185–189.

Brake, S. C., Sager, D. J., Sullivan, R., & Hofer, M. A. (1982). The role of intra-oral and gastrointestinal cues in the control of sucking and milk consumption in rat pups. *Developmental Psychobiology, 15*, 529–541.

Brake, S. C., Shair, H., & Hofer, M. A. (1988). Exploiting the nursing niche: The infant's sucking and feeding in the context of the mother-infant interaction. In E. M. Blass (Ed.), *Handbook of behavioral neurobiology*, Vol. 9. New York: Plenum.

Brake, S. C., Tavana, S., & Myers, M. M. (1986). A method for recording and analyzing intra-oral negative pressure in suckling rat pups. *Physiology and Behavior, 36,* 575–578.

Brake, S. C., Wolfson, V., & Hofer, M. A. (1979). Electromyographic patterns associated with nonnutritive sucking in 11–13-day-old rat pups. *Journal of Comparative and Physiological Psychology, 93,* 760–770.

Cramer, C. P., & Blass, E. M. (1983). Mechanisms of control of milk intake in suckling rats. *American Journal of Physiology, 245,* 154–159.

Cramer, C. P., Blass, E. M., & Hall, W. G. (1980). The ontogeny of nipple-shifting behavior in albino rats: Mechanisms of control and possible significance. *Developmental Psychobiology, 13,* 165–180.

Drewett, R. F. (1983). Sucking, milk synthesis, and milk ejection in the Norway rat. In R. W. Elwood (Ed.), *Parental behavior in rodents.* New York: Wiley.

Drewett, R. F., Statham, C., & Wakerly, J. B. (1974). A quantitative analysis of the feeding behavior of suckling rats. *Animal Behaviour, 22,* 907–913.

Drewett, R. F., & Trew, A. M. (1978). The milk ejection of the rat, as a stimulus and a response to the litter. *Animal Behaviour, 26,* 982–987.

Epstein, A. N. (1986). The ontogeny of ingestive behaviors: Control of milk intake by suckling rats and the emergence of feeding and drinking at weaning. In R. Ritter, S. Ritter, & C. D. Barnes (Eds.), *Neural and humoral controls of food intake.* New York: Academic Press.

Galef, B. G., & Henderson, P. W. (1972). Mother's milk: A determinant of the feeding preferences of weanling rat pups. *Journal of Comparative and Physiogical Psychology, 78,* 213–219.

Gubernick, D. J., & Alberts, J. R. (1984). A specialization of taste aversion learning during suckling and its weaning-associated transformation. *Developmental Psychobiology, 17,* 613–628.

Hall, W. G., Hudson, R., & Brake, S. C. (1988). Terminology for use in investigations of nursing and suckling. *Developmental Psychobiology, 21,* 89–91.

Hall, W. G., & Rosenblatt, J. S. (1977). Suckling behavior and intake control in the developing rat. *Journal of Comparative and Physiological Psychology, 91,* 1232–1247.

Hall, W. G., & Rosenblatt, J. S. (1978). Development of nutritional controls of food intake in suckling rat pups. *Behavioral Biology, 24,* 413–427.

Hofer, M. A., Fisher, A., & Shair, H. (1981). Effects of intraorbital nerve section on survival, growth, and suckling behaviors of developing rats. *Journal of Comparative and Physiological Psychology, 95,* 123–133.

Kehoe, P., & Blass, E. M. (1986). Conditioned aversions and their memories in 5-day-old rats during suckling. *Journal of Experimental Psychology: Animal Behavior Processes, 12,* 40–47.

Kenny, J. T., Stoloff, M. L., Bruno, J. P., & Blass, E. M. (1979). Ontogeny of preference for nutritive over nonnutritive suckling in albino rats. *Journal of Comparative and Physiological Psychology, 93,* 752–759.

Leon, M., Adels, L., & Coopersmith, R. (1985). Thermal limitation of mother-young contact in Norway rats. *Developmental Psychobiology, 18,* 85–105.

Leon, M., Croskerry, P. G., & Smith, G. K. (1978). Thermal control of mother-infant contact in rats. *Physiology and Behavior, 21,* 793–811.

Lincoln, D. W. (1983). Physiological mechanisms governing the transfer of milk from mother to young. In L. A. Rosenblum & H. Moltz (Eds.), *Symbiosis in parent-offspring interactions.* New York: Plenum.

Loeb, G., & Gans, C. (1986). *Electromyography for experimentalists.* Chicago: University of Chicago Press.

Pedersen, P., Williams, C. L. & Blass, E. M. (1982). Activation and odor conditioning of suckling behavior in 3-day-old albino rats. *Journal of Experimental Psychology: Animal Behavior Processes, 8,* 329–341.

Pelchat, M. L., & Brake, S. C. (1982). Aromatic associations: Studies of olfactory conditioning in suckling rat pups. Paper presented at the annual meeting of the International Society for Developmental Psychobiology, Minneapolis, MN.

Pelchat, M. L., & Brake, S. C. (1987). Sapid savvy in sucklings: The effect of quinine hydrochloride on intraoral negative pressure and intake by 11–13-day-old rat pups. *Developmental Psychobiology, 20,* 261–275.

Schanberg, S. M., Evonuik, G., & Kuhn, C. M. (1984). Tactile and nutritional aspects of maternal care: Specific regulators of neuroendocrine function and cellular development. *Proceedings of the Society of Experimental Biology and Medicine, 175,* 135–146.

Shair, H., Brake, S. C., & Hofer, M. A. (1984). Suckling in the rat: Evidence for patterned behavior during sleep. *Behavioral Neuroscience, 96,* 366–370.

Singh, P., & Hofer, M. A. (1978). Oxytocin reinstates maternal olfactory cues for nipple orientation and attachment in rat pups. *Physiology and Behavior, 20,* 385–389.

Sullivan, R. M., Hofer, M. A., & Brake, S. C. (1986). Olfactory-guided orientation in neonatal rats is enhanced by a conditioned change in behavioral state. *Developmental Psychobiology, 19,* 615–623.

Teicher, M. H., & Blass, E. M. (1976). Suckling in newborn rats: Eliminated by nipple lavage, reinstated by pup saliva. *Science, 193,* 422–425.

Teicher, M. H., & Blass, E. M. (1977). First suckling response of the newborn albino rat: The roles of olfaction and amniotic fluid. *Science, 198,* 635–636.

Voloschin, L. M., & Tramezzani, J. H. (1979). Milk ejection reflex linked to slow wave sleep in nursing rats. *Endocrinology, 105,* 1202–1207.

Wakerley, J. B., & Drewett, R. F. (1975). Pattern of sucking in the infant rat during spontaneous milk ejection. *Physiology and Behavior, 15,* 277–281.

4

Physiological Recording During Normal Mother-Infant Interactions

HARRY N. SHAIR

At the beginning of my research career, now more than 20 years ago, I never would have guessed the number of things I would need to unlearn. In this chapter I describe studies of two phenomena in which a particular approach led to an unlearning that was as important as the learning that had occurred. First, it seemed clear to me (and others, e.g., Kleitman, 1963) that sleep/wake state regulation in young infants was fairly simple. Infants should be awake while hungry and fall asleep when well fed. The second oversimplified assumption was that blood pressure should be homeostatically maintained at a constant level by variations in cardiac output and rate. The techniques and experiments detailed in this chapter helped demonstrate that the exceptions to these rules are as interesting and informative as the rules themselves.

What will be described in this chapter are methods for recording physiological variables from infant rats while those pups are engaged in normal interactions with dam and littermates. The techniques themselves are not orginal. They are adaptations of the electrophysiological methods known and used in psychobiological and psychophysiological laboratories throughout the country. It was the combination of these techniques with an old experimental principle—studying organisms in as natural a state as possible—that allowed me to make progress in this field. I will first try to convince you that there are good reasons to attempt this seminaturalistic kind of study. Next, I will describe the methods involved and some sample results. I hope these illustrations and the related changing concepts will be further evidence of the importance of these kinds of studies.

In almost every experimental design or basic psychology course, the differences between observational and manipulational experiments are described and discussed. One experimental approach frequently recommended is to observe phenomena in a natural setting (in the wild, for ethologists) and then try to work out the controls of these phenomena in the laboratory (see also Galef, chap. 7, this volume). The problem with starting in the laboratory is that the range of behavior shown is too limited and may be very different from the behavior shown in the wild. The problem with staying in the wild is that there are too many uncontrolled variables.

The combination of observation and manipulation/wild and lab is familiar. Yet I believe this useful and powerful method is underutilized and may be especially helpful in studying young organisms for the following reasons.

47

1. As noted above, behavior studied under artificial or controlled conditions may not accurately reflect the range of behavior seen in the wild. The two studies on sleep/wake states and blood pressure changes which I describe in some detail later provide examples of surprising results uncovered by studying subjects in as natural a setting as possible. A related issue is that of a "false positive." Behaviors found in laboratory settings either do not occur in nature or do not occur in the settings the experimenter assumes. For example, since experiments showed that rat dams are more likely to retrieve pups that are emitting ultrasonic vocalizations (Allin & Banks, 1972; Smotherman, Bell, Hershberger, & Coover, 1978) and because it is not yet clear what mechanism causes dams to return and nurse their litters, I and others believed that the ultrasounding of hungry pups brought the dams back to begin nursing bouts. However, we disproved this hypothesis by listening to pups in their home cage nest piles for long periods of time and finding no increase in ultrasound prior to the dam's return (Hofer & Shair, 1978).

2. The behaviors of interest may be controlled by or influenced by factors in the natural environment that are not obvious to the experimenter. For example, temperature effects are critical in studies of sleep/wake state behavior, feeding behavior, and learning behavior in the young organism (see Satinoff, chap. 10, this volume). In another example, taste- and odor-aversion learning have been demonstrated in rat pups under many conditions, even in utero (Gemberling, Domjan, & Amsel, 1980; Rudy & Cheatle, 1977; Smotherman, 1982; Stickrod, Kimble, & Smotherman, 1982). But there is the intriguing study of Martin and Alberts (1979) demonstrating that pups attached to the teat of their dam are unable to learn a taste aversion. This classic example is hard to explain physiologically but easy to understand in evolutionary terms.

3. The physiology and behavior of young organisms depend upon environmental input to such a large extent (Hofer, 1978) that removing the experimental subject from its normal environment may quickly have consequences different from those in the adult. In other words, there may be no true baseline of behavior away from the natural setting. An example of this kind of phenomenon is the work of Schanberg and Kuhn's group that shows how rapidly the levels of ornithine decarboxylase (ODC) and growth hormone (GH) fall after the rat pup is removed from its mother even when left in the home cage with its littermates. The levels begin to fall within 20 minutes (Schanberg, Evoniuk, & Kuhn, 1984; Butler & Schanberg, 1977).

4. Removal of the rapidly developing organism from its natural environment may also have long-term consequences. Not only might the mother and littermates treat the returned pups differently but the pups themselves may be changed on any number of measures. There are many examples of this kind of effect. Early-weaned rats remain supersensitive to gastric ulcer production for several months but later become more resistant to ulceration (Ackerman, Hofer, & Weiner, 1975). Nutrient deprivation of rat pups often produces permanent effects on body weight and brain development (see Diaz, chap. 16, this volume; Dobbing, 1976). The changes in ODC and GH mentioned above also have longer-term consequences. After a period of separation, GH no longer induces ODC production (Kuhn, Butler, & Schanberg, 1978; Kuhn, Evoniuk, & Schanberg, 1979).

5. The next point I would like to overstate, for argument's sake. For species like the rat, in which the young are cached, the "natural environment" consists of the nest

area, littermates, and interactions with the nursing dam. The mother acts as a buffer between the pups and the rest of the environment, both in the wild and in the laboratory setting. Undoubtedly she is affected by the limited enviornment of the laboratory, but the pups are less affected because of their immature sensory-motor capabilities and by the fact that so much of their physiology and behavior is directed toward the dam. One might even go so far as to say that the inherited environment for the young pup, the environment of evolutionary adaptiveness, is the mother-infant relationship.

SLEEP/WAKE STATES AND FEEDING

There are many reasons to believe that young infants are awake when hungry and go to sleep after feeding. Some of the reasons are well founded, based on personal observation and research data. However, other reasons are probably based on untested assumptions.

Several studies have shown that, for adults, there is a relationship between sleep/wake states and feeding. This has been demonstrated by experiments that changed the size, composition, or timing of meals and observed changes in sleep/wake state behavior (Danguir, Nicolaidis, & Gerard, 1979; Gaensbauer & Emde, 1973; Mouret & Bobillier, 1971). Total food deprivation causes the most dramatic change. In general, the amount of sleep decreases dramatically following such deprivation (Danguir & Nicolaidis, 1979; Hockman, 1964; Jacobs & McGinty, 1971). Further results have demonstrated that giving food either intragastrically or intravenously can increase the amount of sleep shown by various organisms (Danguir & Nicolaidis, 1980; Rosen, Davis, & LaDove, 1971; Rubenstein & Sonnenschein, 1971; Sudakov, 1965).

Similar relationships have been seen in infants. For example, infants are more likely to be asleep following a feeding than either prior to that feeding or following a sham feeding (Wolff, 1972). The composition of the baby formula can alter sleep latency (Yogman & Zeisel, 1983). However, with adults I would have never made the assumption that the relationship between feeding and sleep/wake states was a simple one. Yes, I know that I tend to fall asleep after a heavy lunch, but I also know that other factors influence my sleep/wake behavior (e.g., whether I am listening to a lecture in a warm room). It may have been the unconscious assumption that infants are simpler than adults that led me to believe that the relationship between sleep/wake states and feeding must be simpler as well. After all, infants really don't do anything but eat and sleep, do they?

Kleitman (1963) hypothesized that for very young infants, the only wakefulness was that of necessity based on hunger. Older infants and organisms had wakefulness of choice. With a species like the rat, it may be especially easy to make these kinds of assumptions. The nursing situation appears to be an active one, with the pups treadling, nipple switching, and stretching in response to milk ejections. The mother steps on them, licks them, and otherwise stimulates them. In the period between nursing bouts, the dam leaves young pups in a nest pile. There is some movement within the pile as pups wriggle and burrow, but the appearance is one of much less activity for rats younger than 2 weeks old. Most polygraphic studies of the pups' state behavior tended to reinforce my assumption. Young pups have generally been studied away from the dam and been found to be asleep a large portion of the time (Gramsbergen, 1976; Gramsbergen, Schwartze, & Prechtl, 1970; Hofer, 1976; Hofer & Shair, 1982; Jouvet-Mounier, Astic, & LaCote, 1970). Since pups sleep between nursing bouts, I assumed they must be awake during the time attached to the teat.

Yet some intriguing hints from our own studies and from the literature suggested that the feeding/state relationship was not as simple as it seemed. When activity patterns of young rats were actually studied, both during nursing bouts and in the interbout interval, Hofer and Grabie (1971) found great similarity in the timing of activity under the two conditions. A series of anecdotal remarks in studies by Emde and Metcalf (1970), Jouvet-Mounier et al. (1970), and Ruckebusch (1972) suggested that infants of some species might actually sleep while attached to the teat, especially toward the end of the nursing bout. I had noticed a similar finding in observing my daughter. She often seemed to fall asleep while being nursed by her mother. She also showed very intriguing behavior when using a bottle. She appeared to be asleep, yet when the bottle would droop and milk no longer flowed, she adjusted the bottle without appearing to wake. For these reasons as well as others, investigation of sleep states during the actual nursing behavior of the young animal seemed worthwhile.

Methodology

In planning to study physiological behaviors in young organisms that are kept with their dam, certain critical principles need to be followed when designing the apparatus and the experiment. These principles are true not only if you plan to test the pups in the presence of the dam, but even if you just plan to return the pups to the dam after surgery and prior to testing.

1. Streamline the apparatus. Try to minimize any disadvantage that the test pup might be under in the interactions among littermates and with the mother. Streamlining will also make it less likely that the mother will attempt to destroy the apparatus (see Figure 4.1A).

2. Make the apparatus difficult to destroy. Make it sturdy, if that can be done without interfering with the streamlining. Bury it under the skin. Cut it short. Put collodion or some other flexible glue over it.

3. Disguise the smell. Always return the pup to the nest and littermates for a considerable period of time prior to returning the mother to the litter. The pup will then have time to recover the proper smell. It may be worth handling the entire litter or implanting more than one pup so the mother is less likely to concentrate her attention on a single pup. The other implants may be shams.

4. Make the pup's life easier. If the operation is a long one, give the pup an intubation feeding prior to returning it to the litter. Make sure the pup is fully recovered in terms of temperature and mobility. Reduce the competition for the pup by removing some of the littermates, if such action is unlikely to change the variables of the mother-infant relationship that you wish to study.

5. Make the dam as comfortable as possible. If you plan to record during the mother-infant interaction, accustom the dam to the place in which the study will occur as well as to the presence of the observer and the experimental apparatus.

The application of these principles to the techniques used in our laboratory for studying sleep/wake states in 2-week-old rat pups have been described previously (Hofer, 1976; Hofer & Shair, 1982; Shair, Brake, & Hofer, 1984). These techniques

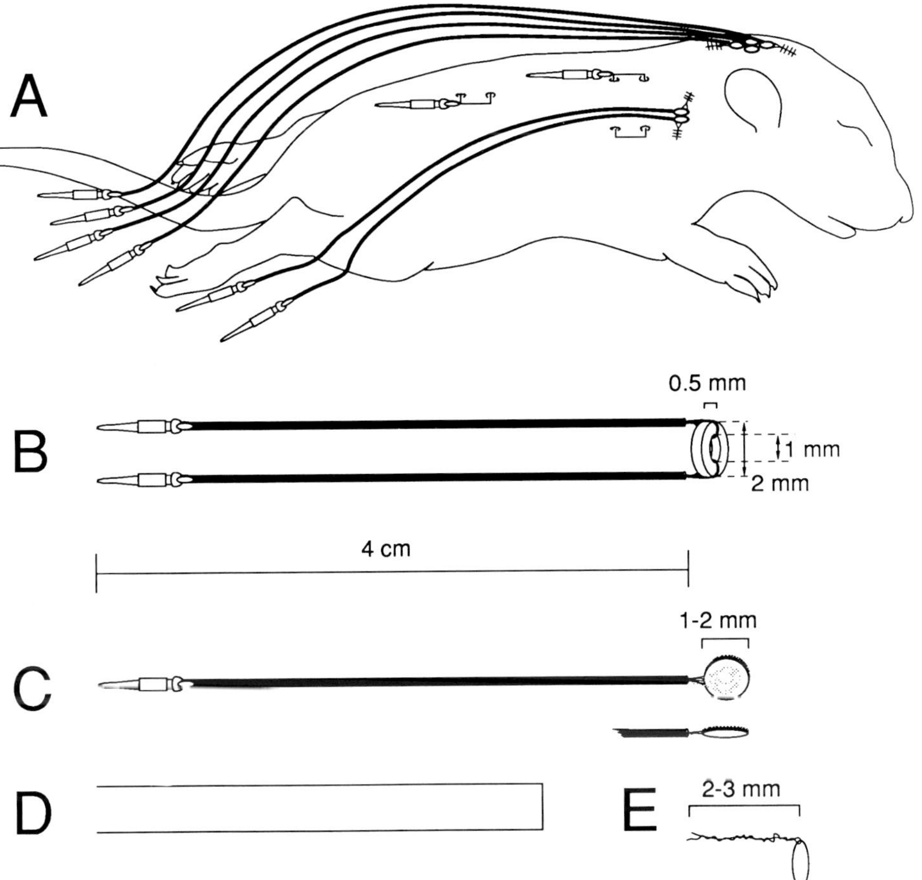

Figure 4.1A. Illustration of 2-week-old rat pup after implantation of electrodes necessary for monitoring sleep/wake state and sucking behaviors. (**B**) Electrocorticogram electrodes. (**C**) Electromyogram electrode. (**D**) Electrocardiogram and respiration electrode. (**E**) Neck electromyogram electrode anchor.

were adapted from those used by others at various ages in the rat (Almli, Henault, Velozo, & Morgane, 1985; Jouvet-Mounier et al., 1970; Mirmiran & Corner, 1982) and in other species (Hoppenbrouwers & Sterman, 1975; Jouvet-Mounier et al., 1970; Pauley & Reite, 1981; Shimizu & Himwich, 1968). The techniques used for measuring sucking behavior have also been described (Brake, Wolfson, & Hofer, 1979). The next sections will expand those descriptions by emphasizing the ways the techniques were adapted in order to meet the critical principles described above.

Electrode Construction

EEG

The cortical electrode (Fig. 4.1B) is made from a miniature machine-tooled washer (Delrin plastic, Dupont) 2 mm in diameter, designed to hold separate the bipolar recording wires, which will rest on the dura over the cortex. A good plastics fabricating company willing to do small orders can supply these pieces, cutting washers from a small tube of Delrin. Attached on opposite sides of the washer are two lengths of silver wire (Cooner Wire Company, AS632) from which insulation has been removed at the ends attached to the washer. A subminiature male Reliatac connector (Amphenol, Part #220-P02, Allied Electronics) is crimped onto the free end. To reduce shearing, crimp the Amphenol onto the insulation as well as onto bare wire. The length of the wire depends on the experiment to be done. In those experiments in which we recorded from the pup with littermates and dam, the wires were 10 cm long.

EMG

The digastric and neck EMG electrodes (Fig. 4.1C) are made from the same silver wire with insulation removed at both ends. At the end to be implanted, 1 cm of uninsulated wire is twisted into a tiny loop. A bead of solder is attached to the loop and flattened to make a broad platelike surface (diameter 1–2 mm). An EMG anchor must also be constructed. We used a piece of uninsulated 30-gauge silver wire (Medwire Corp.) twisted into a small loop and bent at a 90° angle (see diagram, Figure 4.1E).

EKG/Respiratory

Two 30-gauge silver-wire loop electrodes (Figure 4.1D) are placed subcutaneously across the dorsal-lateral thorax, as described in many previous publications (Hofer, 1973; Hofer & Reiser, 1969). Each electrode is approximately 3 cm long, but only a 3–4 mm twisted end protrudes, lying flat along the spine in a caudal direction. These EKG electrodes have Amphenol connectors crimped onto them as well.

Electrode Placement

Patience is the critical word in doing the electrode placement. At each stage the surgeon must be very sure that both the skull and the electrode cement are dry. I find it helpful to bring a book with me during surgery to stop me from rushing.

The rat is anesthetized with methyoxyflurane (Metofane, Pitman-Moore) and placed in a stereotaxic headset with ear bars filed down to fit the ear canal of a 2-week-old rat pup. During the operation, body temperature is maintained by placing a heating pad under the stereotaxic. Metofane is administered throughout the operation to a miniature face mask made of cotton encased in wire that covers the pup's nose. A midline longitudinal incision of 1–2 cm is made through the skin of the head. The skin is retracted, and the space between the skull and skin packed with miniaturized cotton dental pellets (Richmond Dental Cotton Co.). These pellets serve both to keep the skin retracted and to dry the operation area. The membranes on the skull are scraped to the side with a scalpel blade, and the skull is allowed to dry. We use a hairdryer without heat to speed drying. Remember, patience! I don't recommend using bone wax to stop bleeding since it tends to prevent the dental cement from gripping the skull later in the operation. When the bleeding has stopped and the skull is dry (appearance should be whitish), electrode placement can begin.

The cortical electrodes can be placed on either side of the skull, between bregma and lambda, 1–2 mm from the midline. The exact location is not critical, but care should be taken to maintain adequate distance from the skull sutures in order to minimize bleeding. Using a dental drill (Teledyne Emesco, bit size: approximately

0.75 mm), the surgeon drills a small burr hole just large enough to accommodate the EEG washer. The bone may also be worn down with the drill until it is thin enough to pick away with a forceps. The EEG washer is inserted under the skull with the two wire tails extending caudally. The washer should be kept as horizontal as possible during the insertion process. It and the electrodes rest on the dura mater.

On the other side of the midline and approximately 2 mm caudal to the EEG burr hole, another burr hole is drilled for the EMG anchor. This should be done at the same time as the EEG placement. The loop of the anchor is inserted under the edge of the skull, resting on the dura. The twisted tail extends above the skull 2–3 mm.

This is the second point at which to exercise patience. Bleeding has to have stopped entirely prior to application of the dental cement. Slight tamping with small pieces of cotton can help to control the bleeding. The cortical electrode is cemented to the skull by means of a cranioplastic cement (Plastics One, Inc.). My coeditors suggest that "crazy" glue or resin might be as good or better (see Barr, chap. 19, and Hofer, chap. 2, this volume). The cement should be applied in a very small amount just over the electrode and allowed to dry. Gradually build up more layers of cement to cover a final area of approximately 5 mm in diameter and 4–5 mm in height. Care should be taken to confine cement to the dry scraped skull away from the tissue. Seepage of blood under the cement will prevent proper adhesion. Only the cortical electrode is cemented at this time. The cement should be allowed to harden for 10–15 min before beginning the EMG implant.

The EMG electrodes are implanted bilaterally on the nuchal musculature which inserts along the occipital crest. The flattened solder portion of the electrodes is sewn directly onto the muscle. The rest of each electrode is shaped rostrally to lie flat on the skull, around the EMG anchor, and to exit caudally along the rat's dorsum. The EMG electrodes and EMG anchor are cemented together onto the skull by means of the cranioplastic cement. When the cement is dry the cotton padding is removed, and the skin incision is closed around the implant with sutures. It is frequently possible to cover the head implant almost entirely.

The digastric EMG electrodes are implanted in a way similar to the nuchal electrodes. The digastric muscle is cleared and the plates sewn directly onto the muscle. The tail of the electrode is led out under the skin to the back of the rat and tied down with a suture covered with a bit of collodion (Fisher Scientific). Another method of recording sucking activity in the rat pup has now been described by Brake (see Brake, chap. 3, this volume; Brake, Tavana, & Myers, 1986).

Prior to returning the pup to its home cage, remove the dam. The implanted pup should be placed with the litter pile to regain the proper odor, as well as temperature if any was lost during surgery. We feed the implanted pup 1 ml bovine milk by intubation after it recovers from anesthesia. Two to 3 littermates are removed from the litter in order to reduce the competition the pup would face in nursing during the postoperative period. All testing is done at least 1½ days postsurgery. If pups have not gained weight during that postoperative period, they are dropped from the study.

Test Setting

In any behavioral observation of litter and dam, the animals must be accustomed to the observational setting prior to the beginning of the test period. This necessity is especially true when recording physiologic variables, since the introduction of wires and apparatus will also upset the dam. In my experiments, the cage containing litter and dam were placed in the testing room within 2 to 3 days postpartum. The litter and dam became accustomed to researchers coming in and out of the room and to the testing and observation of other litters in that room as well. An alternative strategy is to keep the dam and littermates in a totally shielded area that can be observed without its being entered. For certain experiments this procedure may be better.

The fact that we do not test the animal until approximately 2 days postsurgery allows dam and littermates time to become accustomed to the wires attached to the experimental pup. Also, it is possible to handle pregnant females prior to parturition, which makes them more docile in response to later intervention. There are obvious dangers to such handling, since its influence on the mother-pup interaction has not yet been studied, and influences on offspring behavior with similar prenatal procedures have been reported (Ader & Conklin, 1963).

Test Day

On the day of testing, 15 minutes prior to the test session, the dam was removed from the cage and the implanted pup's weight and temperature were taken. We used a Grass Model 7 polygraph (Grass Instrument Co.) for these physiological recordings. If the pup had gained weight, the animal was connected to the shielded polygraph leads (Alpha Wire, 4/c /3327/40 PVC, Allied Electronics) using Reliatac connectors and the leads buried under the nest shavings. We have had equal success either bringing the leads out under the shavings through a hole drilled in the side of the cage or bringing the leads up along one corner of the cage and covering them well with tape. After the dam is returned, care must be taken to watch her closely during the first 15 minutes of the recording session. Only during this time is she likely to attack the electrodes and leads. It is sometimes necessary to distract her attention from the electrodes by touching the top of the cage or occasionally pushing her head away from the electrodes with a long probe. After this initial adaptation period, the dam is very unlikely to pay attention to the leads during the rest of the recording session. Recording sessions should be long enough to allow for elimination of this first 15-minute period from the data analysis, if necessary.

The leads buried under the shavings should be long enough to allow the test pup to travel from one side of the cage to the other. In the optimal situation, the pup remains in the nest, and most of the leads stay buried. If the pups start to follow the mother around the cage, we found that the 12–14-day-old pups were capable of pulling the wires along with them without much problem. If, as occasionally happened, the leads got tangled as the pup moved around the cage, the dam might again be likely to show interest in them. It may be necessary to reach in and untangle the pup on rare occasions.

Results and Changing Concepts

Having described the complicated procedures necessary to get the recordings while the mother comes and goes from the litter, we felt we had to assure ourselves that the behaviors exhibited by dam and litter were close to normal. There are at least three complementary ways to approach this problem. The first is to compare the behavioral data obtained with data from similar uninstrumented observation sessions, either from your own experience or from the literature if available. The second technique is to compare the behavior of the implanted pup with its own littermates if such secondary observations are possible. A third technique is to look at changes in the behavioral interaction over time of the test session. If the interventions are disruptive, the mother-infant interactions may adapt to that intervention and show predictable changes over time.

By means of these techniques, our results give strong evidence that there was very little disruption of the normal interaction. During the 2-hour recording sesssion, the number of nursing bouts observed, the average time the pup was attached to the teat, the number of nipple switches, the latency to first milk ejection, the intermilk-ejection interval, and the number of milk ejections observed all agreed with data from the literature (Grota & Ader, 1969; Hall, Cramer, & Blass, 1975; Leon, Adels, & Coopersmith, 1985; Lincoln, Hill, & Wakerley, 1973). There were no obvious differences between the behavior of implanted pups and their nonimplanted littermates. The implanted pups were able to attach to the teat, treadle, and nipple switch with no apparent difficulty. In looking at the first versus the second half of the recording session, the interactions of dams and pups and the number of milk ejections that occurred did not show major changes, as if adapting to disturbance.

Having, I hope, convinced you that the recording situation approximates natural conditions, I will briefly describe some of the results that make doing this difficult kind of study worthwhile.

Most of the basic findings of this work can be illustrated in a typical example of a nursing episode. I hope that in looking at Figure 4.2 you can imagine my feelings and surprise as I observed the first litter during this normal nursing situation. As I watched both cage and polygraph, the test pup was in the nest, alternating among the three sleep/wake states. The mother came near the litter and began to lick and stimulate the pups. The test pup attached to a teat, sucking vigorously during the period of attachment. Then, prior to any milk ejections from the mother, the pup rapidly fell asleep. This was my first real inkling that the relationship between sleep/wake states and feeding for infants was not the simple notion I had previously entertained. The second surprise for me was that sucking behavior continued while the pup was a asleep. The pup was always asleep prior to a milk ejection. In fact, each milk ejection from the dam woke the pup, but the pup returned to sleep rapidly.

The impressions given from this example were validated statistically (Shair et al., 1984). For 2-week-old pups, the amount of time spent in each arousal state did not depend on whether they were attached to a teat or not. The percentage of time spent awake or in slow-wave sleep or paradoxical sleep did not differ under the two conditions. Also, there was no increasing amount of time asleep as the nursing bout progressed. Thus, any simple relationship between sleep/wake state and satiety has been shown to be untrue, at least for rat pups of this age. Second, we found characteristic levels of sucking behavior that occurred in each sleep/wake state (see Figure 4.3A). These rates of sucking in each state differed statistically. A comparison of the first and second halves of each nursing bout showed that there were no changes in rates between the two halves. Thus, as has been suggested by others, these data indicate a relationship between activity levels and sleep/wake states.

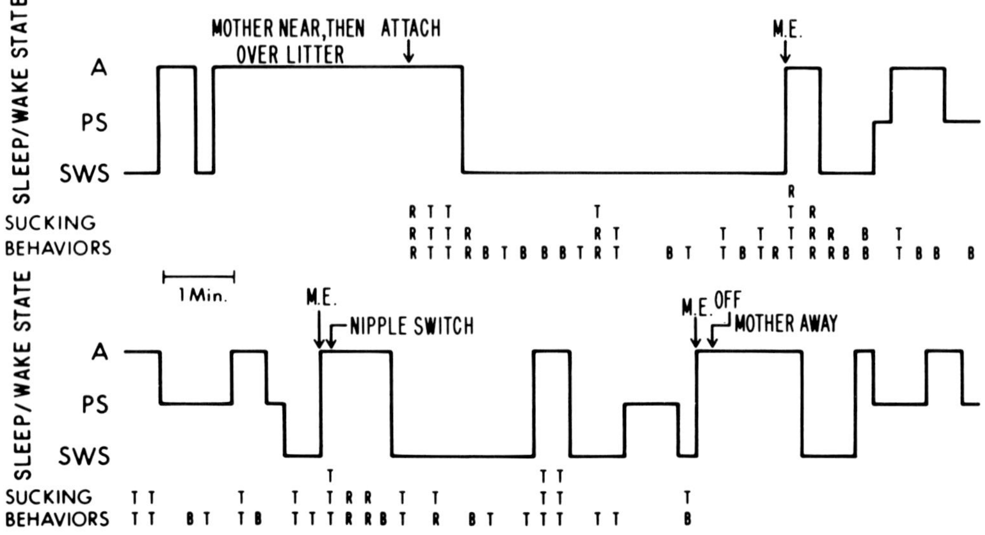

NURSING EPISODE EXAMPLE: Normal Mother

Figure 4.2. Sleep/wake data and sucking behavior during a typical nursing episode. The height of the top line indicates the state of the pup at any given time. After nipple attachment, the type and amount of sucking during each 15-s interval is shown by the letters below the state tracing. Each letter stands for 1–5 s of that sucking type. Thus, three R's indicate 11–15 s of rhythmic sucking. Other behaviors of interest are noted above the sleep/wake state tracing. (A) awake, (PS) paradoxical sleep, (SWS) slow-wave sleep, (R) rhythmic sucking, (B) bursts, (T) treadles, (M.E.) milk ejection. (From "Suckling in the Rat: Evidence for Patterned Behavior During Sleep" by H. Shair, S. Brake, and M. Hofer, 1984, *Behavioral Neuroscience*, 98, pp. 366–370. Copyright by the American Psychological Association. Reprinted by permission of the publisher.)

A separation experiment has been performed to examine further the relationship between activity rates and states. The techniques described above worked perfectly well in recording pups after a period of 22 h of separation. The dam was maintained with 3 pups during this time. One reason for this was to reduce the dam's activation when the instrumented pup was returned to her. Upon reunion, as had been predicted in the work of Brake (Brake, Sullivan, Sager, & Hofer, 1982; Brake et al., 1979), deprived pups demonstrated increased levels of sucking. We also found that the sleep/wake states shown by the pups also had been changed in the direction predicted by work on pups recorded in isolation (Hofer, 1976; Hofer & Shair, 1982). The amount of time in the awake state was increased while the amount of time in paradoxical sleep was decreased. The time in slow-wave sleep increased but not statistically significantly. The exciting new finding of this study (see Figure 4.3B) concerned the relationship between state and sucking behaviors. There was no increase in rate of any sucking type in any sleep/wake state. The only changes in sucking rate in any state were decreases in the rate of bursts during both awake and paradoxical sleep. Thus, rat pups may respond to deprivation not by increasing levels of motivated behavior in all states but by switching to those states with the highest levels of behavior. Further

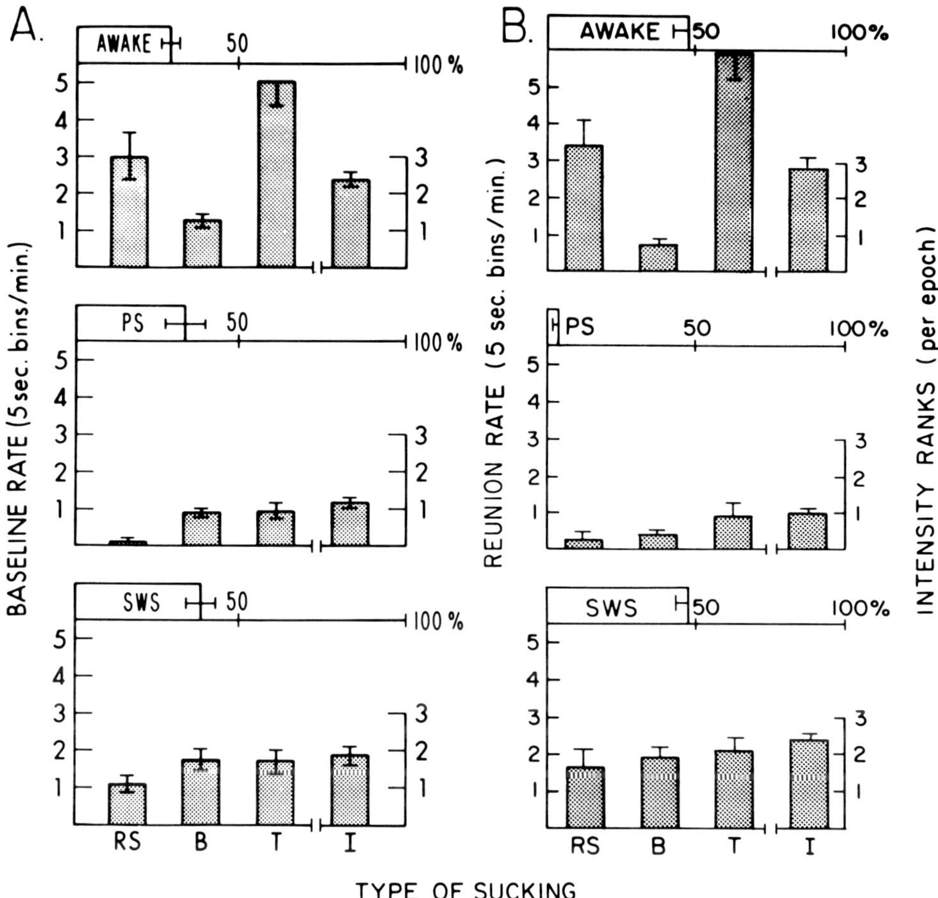

Figure 4.3. The rate of sucking behaviors and the percentage of time in each sleep/wake state during nursing on Baseline and Reunion days. The horizontal bars represent the mean percentage of total recording time spent in each state. The height of the three left-hand vertical bars gives the mean rate of sucking for each kind of sucking. The bar to the right measures the amplitude of the digastric EMG recording when no sucking is occurring (an index of muscle tone involved in maintaining a seal on the nipple). (From "Suckling in the Rat: Evidence for Patterned Behavior During Sleep" by H. Shair, S. Brake, and M. Hofer, 1984, *Behavioral Neuroscience, 98*, pp. 366–370. Copyright by the American Psychological Association. Reprinted by permission of the publisher.)

analysis of these data—too detailed to report here—supports the general idea that there is a relative stability of sucking type within any given sleep/wake state. I do not mean to imply that there is a central pattern generator rigidly emitting the treadles at a given rate during a given sleep/wake state. Rather, there is a general level of treadles that is seen during this sleep/wake state that differs from the levels in other states. In fact, my data do show that the rate of sucking while awake does increase in the periods immediately following milk ejection, but this rise is a response to a change in the environment (i.e., milk letdown) and not dependent on deprivation condition.

The results described here have changed my thinking radically in a number of ways. First, the idea that there is a simple relationship between sleep/wake states and feeding in young organisms is no longer tenable. Undoubtedly, hunger influences sleep/wake states, but many other factors also have an influence on the infant, as we knew to be true for adults. These data have also made me consider more favorably the idea that states may be one way of timing many behavioral and physiological events. This is certainly not a new idea with me. Many kinds of behaviors have shown state-linked frequencies. For example, human rhythmic mouthing is highest during slow-wave sleep; reflex smiles are highest during paradoxical sleep (Korner, 1969). The regulation of the cardiovascular system changes during the different states (Combs, 1982). In fact, the amount of heart rate variability, which is largest during paradoxical sleep, has been used as a way of monitoring state in fetuses (Junge, 1979). Michael Chase (1972) has demonstrated that the amplitude of reflex responses of the jaw muscle of the cat is modulated differently, depending upon state. Rate of sucking itself has previously been linked to the behavioral state of the organism (Kaye, 1966; Wolff, 1972). Both reports are concerned only with nonnutritive sucking and use somewhat different definitions of state, yet their observations agree favorably with the results I have obtained.

Finally, the most intriguing idea to me is the suggestion that changes in sleep/wake state regulation may be one way in which motivational levels are expressed. Assume that the rat pups, upon reunion after deprivation, have high motivation to get milk and suck. The evidence for this assumption is the increased level of sucking (Brake et al., 1979) and of intake (e.g., Houpt & Houpt, 1975). The rat pups could respond to deprivation by showing increased levels of sucking in all states, or just while awake. For example, studying the hypothalamus of cats, Jacobs, Harper, and McGinty (1970) found that the firing rate of certain cells was influenced by the level of hunger during all states recorded. Yet in my experiment, the rate of behavior within each state was independent of deprivation. Rat pups responded to deprivation by shifting sleep/wake percentages to those states with the highest levels of sucking.

BLOOD PRESSURE REGULATION

Our studies of the development of blood pressure regulation provide a second illustration of the importance of recording physiological variables in the natural environment. My earlier oversimplified view of the homeostatic control of blood pressure was that variations in heart rate and cardiac output are used to maintain blood pressure at a constant level. Even a careful reading of the work of W. Bradford Cannon (1939), who coined the term *homeostasis*, would have demonstrated the fallacy of the idea of a constant level of blood pressure. Homeostasis implies only that blood pressure is maintained within certain upper and lower limits. However, as I will now report, there is a normal, often repeated, situation in which the blood pressure increases far beyond what are thought to be those limits.

We were led to these studies of blood pressure levels by the work of Michael Myers (see chap. 1, this volume; Myers, Brunelli, Shair, Squire, & Hofer, 1989; Myers, Brunelli, Squire, Shindledecker, & Hofer, 1989). Myers and his colleagues studied mother-infant interactions in rats. They demonstrated that certain behaviors seen in those interactions predicted the adult blood pressure of the offspring. The frequency that the dam was observed in the arched nursing posture during the 3-week preweanling period was one of the factors that correlated most strongly with later blood pressure levels. This arched nursing posture often accompanies the marked behavioral activation shown by rat pups at the time of milk ejection. We hypothesized that a sympathetic nervous system surge occurring in these pups with each milk ejection might have long-term effects on the central cardiovascular control mechanisms or structure of the blood vessels themselves. It seemed possible that there might be some sort of blood pressure change, and accompanying sympathetic surge at the time of each milk ejection, since other investigators have described blood pressure increases during feeding (Bloom, Edwards, Hardy, Malinowska, & Silver, 1975; Cohen, Fifer, Better, Dweck, & Myers, 1987; Gupta & Scopes, 1965; Harding et al., 1978; LeDoux, Sakaguchi, & Reis, 1982; Matsukawa & Ninomiya, 1985). The behavioral studies of Henry (Henry, Meehan, & Stephens, 1967; Henry & Stephens, 1977) in connection with the anatomical studies of Folkow (Folkow, Grimby, & Thulesius, 1958; Folkow, Hallback, Lundren, Sivertsson, & Weiss, 1973) provided a model for this hypothesis—an example in which long-term social interactions in mice led to hypertension, as a result of structural changes in the blood vessels.

Method

The techniques for recording direct blood pressure from 2-week-old rat pups were adapted from the work of earlier physiologists (DiCara, Pappas, & Pointer, 1969; Jansons & Mosier, 1986; Yoburn, Morales, & Inturrisi, 1984), have been reported in papers from our laboratory (Shair, Brake, Hofer, & Myers, 1986; Shear, Brunelli, & Hofer, 1983), and share many similarities with the techniques for sleep/wake studies described above. However, in none of the earlier reports were active animals studied during social interactions. I will sketch the procedure used here and emphasize a few tricks in the surgery and recording situation that have not been reported elsewhere.

The key to recording blood pressure in these small freely moving animals is to use tubing that is flexible enough to be implanted without tearing the artery when the animal moves, but not so distensible that the pulse pressure force is unduly damped in transmission. Furthermore, the tubing should be small enough to go into the artery but large enough to transmit the pulse waves and to reduce the likelihood of clotting. Combining two types of tubing provides the qualities needed. Silastic tubing #602-105 (Dow-Corning Corp.) provides fine, flexible tubing for insertion into the artery. Polyethylene tubing PE50 (Clay Adams) is stiffer and wider, which prevents damping of the pressure in transmission to the recorder.

Four to 5 cm of Silastic tubing, dipped into ether to cause it to expand, is then forced at least a centimeter over the end of the PE50 tubing. As the ether evaporates, the Silastic will form an extremely tight seal to the PE50. The PE50 must be drawn out slightly at the end so that the Silastic is able to go over it. We have found that if you cut the PE50 tubing on a bevel to aid connection there is more likelihood of clotting within the cannula. Therefore, the proximal end of the PE50 tubing must be cut perpendicularly across its longitudinal axis. Too great a taper for the PE50 tubing will reduce the transmission of the pulse pressure.

The cannula is filled with heparin (1,000 USP U/ml, Elkins-Sinn, Inc.), and the distal PE50 end is plugged with a male mini-Amphenol connector that fits it snugly.

One cm of air should be left next to the Amphenol connector. This air bubble will allow visualization of the pulse after insertion into the artery. Care needs to be taken not to infuse any heparin into the pup.

We have tried other kinds of cannula tubing such as micro-renathane (Braintree Scientific, Inc.), and we have tried TD Mac heparin (Braintree Scientific, Inc.) in an attempt to reduce the amount of clotting that occurred. Both techniques worked well, but neither improved our percentage of success over the techniques described above.

The length of the PE50 portion of the cannula should vary, depending on the kind of experiment. If the pup is to be separated from its mother, make the cannula approximately 10 cm in length to allow easy attachment of the connecting tubing. If the pup is to be returned to the mother, the PE50 tubing must be cut short, approximately 3–4 cm.

The pup to be tested is anesthetized with inhaled Metofane and operated on with the use of microsurgical instruments and a stereo-dissecting microscope (see Hofer, chap. 2, this volume). Either the left or right common carotid artery can be implanted, whichever is more convenient to the surgeon. The artery is exposed and ligated just caudal to its bifurcation to prevent back flow of blood when the artery is later opened. Leave 2-cm-long thread ends, to be used later (see below). The cannula is passed under the skin from the back of the animal forward under the jaw to the anterior neck. Cut the Silastic tubing on a slight bevel at a point that, after insertion into the carotid, will allow just enough of the flexible tubing to reach the exit wound on the back. The surgeon then clamps the artery as close as possible to the clavicle, under which it disappears, using a microvascular clamp (Roboz Surgical Instrument Co.). Place a loose ligature around the artery just anterior to the clamp. A small cut is made in the artery close to the ligation, and the heparin-filled cannula is inserted into the artery. This is the trickiest part of the surgery and involves moving the cannula down, but not pushing so hard as to tear the artery, while trying to ease the artery up around the outside of the cannula using microforceps. The loose ligature is then tightened just enough to hold the cannula in place, and the clamp is opened. Push the cannula a few more millimeters into the artery. Make sure a pulse is visible at the air bubble, then secure the caudal ligature. The catheter is anchored with the 2 cm ends of the rostral carotid ligature as well.

The joint between the Silastic and the PE50 tubing is secured at the point of exit from the skin with collodion and a suture wrapped just distal to the joint and anchored to the skin. Leave approximately 1 cm of Silastic within the potential space under the skin to prevent the torque of the animal's twisting body from pulling the cannula out of the carotid. For nondeprived animals, the PE50 tubing is cut short and covered with more collodion and an extra suture to prevent the mother from biting it. As is usual, we return the pups to their home cages with the dam absent for a period of time to allow the pups to recover and to regain the smell of the littermates and home cage nest area. It is also possible that the smells of the operated pup are diffused to the nonoperated pups, and this may also work to prevent the mother from directing too much attention toward the operated pup.

Pups will recover from anesthesia and can be recorded as early as 3–4 hours after the operation. When the pups are recorded a day or two later, the cannulas may be clogged slightly. In this case, attach a syringe using a 23-gauge needle filled with a solution of 10% heparin in 5% dextrose in water and try gently to withdraw the clot. You may be forced to infuse fluid in order to break up the clot.

On the day of testing, we remove the mother from the home cage. The implanted pup is connected to the pressure transducer (Spectramed, P 23) with a length of 10% heparin-filled PE50 tubing connected to the PE50 cannula in the pup by a short connector of Silastic #602-155 into which PE50 fits tightly. Heart rate leads are also connected to the electrodes on the back of the pup by means of mini-Amphenol connectors. The cannula extension and heart rate leads are buried under home cage shavings, as in the sleep studies. After a baseline recording period, the mother is placed in the cage with her litter. Once again, as with the sleep/wake studies, it is this period just after return of the mother that is the most critical for the experiment. The methods for behavioral observations, polygraph scoring, and data analysis have been previously reported (Shair et al., 1986).

Using the techniques described above, we were able to record blood pressure and heart rate of 2-week-old pups during normal mother-infant interaction. These experiments were not easy to do. Our success rate, considering surgery, problems of clotting, and getting through the mother-infant interaction successfully, was approximately 60–70%. The results have justified the effort.

Results and Changing Concepts

We found that there was a large increase in blood pressure lasting 8–9 s that occurred coincident with the milk ejections given by the mother. As described (Shair et al., 1984; Shair et al., 1986), these milk ejections were defined behaviorally. The blood pressure increase was from 30 to 50% over the levels between milk ejections, depending on whether the pup had been previously deprived or not (see examples in Figure 4.4). Intriguingly, the heart rate occasionally showed a compensatory decrease at the same time as the blood pressure rose, presumably due to baroreceptor effects. But, just as frequently, there was no such response. In fact, the blood pressure responses of baroreceptor-denervated animals did not differ from normals (unpublished observa-

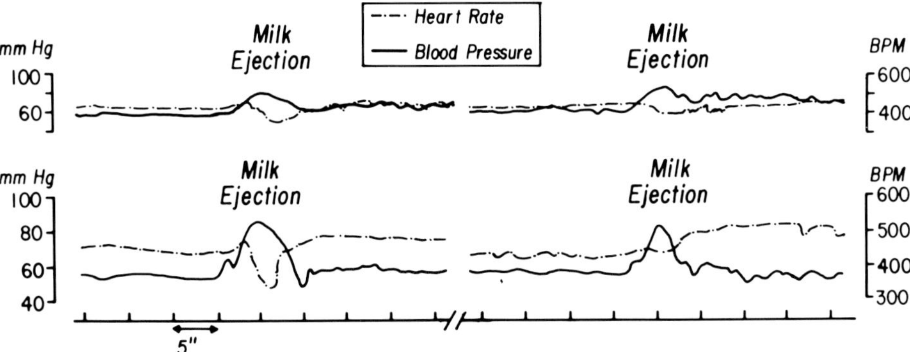

Figure 4.4. Examples of blood pressure and heart rate recordings of 2-week-old rat pups during naturally occurring nursing bouts. The top two tracings show results during milk ejections from a single nursing bout of a nondeprived pup. The bottom two records are from a single nursing bout of a deprived pup. (From "Blood Pressure Responses to Milk Ejection in the Young Rat" by H. Shair, S. Brake, and M. Myers, 1986, *Physiology and Behavior, 37*, pp. 171–176. Copyright by Pergamon Press. Reprinted by permission of the publisher.)

tions), possibly indicating some central "gating" of baroreceptor control during milk ejections.

These blood pressure rises seen during milk ejections were much greater than any blood pressure rises observed at any other point during the mother-infant interactions. Even when the pups were burrowing under the mother, scrambling to attach to the nipple, and treadling with the forepaws against the mother's ventrum during the periods between milk ejections, we found no comparable rises.

This phenomenon may explain the finding of Myers's group (Myers Brunelli, Shair, Squire, & Hofer, 1989; Myers, Brunelli, Squire, Shindledecker, & Hofer, 1989) relating the frequency of an early experience and later adult blood pressure levels. The most interesting hypothesis, to us, is that the blood pressure rise produced in response to milk ejection is caused by a sympathetic surge in the nervous system of the pups. This sympathetic nervous system surge, repeated anywhere from 600 to 1,600 times during the 3-week preweanling period of the pups, may cause permanent changes in the arteries of the adult rat, especially in those pups receiving the highest number of ejections.

Now that we have discovered this blood pressure response to milk ejections and have become aware of some of the parameters involved in its control in the natural setting (e.g., its greater magnitude in deprived animals, its coincidence with the stretch response), we are in a position to investigate further issues. We can take advantage of the fact that pups will attach to and nurse from an anesthetized dam. Milk ejections can be elicited periodically using i.v. oxytocin infusions, or fluid can be infused into the pup using a tongue or gastric cannula (see Diaz, chap. 16, and Phifer, chap. 11, this volume). In this way we can now perform various interventions on the pup that would not be possible in the normal environment.

The first question we have begun to study concerns the afferent signal for the blood pressure response. Preliminary results suggest that the signal originates in the mouth, not the stomach or intestines, and that fluid in the mouth is the trigger. Olfactory cues from the dam or changes in either conformation or temperature of the teat do not appear necessary. Mapping of the complete afferent, central nervous system and efferent circuit is one of our goals. Perhaps during this process we will also learn more about the controls of the baroreceptor "gating" described above.

A second issue involves understanding the possible function of the blood pressure rise. Does it have some feedback effect on behavioral state? Is the cardiovascular system shunting blood to the intestines to aid digestion? The reverse could also be possible; the splanchnic bed may be constricted in order to prevent too rapid food absorption. These kinds of questions are raised only when the whole animal is being studied. The complexity of interactions within an organism is so great that researchers studying a particular system often attempt to eliminate or ignore influences from the rest of the animal. Although a narrow focus can be necessary, that focus should be part of a broad perspective that includes many systems, the whole animal, and the animal's developmental history and environment.

A third area we are studying is the long-term effects of this repeated response. The idea that stimulation has not only immediate phasic effects but longer-term tonic influences was best characterized at a conference organized by Bernice Wenzel and Philip Zeigler (1977) called "Tonic Functions of Sensory Systems." This concept may be especially relevant for developing organisms. Yet it remains relatively unexplored and should be an important future direction for our field. In the example I have described, the effects of repetitive stimulation may have other results beyond the setting of adult blood pressure that is suggested by the work of Myers et al. (Myers, Brunelli, Shair, Squire, & Hofer, 1989; Myers, Brunelli, Squire, Shindledecker, & Hofer, 1989). Another possible effect is as a reinforcer of learning. It has been suggested that stimulation over the baroreceptor pathways may be reinforcing (Dworkin, Filewich, Miller, Craigmyle, & Pickering, 1979; Randich & Maixner, 1984). It is possible that part of the reason that young organisms become attached to their dam is repetitive baroreceptor stimulation in situations of great arousal when attached to the mother's teats. It is also possible that this phenomenon can help explain why young rats prefer smells and food associated with their dam. The work of Sullivan (Sullivan, Brake, Hofer, & Williams, 1986; Sullivan, Hofer, & Brake, 1986) on the influence of arousal levels in learning is of relevance here.

CONCLUSIONS

In the introduction to this chapter I described how much I have unlearned in the last 20 years. I conclude by discussing briefly what I believe I have learned from these two series of experiments. The first thing I learned is that my teachers were smarter than I realized. What I was taught about the power of certain experimental methods still holds true, at least in my experience. Using these methods, I was able to discover two phenomena that might have been unsuspected in more experimentally controlled kinds of observations. It was only allowing the animals the freedom to interact in (semi)naturalistic ways that provided the opportunity for discovery:

Next, I hope and believe that my experiments can help answer some of the traditional questions that developmental psychobiology has tried to ask. Two such questions are: How are responses to changes in life events controlled? How is the timing of ongoing events controlled? My sleep/wake studies suggest that a novel mechanism that may help in understanding these issues. It is possible that changes in sleep/wake states may be one of the factors that control the timing of ongoing events. And, perhaps more important, changes in sleep/wake states may be a mechanism that is used in controlling responses to deprivation and motivational levels. Of course, other people have had similar ideas (Kaye, 1966; Wolff, 1972), but I believe my work adds to theirs in new and, it is to be hoped, important ways.

A fundamental issue for developmental psychobiology has been the question of how early environmental changes cause differences in later behavior and physiology. Since the classic handling studies (Denenberg & Zarrow, 1971; Levine, 1975) showed that early effects can have such long-term influences, developmental psychobiologists have been searching for mechanisms to explain these consequences. I hope that my blood pressure studies, in conjunction with the work of others, and the long-term observational studies of Michael Myers can provide at least one mechanism that can be studied in detail.

ACKNOWLEDGMENTS

This research was supported by a project grant to Myron Hofer, M.D., from the National Institute of Mental Health.

REFERENCES

Ackerman, S., Hofer, M., & Weiner, H. (1975). Age at maternal separation and gastric erosion susceptibility in the rat. *Psychosomatic Medicine 37*, 180–184.

Ader, R., & Conklin, P. (1963). Handling of pregnant rats: Effects on emotionality of their offspring. *Science, 142*, 411–412.

Allin, J., & Banks, E. (1972). Functional aspects of ultrasound production by infant albino rats. *Animal Behavior, 20*, 175–185.

Almli, C., Henault, M., Velozo, C., & Morgane, P. (1985). Ontogeny of electrical activity of main olfactory bulb in freely-moving normal and malnourished rats. *Developmental Brain Research, 18*, 1–11.

Bloom, S., Edwards, A., Hardy, R., Malinowska, K., & Silver, M. (1975). Cardiovascular and endocrine responses to feeding in the young calf. *Journal of Physiology, 254*, 135–155.

Brake, S., Sullivan, R., Sager, D., & Hofer, M. (1982). Short- and long-term effects of various milk delivery contingencies on sucking and nipple attachment in rat pups. *Developmental Psychobiology, 15*, 543–556.

Brake, S., Tavana, S., & Myers, M. (1986). A method for detecting and analyzing intra-oral negative pressure in suckling rat pups. *Physiology and Behavior, 36*, 575–578.

Brake, S., Wolfson, V., & Hofer, M. (1979). Electromyographic patterns associated with non-nutritive sucking in 11–13-day-old rat pups. *Journal of Comparative and Physiological Psychology, 93*, 760–770.

Butler, S., & Schanberg, S. (1977). Effect of maternal deprivation on polyamine metabolism in preweanling rat brain and heart. *Life Science, 21*, 877–884.

Cannon, W. B. (1939). *The wisdom of the body* (2nd ed.). New York: Norton.

Chase, M. (1972). Patterns of reflex excitability during the ontogenesis of sleep and wakefulness. In C. Clemente, D. Purpura, & F. Mayer (Eds.), *Sleep and the maturing nervous system* (pp. 253–287). New York: Academic Press.

Cohen, M., Fifer, W., Better, D., Dweck, A., & Myers, M. (1987). Elevation of blood pressure during feeding in newborns. *Pediatric Research Abstracts, 21*, (*Part 2*), 383A.

Combs, C. (1982). Behavioral modulation of arterial baroreflexes. In O. Smith, R. Galosy, & S. Weiss (Eds.), *Circulation, neurobiology and behavior* (pp. 185–199). New York: Elsevier.

Danguir, J., & Nicolaidis, S. (1979). Intravenous infusions of nutrients and sleep in the rat: An ischymetric sleep regulation hypothesis. *American Journal of Physiology, 238*, E307–E312.

Danguir, J., Nicolaidis, S., & Gerard, H. (1980). Dependence of sleep on nutrients' availability. *Physiology and Behavior, 22*, 735–740.

Denenberg, V. H., and Zarrow, M. X. (1971). Effects of handling in infancy upon adult behavior and adrenocortical activity: Suggestions for a neuroendocrine mechanism. In D. H. Walcher & D. L. Peters (Eds.), *Early childhood: The development of self-regulatory mechanisms* (pp. 39–71). New York: Academic Press.

DiCara, L., Pappas, B., & Pointer, F. (1969). A technique for chronic recording of systemic arterial blood pressure in the unrestrained rat. *Behavioral Research Methods and Instruments, 1*, 221–223.

Dobbing, J. (1976). Vulnerable periods in brain growth and somatic growth. In D. F. Roberts & A. M. Thomson (Eds.), *The biology of human fetal growth* (pp. 137–147). London: Taylor and Francis, Ltd.

Dworkin, B., Filewich, R., Miller, N., Craigmyle, N., & Pickering, T. (1979). Baroreceptor activation reduces reactivity to noxious stimulation: Implications for hypertension. *Science, 21*, 1299–1391.

Emde, R., & Metcalf, D. (1970). An electroencephalographic study of behavioral rapid eye movement states in the human newborn. *Journal of Nervous and Mental Disease, 150*, 376–386.

Folkow, B. Grimby, G., & Thulesius, O. (1958). Adaptive structural changes of the vascular walls in hypertension and their relation to the control of the peripheral resistance. *Acta Physiologica Scandanavia, 44*, 255–272.

Folkow, B., Hallback, M., Lundren, Y., Sivertsson, R., & Weiss, L. (1973). Importance of adaptive changes in vascular design for establishment of primary hypertension, studied in man and in spontaneously hypertensive rats. *Circulation Research, 32/33*, Suppl. 1, 2–13.

Folkow, B., & Neil, E. (1971). *Circulation*. London: Oxford University Press.

Gaensbauer, T., & Emde, R. (1973). Wakefulness and feeding in human newborns. *Archives of General Psychiatry, 28*, 894–897.

Gemberling, G., Domjan, M., & Amsel, A. (1980). Aversion learning in 5-day-old rats: Taste-toxiosis and texture-shock aversion learning. *Journal of Comparative and Physiological Psychology, 94*, 734–745.

Gramsbergen, A. (1976). The development of the EEG in the rat. *Developmental Psychobiology, 9*, 501–515.

Gramsbergen, A., Schwartze, P., & Prechtl, H. (1970). The postnatal development of behavioral states in the rat. *Developmental Psychobiology, 3*, 267–280.

Grota, L., & Ader, R. (1969). Continuous recording of maternal behavior in *Rattus norvegicus*. *Animal Behaviour, 17*, 722–729.

Gupta, J., & Scopes, J. (1965). Observations on blood pressure in newborn infants. *Archives of Disease in Childhood, 40*, 637–644.

Hall, W., Cramer, C., & Blass, E. (1975). Developmental changes in suckling of rat pups. *Nature, 258*, 318–320.

Harding, R., Johnson, P., McClelland, M., McLeod, C., Whyte, P., & Wilkinson, A. (1978). Respiratory and cardiovascular responses to feeding in lambs. *Journal of Physiology* (London), *275*, 40P–41P.

Henry, J., Meehan, J., & Stephens, P. (1967). The use of psycho-social stimuli to induce prolonged systolic hypertension in mice. *Psychosomatic Medicine, 29*, 408–432.

Henry, J., & Stephens, P. (1977). *Stress, health and the social environment*. New York: Springer.

Hockman, C. (1964). EEG and behavioral effects of food deprivation in the albino rat. *Electroencephalography and Clinical Neurophysiology, 17*, 420–427.

Hofer, M. (1973). The effects of brief maternal separations on behavior and heart rate of two week old rat pups. *Physiology and Behavior, 10*, 423–427.

Hofer, M. (1976). The organization of sleep and wakefulness after maternal separation in young rats. *Developmental Psychobiology, 9*, 189–205.

Hofer, M. (1978). Hidden regulatory processes in early social relationships. In P. Bateson & P. Klopfer, (Eds.), *Perspectives in ethology* (Vol. III, pp. 135–166). New York. Plenum.

Hofer, M., & Grabie, M. (1971). Cardiorespiratory regulation and activity patterns of rat pups studied with their mothers during the nursing cycle. *Developmental Psychobiology, 4*, 169–180.

Hofer, M., & Reiser, M. (1969). The development of cardiac rate regulation in preweanling rats. *Psychosomatic Medicine, 31*, 372–388.

Hofer, M., & Shair, H. (1978). Ultrasonic vocalization during social interaction and isolation in 2-week-old rats. *Developmental Psychobiology, 11*, 495–504.

Hofer, M., & Shair, H. (1982). Control of sleep/wake states in the infant rat by features of the mother-infant relationship. *Developmental Psychobiology, 15*, 229-243.

Hoppenbrouwers, T., & Sterman, M. (1975). Development of sleep state patterns in the kitten. *Experimental Neurology, 49*, 822–838.

Houpt, K., & Houpt, T. (1975). Effects of gastic loads and food deprivation on subsequent food intake in sucking rats. *Journal of Comparative and Physiological Psychology, 88*, 764–772,

Jacobs, B., Harper, R., & McGinty, D. (1970). Neuronal coding of motivational level during sleep. *Physiology and Behavior, 5*, 1139–1143.

Jacobs, B., & McGinty, D. (1971). Effects of food deprivation on sleep and wakefulness in the rat. *Experimental Neurology, 30*, 212–222.

Jansons, R., & Mosier, H. Jr. (1986). A simplified method for chronic portal vein cannulation in the rat. *Physiology and Behavior, 38*, 739–741.

Jouvet-Mounier, D., Astic, L., & LaCote, D. (1970). Ontogenesis of the states of sleep in the rat, cat, and guinea pig in the first postnatal month. *Developmental Psychobiology, 2*, 216–239.

Junge, H. (1979). Behavioral states and state related heart rate and motor activity patterns in the newborn infant and the fetus antepartum—A comparative study. I. Techniques, illustrations of recordings, and gernal results. *Journal of Perinatal Medicine, 7*, 85–107.

Kaye, H. (1966). The effects of feeding and tonal stimulation on non-nutritive sucking in the human newborn. *Journal of Experimental Child Psychiatry, 3*, 131–145.

Kleitman, N. (1963). *Sleep and wakefulness*. Chicago: University of Chicago Press.

Korner, A. (1969). Neonatal startles, smiles, erections, and reflex sucks as related to state, sex, and individuality. *Child Development, 40*, 1028–1053.

Kuhn, C., Butler, S., and Schanberg, S. (1978). Selective depression of serum growth hormone during maternal deprivation in rat pups. *Science, 201*, 1034–1036.

Kuhn, C., Evoniuk, G., & Schanberg, S. (1979). Loss of tissue sensitivity to growth hormone during maternal deprivation in rat pups. *Life Science, 25*, 2089–2097.

LeDoux, J. E., Sakaguchi, A., & Reis, D. G. (1982). Behaviorally selective cardiovascular hyperreactivity in spontaneously hypertensive rats: Evidence for hypoemotionality and increased appetitive motivation. *Hypertension, 4*, 853–863.

Leon, M., Adels, L., & Coopersmith, R. (1985). Thermal limitation of mother-young contact in Norway rats. *Developmental Psychobiology, 18*, 85–105.

Levine, S. (1975). Psychosocial factors in growth and development. In L. Levi (Ed.), *Society, stress and disease* (Vol. 2, pp. 43–50). New York: Oxford University Press.

Lincoln, D., Hill, A., & Wakerley, J. (1973). The milk-ejection reflex of the rat: An intermittent function not abolished by surgical levels of anesthesia. *Journal of Endocrinology, 57*, 459–476.

Martin, L., & Alberts, J. (1979). Taste aversions to mother's milk: The age-related role of nursing in acquisition and expression of a learned association. *Journal of Comparative and Physiological Psychology, 93*, 430–445.

Matsukawa, K., & Ninomiya, I. (1985). Transient responses of heart rate, arterial pressure, and head movement at the beginning of eating in awake cats. *Japanese Journal of Physiology, 35*, 599–611.

Mirmiran, M., & Corner, M. (1982). Neuronal discharge patterns in the occipital cortex of developing rats during active and quiet sleep. *Developmental Brain Research, 3*, 37–48.

Mouret, J., and Bobillier, P. (1971). Diurnal rhythms of sleep in the rat: Augmentation of paradoxical sleep following alterations of the feeding schedule. *International Journal of Neuroscience, 2*, 265–270.

Myers, M., Brunelli, S., Shair, H., Squire, J., & Hofer, M. (1989). Relationships between maternal behavior of SHR and WKY dams and adult blood pressures of cross-fostered F1 pups. *Developmental Psychobiology, 22*, 29–53.

Myers, M., Brunelli, S., Squire, J., Shindledecker, R., & Hofer, M. (1989). Maternal behavior of SHR rats and its relationship to offspring blood pressures. *Developmental Psychobiology, 22*, 29–53.

Pauley, J., & Reite, M. (1981). A microminiature hybrid multichannel implantable biotelemetry system. *Biotelemetry Patient Monitor, 8*, 163–172.

Randich, A., & Maixner, W. (1984). Interactions between cardiovascular and pain regulatory systems. *Neuroscience and Biobehavioral Reviews, 8*, 343–367.

Rosen, A., Davis, J., & LaDove, R. (1971). Electrocortical activity: Modification by food ingestion and a humoral satiety factor. *Communications in Behavioral Biology, 6*, 323–327.

Rubenstein, E., & Sonnenschein, R. (1971). Sleep cycles and feeding behavior in the cat: Role of gastrointestinal hormones. *Acta Cientifica Venezolana (Supplementum), 22*, 125–128.

Ruckebusch, Y. (1972). Development of sleep and wakefulness in the foetal lamb. *Electroencephalography and Clinical Neurophysiology, 32*, 119–128.

Rudy, J., & Cheatle, M. (1977). Odor-aversion learning in neonatal rats. *Science, 198*, 845–846.

Schanberg, S., Evoniuk, G., & Kuhn, C. (1984). Tactile and nutritional aspects of maternal care: Specific regulators of neuroendocrine function and cellular development. *Proceedings of the Society of Experimental Biology and Medicine, 175*, 135–146.

Shair, H., Brake, S., & Hofer, M. (1984). Suckling in the rat: Evidence for patterned behavior during sleep. *Behavioral Neuroscience, 98*, 366–370.

Shair, H., Brake, S., Hofer, M., & Myers, M. (1986). Blood pressure responses to milk ejection in the young rat. *Physiology and Behavior, 37*, 171–176.

Shear, M., Brunelli, S., & Hofer, M. (1983). The effects of maternal deprivation and refeeding on the blood pressure of infant rats. *Psychosomatic Medicine, 45*, 3–9.

Shimizu, A., & Himwich, H. (1968). The ontogeny of sleep in kittens and young rabbits. *Electroencephalography and Clinical Neurophysiology, 24*, 307–318.

Smotherman, W. (1982). Odor aversion learning by the rat fetus. *Physiology and Behavior, 29*, 769–771.

Smotherman, W., Bell, R., Hershberger, W., & Coover, G. (1978). Orientation to rat pup cues: Effects of maternal experiential history. *Animal Behaviour, 26*, 265–273.

Stickrod, G., Kimble, D., & Smotherman, W. (1982). In utero taste/odor aversion conditioning in the rat. *Physiology and Behavior, 28*, 5–7.

Sudakov, K. (1965). The electroencephalogram and neurohumoral mechanisms of satiety. *Bulletin of Experimental Biology and Medicine, 59*, 103–106.

Sullivan, R., Brake, S., Hofer, M., & Williams, C. (1986). Huddling and independent feeding of neonatal rats can be facilitated by a conditioned change in behavioral state. *Developmental Psychobiology, 19*, 625–635.

Sullivan, R., Hofer, M., & Brake, S. (1986). Olfactory-guided orientation in neonatal rats is enhanced by a conditioned change in behavioral state. *Developmental Psychobiology, 19*, 615–623.

Wenzel, B., & Zeigler, H. P. (Eds.). (1977). *Tonic functions of sensory systems. Annals of the New York Academy of Sciences, 290*, 1–435.

Wolff, P. (1972). The interaction of state and non-nutritive sucking. In *Third symposium on oral sensation and perception: The mouth of the infant*. Springfield, IL: Thomas Publishing Company.

Yoburn, B., Morales, R., & Inturrisi, C. (1984). Chronic vascular catheterization in the rat: Comparison of three techniques. *Physiology and Behavior, 33*, 89–94.

Yogman, M., & Zeisel, S. (1983). Diet and sleep patterns in newborn infants. *New England Journal of Medicine, 309*, 1147–1149.

B.
LEARNING

As Spear and Rudy point out, there has been an implicit assumption in the literature that young organisms are either especially poor learners (unable to make connections between reward and stimulus; very quick to forget) or that they are especially good learners (since they are tabula rasa). Neither alternative is true. The young have specific sensory and motor capabilities, as well as social and motivational requirements that must be appreciated for the proper analysis of their behavior. As the authors in this section elegantly demonstrate, young organisms are extremely competent at learning tasks, but often follow different rules than adults. These newly uncovered abilities can be examined both to increase knowledge of the behavioral, social, and neural controls of learning and to illuminate the importance of learning as one of the forces controlling development.

Chapter 5, by Stopfer, Marcus, Nolen, Rankin, and Carew, describes studies that have as their goal the analysis of neural control of behavior. As the title of the chapter states, these researchers have combined a developmental and simple systems approach in that process. This chapter provides an excellent example of a "natural" experiment. The authors use development to disentangle processes that appear at different times during ontogeny. The physiological controls of those processes can thus be identified as the anatomy of the system changes. One of the new concepts that has come from this extremely elegant approach is the fact that dishabituation and sensitization, previously thought to be parts of one underlying facilitatory process, appear at different times during ontogeny and have other features that demonstrate they are actually two separate processes.

Chapter 6, by Spear and Rudy, also covers the concepts of dishabituation and sensitization, among many others. These authors describe a whole series of methods, all of which can be used to study the various processes of learning and memory in rats of different ages. Although designed for rats, these methods can be adapted *with care* to other species as well. Among the questions addressed in this chapter are: What are the rules for learning and memory? Do these rules change at different ages in the young animal? And what can be inferred about the underlying neural processes involved? In a sense, the chapter describes techniques for doing cognitive psychology in developing rats. The question of infantile amnesia is but one area addressed in detail. Since this phenomenon is also found in rats, it is hard to believe that Freud's concept of repression can correctly explain infantile amnesia. In fact, as Spear and Rudy demonstrate, the rules for infantile forgetting appear to be the same as those for adult forgetting.

In Chapter 7, by Galef, the question is not one of neural control, but of social control of behavior. One of the themes of this book is that the methods available to scientists tend to limit or direct the questions asked. Galef's chapter describes an example in which a new way of thinking, a new experimental paradigm, has led to great progress. He demonstrates that once an experimenter becomes aware of the guiding/limiting nature of his paradigm, broad new vistas can be opened up. It is now possible to understand and address the role of social learning in the actual life history of an organism. The social learning area has been put in touch with the biological and behavioral studies of the ethology movement.

A great advantage of comparative studies is that fresh ideas are in effect forced on the researcher. Developmental research shares many of the strengths of comparative work. If one works with a single species at a single age, it can be hard to imagine that behavior could be organized in other fashions, with different rules, with different physiological controls. Along with a fresh understanding of ontogenetic processes, the ideas and rules developed in studying young organisms can lead to new ways to approach learning in adults.

5

Learning and Memory in *Aplysia*: A Combined Developmental and Simple Systems Approach

MARK A. STOPFER, EMILIE A. MARCUS, THOMAS G. NOLEN, CATHARINE H. RANKIN, AND THOMAS J. CAREW

A primary goal of neurobiology is to understand the neural mechanisms underlying behavior. Within this general framework, few issues have attracted more experimental interest than the investigation of the cellular mechanisms responsible for learning and memory. As the work of several contributors to this volume clearly illustrates, a *developmental approach* has been a successful strategy in addressing such questions. The goal of a developmental analysis is to investigate the way in which different learning and memory processes emerge and interrelate during ontogeny. The power of this approach is it can reveal relationships that can be difficult or impossible to establish in the complex circuitry of fully mature adult animals. In recent years this kind of developmental approach has been applied to a variety of forms of learning, most often in altricial mammalian species that are born with relatively immature nervous systems.

A *simple systems approach* to the analysis of learning and memory is another strategy that has proved quite valuable. This strategy typically involves the study of animals whose nervous systems are readily accessible for cellular and biophysical analyses of fundamental neuronal processes. For example, several invertebrate preparations have now been developed in which it is possible to specify the neuronal circuitry involved in learning, and in some cases it has been possible to analyze aspects of the biophysical and molecular mechanisms underlying different forms of learning and memory (Carew & Sahley, 1986; Byrne, 1987; Hawkins, Clark, & Kandel, 1987). In our laboratory we have combined the advantages of a developmental approach with the reductionistic power afforded by a simple systems strategy in our analysis of the development of learning and memory in the marine mollusk *Aplysia californica*.

Aplysia is advantageous for a developmental analysis of learning and memory for several reasons: First, considerable insights have been gained into important aspects of development in *Aplysia*, including the emergence and formation of the major central ganglia (Kriegstein, 1977a), the possible functional significance of specific kinds of neuron-glia interactions (Schacher, Kandel, & Wooley, 1979a,

1979b), the site of origin of identified classes of central neurons (McAllister, Scheller, Kandel, & Axel, 1983; Jacob, 1984), and the ontogeny of different forms of synaptic plasticity (Rayport, 1981; Ohmori, Rayport, & Kandel, 1981; Rayport & Camardo, 1984; Nolen, Marcus, & Carew, 1987; Nolen & Carew, 1988). Second, *Aplysia* has been shown capable of a variety of forms of learning, which range in complexity from nonassociative to associative, and in time course from short term, lasting minutes or hours, to long term, lasting days or weeks. These advantages have enabled significant progress toward elucidating the cellular and molecular mechanisms that contribute to these various forms of learning and memory (Carew, 1987). Third, in *Aplysia* it has been possible to relate a number of identified neurons and neuronal circuits to a variety of discrete behaviors, including such defensive responses as the reflexive withdrawal of mantle organs (the gill and siphon) and the tail, as well as inking, and escape locomotion; and appetitive responses such as feeding and egg laying (Kandel, 1976, 1979).

In recent years we have examined the emergence and assembly of a variety of different forms of learning and memory in *Aplysia*. We have carried out this investigation at a number of levels of analysis. In this chapter we discuss four of them: behavioral, cellular, immunocytochemical, and morphological. In the sections that follow, we illustrate our approach with a single response system, the defensive withdrawal of the siphon. In each section we first describe the rationale for that level of analysis and then, in keeping with the theme of this volume, we describe the principal methodological and conceptual issues of relevance to the analysis.

DEVELOPMENTAL STAGES OF JUVENILE *APLYSIA*

An appreciation of any of the levels of analysis that follow requires a brief introduction to the life cycle of *Aplysia* (Figure 5.1). *Aplysia* development has been characterized by Kriegstein (1977b) as occurring in four phases: *embryonic* (lasting about 10 days), *planktonic* (lasting about 30 days), *metamorphic* (lasting about 3 days), and *juvenile* (lasting about 120 days; see also Rankin & Carew, 1987). The animals become

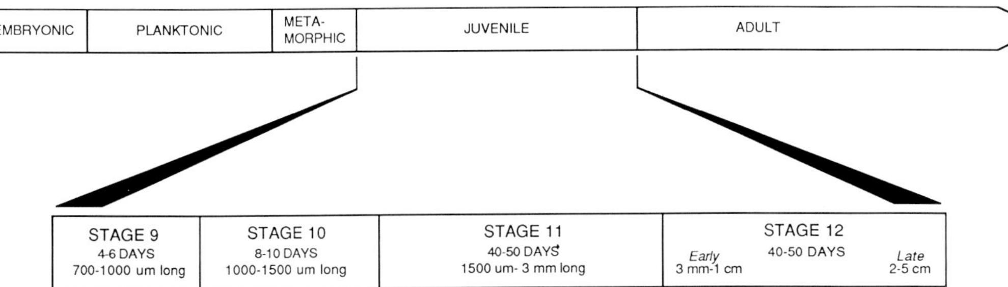

Figure 5.1. Stages in *Aplysia* development. *Aplysia* development can be divided into a series of phases: *embryonic* (lasting about 10 days), *planktonic* (about 30 days), *metamorphic* (about 3 days), *juvenile* (about 120 days), and *adult* (thought to last about 1 year). We have focused upon the juvenile phase, which can be subdivided into four stages, based upon external morphological criteria. The approximate duration of each stage is shown, although this duration can vary substantially with temperature and other rearing conditions.

sexually mature adults at the end of the juvenile stage and are thought to have a lifespan of about 1 year. Each developmental phase can be divided into a series of stages, which are readily identified by discrete external morphological criteria. We began our behavioral analysis in the juvenile phase of development because it is during this phase that the behaviors we are interested in first emerge. The juvenile phase is divided into several stages (Figure 5.1): *Stage 9 Aplysia* are the youngest animals we have studied. They are 700–1,000 μm long, are pink, and lack part of the shell called the *operculum*. However, even at this earliest of juvenile stages, all central ganglia are already essentially in place, and several large neurons can be individually identified. *Stage 10* animals are 1–1.5 mm long and have white spots over the eyes and siphon rim. *Stage 11* animals are 1.5–3 mm long and have the beginnings of rhinophore buds. *Stage 12* animals have a genital groove along the anterior, right side of the body. We have subdivided Stage 12 into early and late substages: *Early Stage 12* (*E12*) animals are 5–10 mm long and *Late Stage 12* (*L12*) animals are 2–5 cm long (Rankin & Carew, 1987).

BEHAVIORAL ANALYSIS

Different Forms of Learning Emerge According to Different Developmental Timetables in *Aplysia*

As mentioned above, in this chapter we restrict our discussion to a single response system in *Aplysia*, the siphon-withdrawal reflex. The siphon is a delicate, spout-shaped structure that serves as an exhalant funnel for seawater circulating over the gill. Stimuli to the siphon or tail cause the animal to briskly withdraw its siphon into the parapodial folds. The siphon-withdrawal reflex is advantageous for study because it is easy to elicit and quantify, and its underlying neural circuitry is relatively well understood.

The preparation used to study the siphon-withdrawal reflex in developing *Aplysia* is shown in Figure 5.2. Animals were viewed through a stereomicroscope fitted with a video camera (JVC S62-U available from any good video supply firm). A timing signal from a video timer (Panasonic Time-Date Generator WJ-810) was superimposed onto the video image. Image and signal were recorded by a videotape recorder (Panasonic NV-8950), which also permitted slow-motion and stop-frame analysis. Animals were restrained above the substrate (under seawater) with three suction micropipettes, one on each parapodium and one on the tail. The reflex was elicited by a brief, reproducible water-jet stimulus to the siphon, delivered through a fourth micropipette by a Picospritzer II (General Valve Corporation) water-jet device. Electric shocks generated by a Grass S88 stimulator could be delivered to the tail via electrodes within the tail-suction micropipette. Reflexes were later quantified by making acetate tracings of the siphon from a high-resolution video monitor (Sony PMV-1270Q). Tracings were made (1) immediately prior to stimulation and (2) at the peak of siphon contraction (Figure 5.2B$_1$ and 5.2B$_2$). These tracings were computer digitized (Bioquant, R&M Biometrics), and the response was quantified (using an IBM PC-XT) as the percentage change in siphon area for each stimulus (Figure 5.2C).

Using these techniques, we examined the emergence of three forms of non-associative learning in *Aplysia*: habituation, dishabituation, and sensitization. Habituation refers to a decrease in response amplitude produced by repeated activation of

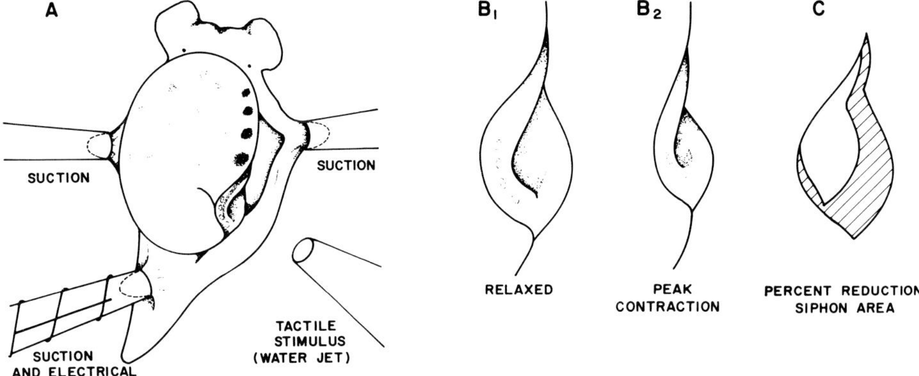

Figure 5.2. Experimental preparation for the study of the siphon withdrawal reflex in juvenile *Aplysia*. (**A**) A juvenile animal is restrained with three suction pipettes. The reflex is evoked with brief tactile stimuli (water-jet) to the siphon. For dishabituation and sensitization training, a series of electric shocks is delivered through the tail pipette. (**B and C**) Quantification of the reflex contraction: (**B₁**) enlarged view of the relaxed siphon 1 s prior to stimulation, (**B₂**) peak of the contraction following tactile stimulation, (**C**) superimposed outlines of **B₁** and **B₂**. The reduction in siphon area (shaded portion) is used as a measure of response amplitude.

the reflex; dishabituation describes the facilitation of an habituated response by the presentation of a strong or novel stimulus; and sensitization refers to the facilitation of a nonhabituated response by a similar novel or noxious stimulus.

We found that habituation is already present in the earliest stage we examined (Stage 9). Dishabituation, however, was absent in Stage 9. Specifically, dishabituation training (consisting of tail shock delivered after the reflex had been habituated with repeated water-jet stimuli) could not induce facilitation of the reflex. Dishabituation did not emerge until about one week later, in Stage 10 (Rankin & Carew, 1987).

Having established that habituation exists as early as Stage 9, a week prior to the emergence of dishabituation we turned our attention to sensitization. Until recently, dishabituation and sensitization had commonly been considered to reflect a single underlying facilitatory process (Groves & Thompson, 1970; Carew, Castellucci, & Kandel, 1971). We addressed the question of whether dishabituation and sensitization reflected a unitary process by determining behaviorally when they emerged during development. Our rational was that if only a single process is involved, then, on-togenetically, both forms of learning must emerge at the same time. Alternatively, if more than one process is involved, it might be possible to separate the processes developmentally. To examine this possibility, Rankin and Carew (1988) analyzed the effects of tail shock on both habituated and nonhabituated responses in developmental stages 11, E12, and L12. In each stage, two groups of animals were examined. One group received dishabituation training: first, an habituating series of mild water-jet stimuli to the siphon and then a tail shock; the other group received sensitization training: first, several baseline water-jet stimuli delivered at an ISI that was too long to produce habituation and then a tail shock identical to that of the dishabituation group. Both groups were then tested for response facilitation resulting from the tail shock. This analysis revealed two principal findings: (1) sensitization emerges *several weeks* after the emergence of dishabituation, between Early and Late Stage 12 of juvenile development; and (2) tail shock actually produces *inhibition* of the reflex response in young animals prior to the development of sensitization.

In a recent series of experiments Rankin and Carew (1989) further examined this inhibitory process in stages 11 and E12 and found that its magnitude (measured as the degree to which it could decrement siphon withdrawal) increased directly as a function of tail-shock intensity. Interestingly, the magnitude of dishabituation showed an *inverse* relationship: the stronger the tail shock, the *weaker* the dishabituation. These observations suggest that the inhibitory process can compete with the expression of the faciliatory process of dishabituation.

Having behaviorally dissociated three forms of behavioral plasticity (dishabituation, sensitization, and inhibition) on the basis of their different developmental timetables, it was of interest to determine whether these same processes could be behaviorally dissociated in the adult. Recent work by Marcus, Nolen, Rankin, and Carew (1988) shows that these processes can indeed be dissociated in adult *Aplysia* on the basis of two independent criteria: (1) their time of onset and (2) their stimulus-intensity requirements.

In summary, our behavioral analysis revealed that habituation, sensitization, and inhibition can be dissociated in juvenile *Aplysia* on the basis of their ontogenetic timetables. The developmental separation of dishabituation and sensitization (by at least 60 days) was especially striking and suggested that these two forms of learning might indeed reflect distinct cellular mechanisms (see also Hochner, Braha, Klein, & Kandel, 1986; Hochner, Klein, Schacher, & Kandel, 1986). Finally, our ability to dissociate developmentally different forms of learning prompted us to examine whether these same forms of learning could be dissociated in adult animals. We found that they could, both on the basis of their time of onset and their stimulus requirements. Thus, a developmental analysis provided important insights not only into the ontogeny of learning but also into different forms of learning expressed in adult animals.

CELLULAR ANALYSIS

The behavioral studies described above revealed that *Aplysia*'s ability to learn and remember reflect separate, interacting processes that develop gradually during the course of juvenile life. These observations led us to investigate the cellular mechanisms responsible for this developmental dissociation. A first step in this analysis required establishing that cellular analogs of these forms of learning (modified neural responses observed in reduced preparations) could be identified and analyzed in the CNS of juvenile *Aplysia* at the same stages of development that those forms of learning were first behaviorally expressed. The neural circuit that controls the siphon-withdrawal reflex has not yet been delineated in juveniles, but has been localized to the abdominal ganglion (Rankin & Carew, 1987). However, the giant motor neuron R2 (Figure 5.3), which receives afferent input from the siphon, can provide a cellular vantage point to monitor plastic changes that emerge in the reflex pathway for siphon withdrawal. This neuron (also located in the abdominal ganglion) is readily identifiable throughout juvenile development.

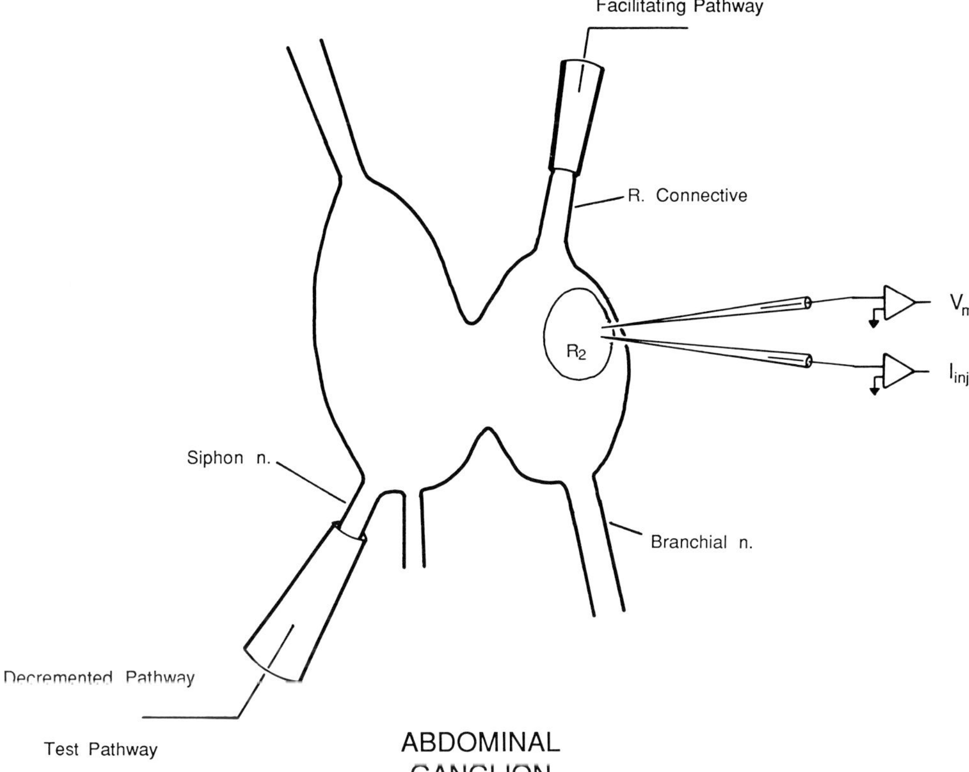

Figure 5.3. Physiological preparation for the examination of the development of cellular analogs of dishabituation and sensitization. Intracellular electrodes were used to record the membrane potential (V_m) of the identified motor neuron R2 in the abdominal ganglion of Stage 12 animals. In Late Stage 12 animals, a second electrode (I_{inj}) was used to hyperpolarize the cell. In Early and Middle Stage 12 animals hyperpolarizing current was passed through the recording electrode by means of a bridge circuit. Afferent input from the siphon was activated by a brief shock to the siphon nerve, which evoked a complex EPSP in R2. Input from the tail pathway was activated by a train of brief shocks to the right connective.

Since our behavioral studies indicated that sensitization emerged during Stage 12 of juvenile development, Nolen and Carew (1988) focused on this stage for their studies of the emergence of the cellular analog of sensitization. Animals were anesthetized with an injection of $MgCl_2$ in a concentration roughly five times that normally found in the hemolymph. (The effect of a high Mg^{++} solution is to block all chemical synaptic transmission.) Animals were then pinned ventral side up to the bottom of a Sylgard-coated petri dish. The foot was bisected along the midline, and the abdominal ganglion was removed and transferred to a Sylgard-coated recording dish containing normal seawater (which is similar in basic composition to *Aplysia* hemolymph).

To mimic water-jet stimuli to the siphon skin, brief electrical pulses were delivered to the siphon nerve through glass suction electrodes (see Figure 3.5), thus activating afferent axons from sensory neurons, which, through interneurons, pro duce a complex excitatory postsynaptic potential (EPSP) in R2. Tail shock was mimicked by delivering stronger electrical pulses to the right pleural-abdominal connective. Intracellular recordings were made from R2 using standard electrophysiology techniques. Subthreshold EPSP records were simultaneously displayed on a dual-beam oscilloscope (Tektronix 5111 A) and recorded on FM tape (Teac 1230 tape deck or Vetter model 2D FM recording adaptor), as well as on a chart recorder (Gould 2400S) (see Nolen & Carew, 1988, for details).

Two stimulus paradigms were used: (1) a dishabituation analog, which consisted of 15 presentations of siphon nerve stimuli at an habituating ISI, then a train of shocks to the connective, then 3 more siphon nerve (test) stimuli; and (2) a sensitization analog, consisting of 3 stimuli to the siphon nerve at a nondecrementing ISI, then stimulation of the right connective (using the same stimulation parameters as in the dishabituation analog), and then 3 more siphon nerve stimuli. Thus, the only difference between the two paradigms was that connective (tail) input was superimposed on either a decremented EPSP or a nondecremented EPSP.

In this study, Stage 12 was divided into three substages: Early, Middle, and Late. Consistent with our previous, behavioral results, the cellular analog of *dishabituation* (facilitation of *decremented* EPSPs) was present as early as Early Stage 12. Furthermore, as one would predict from behavioral findings, the analog of *sensitization* was clearly present in Middle and Late Stage 12 animals but not in Early Stage 12 animals. Specifically, connective stimulation in these substages produced significant facilitation of *nondecremented* EPSPs. However, in contrast to the older substages, connective stimulation produced *no* facilitation of nondecremented EPSPs in Early Stage 12; thus, the analog of sensitization was absent in this early stage. In fact, in Early Stage 12 connective stimulation actually produced significant *inhibition* of EPSP amplitude.

In summary, there are two clear parallels between the development of behavioral sensitization and the development of its cellular analog: First, the cellular analog of sensitization (produced by activation of the tail pathway) emerges in the same late juvenile stage of development as does behavioral sensitization (Rankin & Carew, 1987); second, the cellular analysis of sensitization revealed an inhibitory cellular process produced by connective stimulation that parallels the behavioral inhibition produced by tail shock prior to the emergence of behavioral sensitization. Thus, there is a close temporal correspondence between the development and maturation of behavioral sensitization and a cellular representation of that process in the CNS of juvenile *Aplysia*.

IMMUNOCYTOCHEMICAL ANALYSIS

Our behavioral analysis revealed that the capacity for sensitization in the siphon-withdrawal reflex emerges during Stage 12 of development, and our cellular analysis paved the way toward a mechanistic analysis of the processes underlying the expression of this form of learning. One possibility that can be tested is that young juveniles (earlier than Middle Stage 12) lack one or more of the facilitating neurotransmitters necessary for the expression of sensitization. The transmitter serotonin (5-HT) is a particularly interesting candidate because it is known to produce heterosynaptic facilitation in neural circuits underlying siphon withdrawal, as well as in other response systems in adult *Aplysia* (for review, see Carew, 1987).

To examine the developmental expression of serotonin-containing neurons throughout juvenile development, Nolen, Mindell, and Carew (1986) used immunocytochemical techniques to identify 5-HT immunoreactive neurons in stages 9–12. Complete central nervous systems were dissected and then fixed in a paraformaldehyde solution. In order to identify neurons containing 5-HT, the nervous systems were then incubated with a primary antibody raised in rabbits against 5-HT. Next, in order to amplify and visualize the 5-HT reactive neurons, nervous systems were incubated with the secondary antibody (goat antirabbit antisera), which was tagged with a fluorescent marker, either Fluorescein or Rhodamine. The resulting preparations were then analyzed and photographed using a fluorescence microscope (Zeiss IM35).

Using these techniques, Nolen and colleagues (1986) found that, in general, the numbers of 5-HT immunoreactive neurons increased gradually in all ganglia throughout the juvenile phase. By comparing cell position, size, and fluorescence characteristics, they also identified a candidate subpopulation of 5-HT immunoreactive neurons in the cerebral ganglion that has been shown in the adult to produce heterosynaptic facilitation in the neural circuit for siphon withdrawal (Mackey, Hawkins, & Kandel, 1986). In contrast to the cells in other central ganglia, these cerebral neurons appeared to be in place very early, by Stage 10, and showed no further increase in cell numbers through development. Thus, the data do not appear to support the hypothesis that the developmental emergence of 5-HT neurons is responsible for the emergence of sensitization, since candidate facilitatory neurons containing 5-HT are present long before sensitization is first expressed.

A second, peptidergic class of facilitatory neurotransmitters has also been identified in *Aplysia*. One of them, Small Cardioactive Peptide B (SCP$_B$), has been implicated in synaptic facilitation contributing to sensitization in *Aplysia* (Abrams, Bernier, Hawkins, & Kandel, 1984). Thus, it was of interest to examine the developmental expression of the facilitatory neuropeptide SCP$_B$. Toward that end, Harkins, Worrall, Marcus, Nolen, and Carew (1988) recently used immunocytochemical techniques similar to those described above in order to localize SCP$_B$ containing neurons in developing *Aplysia*. They examined animals of two juvenile stages: Stage 11, when sensitization is absent, and Stage 12, when sensitization first emerges. Interestingly, they observed a fivefold increase in the number of SCP$_B$ immunoreactive neurons between Stages 11 and 12, raising the possibility that some aspect of the expression of this neuropeptide may contribute to the emergence of sensitization. It will be of interest to determine the functional significance of this neuropeptide in the context of the development of sensitization.

MORPHOLOGICAL ANALYSIS

Thus far we have described behavioral, cellular, and immunocytochemical studies directed at the development of learning. Another important level of analysis concerns the anatomical substrate upon which the learning is superimposed. One possible antecedent to the expression of sensitization, as well as other forms of learning, may be the assembly of particular neural circuits required for learning. As a first step toward examining this possibility, Cash and Carew (1987, 1989) determined the number of neurons that are present in each of the major central ganglia and estimated the neuropilar volume of representative ganglia at each stage of juvenile development.

Animals were anesthetized and dissected to expose the CNS. In early juvenile stages (9–12E) the nervous systems were fixed (with Bouin's solution) within the animal; in stages 12L and Early Adult the central ganglia were first removed from the animal and pinned in a Sylgard dish prior to fixation. Preparations were dehydrated in ethanol and xylene, embedded in paraffin, and sectioned (at 10 μm) using a rotary microtome. Sections were mounted on slides, rehydrated, and stained with cresyl violet.

Cells were identified as neurons by means of a number of conventional criteria, including their relative large cell bodies, pale coloration, and presence of chromatin bodies (Coggeshall, Kandel, Kupfermann, & Waziri, 1966). Cell counts were made in every section of every ganglion for each animal studied. Because of the possibility of multiple counting of the same cell in different sections, an appropriate correction factor, based on the average cell diameter, was used (Abercrombie, 1946). In addition, neuropile volume was estimated by examining two ganglia: (1) the abdominal ganglion, which contains relatively few neurons, and (2) the pedal ganglion, which contains the largest number of neurons. To estimate neuropilar volume, in each tissue section the perimeter of the neuropile was traced using a camera lucida. The perimeter was digitized, and its area was computed using a Bioquant morphometry program. Neuropilar volume of each ganglion was then estimated by adding the successive areas of the serial sections and multiplying the sum by the thickness of the sections.

Cash and Carew (1987, 1989) found that a striking, nonlinear proliferation of neurons occurs during juvenile development. By far the largest increase in cell numbers occurs from Early to Late Stage 12 (which is the same time sensitization first emerges in both behavioral and cellular preparations); during this time there is more than a doubling of the number of neurons—from 2,900 to 6,700. The neuronal proliferation is system-wide, occurring in each of the central ganglia simultaneously, suggesting the action of a general developmental signal or trigger (perhaps a hormone). In Stage 12 there is also a very large increase in neuropilar volume (150-fold), increasing the opportunity for synaptic interaction. This large proliferation of cells is not explained by changes in body weight, since the animal acquired over 80% of its total complement of Early Adult neurons long (several weeks) before it attained its Early Adult body weight.

The system-wide increase in central neuronal population size raises two important questions: First, where are the neurons coming from? Migration of neurons from the periphery into the CNS in developing *Aplysia* has already been described by McAllister et al. (1983) and Jacob (1984). In both studies, central neurons were found originating in the ectoderm of the body wall and then migrating into the CNS during development. Recent experiments by Hickmott and Carew (1988) support the hypothesis that this form of migration may be an important source of the new neurons observed in Stage 12. They used a [³H]-thymidine labeling technique to localize newly divided cells in two different stages of juvenile development: Stage 11 (before large-scale proliferation of neurons occurs) and Stage 12 (when proliferation is greatest). Examining the abdominal ganglion, they found that proliferative zones in the body wall immediately adjacent to the ganglion were significantly more productive in Stage 12 compared to Stage 11. Moreover, they found that once cells arrived at the abdominal ganglion, they appeared to continue their migration into deeper regions of the ganglion. Thus, the new neurons appear to be generated in the body wall and subsequently migrate to final destinations within the ganglion.

A second important question concerns the functional significance of the rapid proliferation of neurons observed during State 12. It is interesting to note that long before the large increase in neurons occurs, most of the basic behaviors of *Aplysia*, such as feeding, locomotion, inking, and a variety of withdrawal reflexes, are already intact. The only major behaviors not yet in place are reproductive activities, such as mating and egg laying, which do not emerge until weeks or months later in the adult stage. Thus, the neural circuitry underlying much of the behavioral repertoire of *Aplysia* is intact and functional prior to the large-scale addition of neurons in juvenile development. An interesting possibility we are examining is that the primary role of at least some of the newly added neurons might be modulatory, serving to fine tune previously existing sensory and motor systems and permitting, perhaps for the first time, the expression of a variety of forms of behavioral and synaptic plasticity.

SUMMARY AND PERSPECTIVES

A Simple Systems Approach Offers Advantages for Analyzing the Development of Learning

Invertebrate preparations offer a number of advantages for the analysis of developmental processes in general and the development of learning and memory capabilities in particular. For example, such preparations often contain relatively large central neurons that are amenable to both electrophysiological and biochemical analyses by virtue of both their size and accessibility within the CNS. Moreover, in many cases these neurons are readily and repeatedly identifiable in every animal, greatly facilitating the ability to specify the particular role individual neurons play, both in identified neural circuits and in the developmental expression of specific behaviors. Finally, in many invertebrates, including *Aplysia*, the developmental sequences involved in the expression and assembly of identified neuronal circuits within the CNS throughout ontogeny are relatively well characterized.

In our investigation of the development of learning in *Aplysia*, we have taken advantage of some of the features of invertebrate preparations described above. Specifically, we have been able to differentiate behaviorally two forms of response facilitation, dishabituation and sensitization, as they emerge according to different developmental timetables. This analysis also revealed a novel inhibitory process in *Aplysia* that can be masked by sensitization in the adult. In subsequent physiological studies we found that the cellular analogs of these three different forms of behavioral plasticity emerged in close temporal register with their respective behavioral counterparts. Coupled with immunocytochemical and anatomical approaches, we have begun to pinpoint important cellular events that may underlie the development of specific classes of learning, such as sensitization. A general summary of our combined results is shown in Figure 5.4. Based on these results, a cellular analysis in juvenile *Aplysia* now affords the opportunity to investigage the emergence of particular synaptic,

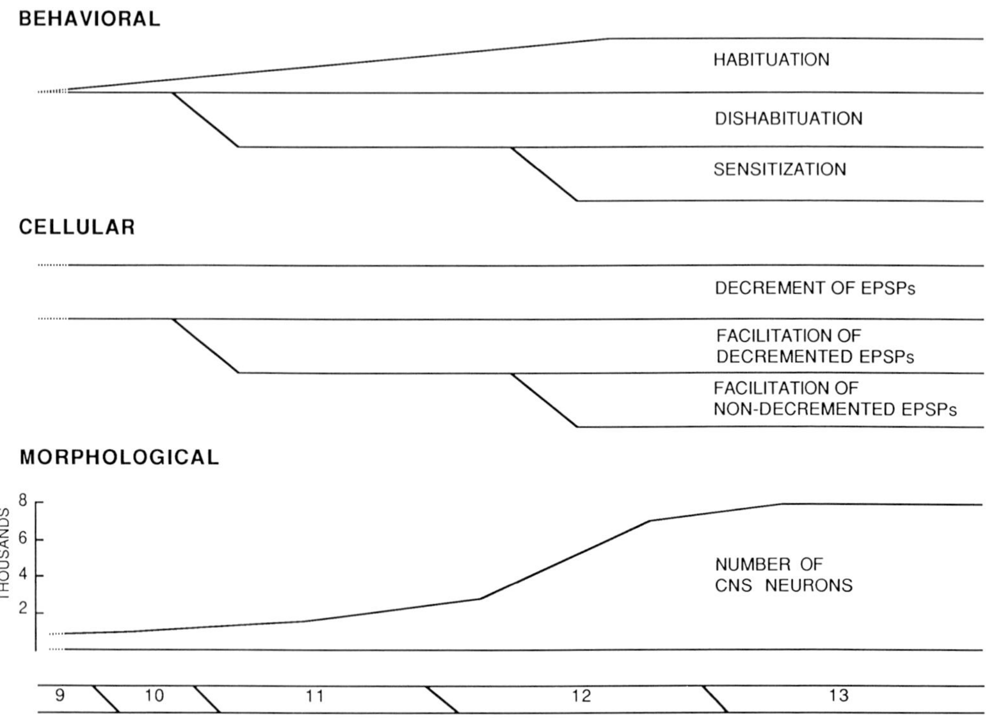

Figure 5.4. Summary of developmental timetables for simple forms of learning in *Aplysia*: behavioral, cellular, and morphological correlates. Several forms of nonassociative learning emerge in the siphon-withdrawal reflex at different times during juvenile development. Habituation and its cellular analog (homosynaptic decrement of EPSPs in R2) emerge by Stage 9; dishabituation and its cellular analog (facilitation of decremented EPSPs in R2) emerge in Stage 10; sensitization and its cellular analog (facilitation of nondecremented EPSPs in R2) emerge much later (approximately 60 days), in Middle to Late Stage 12. During this same developmental stage, the central nervous system exhibits striking neuronal proliferation.

biophysical, and molecular mechanisms that may play critical roles in the developmental expression of particular forms of learning and memory.

Dissociation of Dishabituation, Sensitization, and Inhibition Suggests a Multiprocess View of Nonassociative Learning

The behavioral dissociation of dishabituation, sensitization, and inhibition, which was first revealed by our developmental analysis and subsequently confirmed by studies on adult animals, has important implications for theoretical views of nonassociative learning. Until recently, a commonly held view was that nonassociative processes such as habituation, dishabituation, and sensitization could be explained by a dual-process theory, originally outlined by Groves and Thompson (1970) and further supported by Carew, et al. (1971). The dual-process view involves a single decrement-ing process that gives rise to habituation and a single facilitatory process that gives rise to both dishabituation and sensitization. However, in this review, we have presented evidence in *Aplysia* showing that dishabituation and sensitization differ in three fundamental ways: (1) in terms of the *developmental timetables* (dishabituation emerges 60 days before sensitization emerges); (2) in terms of their *time of onset* (dishabituation is expressed immediately following a shock, whereas sensitization is not expressed until 20–39 min later); and (3) in terms of their *stimulus requirements* (dishabituation is produced by weak tail shock, whereas sensitization is produced by strong tail shock). Collectively, these results show that a simple dual-process view is inadequate to account for the behavioral features of dishabituation and sensitization in *Aplysia*. Therefore, we propose that a multiprocess view, involving multiple facilitatory processes or interactions between facilitatory and inhibitory processes, is necessary to account fully for nonassociative learning.

Current Research Directions

We are currently extending our studies of the development of learning and memory in three principle directions: First, thus far we have focused only on nonassociative learning, but *Aplysia* exhibits several forms of associative learning as well, including differential classical conditioning (Carew, Hawkins, & Kandel, 1983), operant conditioning (Cook & Carew, 1986), and contingency effects (Hawkins, Carew, & Kandel, 1983). It will be extremely interesting to determine the developmental timetables for these different forms of associative learning and compare them to those already established for nonassociative processes.

Second, we have thus far examined only short-term memory for different forms of learning, but *Aplysia* is also capable of exhibiting long-term memory, lasting days and weeks, for several types of learning. Thus, it will be important to analyze the development of long-term memory and compare its timetable to the expression of short-term memory process.

Finally, we have restricted our cellular analysis thus far to cellular analogs utilizing complex EPSPs in the giant neuron R2. It is important to extend these studies to an analysis of identified sensory and motor neurons in the neural circuits underlying the learning we have examined. The cellular and molecular mechanisms underlying a variety of forms of learning and memory in adult *Aplysia* have been extensively investigated and, in some cases, are very well understood. Thus, using a developmental approach in identified neural circuits it may be possible to gain insights into these cellular and molecular mechanisms by studying their developmental expression and assembly.

ACKNOWLEDGMENTS

We are very grateful to the Howard Hughes Medical Institute for its generous support in supplying juvenile *Aplysia*. This work has been supported by a NSF predoctoral fellowship (to E.A.M.), by NIH NRSA grant 5-F32-NS-07480 (to T.G.N.), and by NSF grant BNS8311300, NIMH grant MH41083, and ONR grant N00014-87-K-0381 (to T.J.C.).

REFERENCES

Abercrombie, M. (1946). Estimation of nuclear population from microtome sections. *Anatomy Research, 94*, 239–247.

Abrams, T. W., Bernier, L., Hawkins, R. D., & Kandel, E. R. (1984). Possible roles of Ca^{++} and cAMP in activity-dependent facilitation, a mechanism for associative learning in *Aplysia*. *Society for Neuroscience Abstracts, 10*, 269.

Byrne, J. H. (1987). Cellular analysis of associative learning. *Physiological Reviews, 67*(2), 329–439.

Carew, T. J. (1987). Cellular and molecular advances in the study of learning in *Aplysia*. In J. P. Changeaux & M. Konishi (Eds.), *The neural and molecular basis of learning* (pp. 177–204). New York: Wiley.

Carew, T. J., Castellucci, V. F., & Kandel, E. R. (1971). An analysis of dishabituation and sensitization of the gill-withdrawal reflex in *Aplysia*. *International Journal of Neuroscience, 2*, 79–98.

Carew, T. J., Hawkins, R. D., & Kandel, E. R. (1983). Differential classical conditioning of a defensive withdrawal reflex in *Aplysia californica*. *Science, 219*, 397–400.

Carew, T. J., & Sahley, C. L. (1986). Invertebrate learning and memory: From behavior to molecules. *Annual Review of Neuroscience, 9*, 435–487.

Cash, D., & Carew, T. J. (1987). A quantitative analysis of the development of the CNS in juvenile *Aplysia*. *Society for Neuroscience Abstracts, 13*, 816.

Cash, D., & Carew, T. J. (1989). A quantitative analysis of the development of the central nervous system in juvenile *Aplysia californica*. *Journal of Neurobiology, 20*(1), 25–47.

Coggeshall, R. E., Kandel, E. R., Kupfermann, I., & Waziri, R. (1966). A morphological and functional study on a cluster of identifiable neurosecretory cells in the abdominal ganglion of *Aplysia californica*. *Journal of Cell Biology, 31*, 363–368.

Cook, D. G., & Carew, T. J. (1986). Operant conditioning of head-waving in *Aplysia*. *Proceedings of the National Academy of Science, USA, 83*, 1120–1124.

Groves, P. M., & Thompson, R. F. (1970). Habituation: A dual process theory. *Psychological Review, 77*, 419–450.

Harkins, E. W., Worrall, B. B., Marcus, E. A., Nolen, T. G., & Carew, T. J. (1988). Developmental expression of facilitatory neuropeptides in *Aplysia*. *Society for Neuroscience Abstracts, 14*, 163.

Hawkins, R. D., Carew, T. J., & Kandel, E. R. (1983). Effects of interstimulus interval and contingency on classical conditioning in *Aplysia*. *Society for Neuroscience Abstracts, 9*, 168.

Hawkins, R. D., Clark, G. A., & Kandel, E. R. (1987). Cell biological studies of learning in simple vertebrate and invertebrate systems. In F. Plum (Ed.), *Handbook of physiology, Section I. Higher functions of the nervous system* (Vol. 6, pp. 25–83). Bethesda, MD: American Physiological Society.

Hickmott, P. W., & Carew, T. J. (1988). An autoradiographic analysis of neuronal proliferation in juvenile *Aplysia*. *Society for Neuroscience Abstracts, 14*, 163.

Hochner, B., Braha, O., Klein, M., & Kandel, E. R. (1986). Distinct processes in presynaptic facilitation contribute to sensitization and dishabituation in *Aplysia*: Possible involvement of C kinase in dishabituation. *Society for Neuroscience Abstracts, 12*, 1340.

Hochner, B., Klein, M., Schacher, S., & Kandel, E. R. (1986). Additional component in the cellular mechanism of presynaptic facilitation contributes to behavioral dishabituation in *Aplysia*. *Proceedings of the National Academy of Science, USA, 83*, 8794–8798.

Jacob, M. H. (1984). Neurogenesis in *Aplysia californica* resembles nervous system formation in vertebrates. *Journal of Neuroscience, 4*, 1225–1239.

Kandel, E. R. (1976). *Cellular basis of behavior*. San Francisco: Freeman.

Kandel, E. R. (1979). *Behavioral biology of Aplysia*. San Francisco: Freeman.

Kriegstein, A. R. (1977a). Development of the nervous system of *Aplysia californica*. *Proceedings of the National Academy of Science, USA, 74*, 375–378.

Kriegstein, A. R. (1977b). Stages in the post-hatching development of *Aplysia californica*. *Journal of Experimental Zoology, 199*, 275–288.

Mackey, S. L., Hawkins, R. D., & Kandel, E. R. (1986). Neurons in 5-HT containing region of cerebral ganglia produce facilitation of LE cells in *Aplysia*. *Society for Neuroscience Abstracts, 12*, 1340.

Marcus, E. A., Nolen, T. G., Rankin, C. H., & Carew, T. J. (1988). Behavioral dissociation of dishabituation, sensitization, and inhibition in *Aplysia*. *Science, 241*, 210–213.

McAllister, L. B., Scheller, R. J., Kandel, E. R., & Axel, R: (1983). In situ hybridization to study the origin and fate of identified neurons. *Science, 222*, 800–808.

Nolen, T. G., Marcus, E. A., & Carew, T. J. (1987). Development of learning and memory in *Aplysia*. III. Central neuronal correlates. *Journal of Neuroscience, 7*, 144–153.

Nolen, T. G., & Carew, T. J. (1988). The cellular analog of sensitization emerges at the same time in development as behavioral sensitization in *Aplysia*. *Journal of Neuroscience, 8*, 212–222.

Nolen, T. G., Mindell, J. A., & Carew, T. J. (1986). Development of neurotransmitters implicated in learning in *Aplysia*. *Society for Neuroscience Abstracts, 12*, 399.

Ohmori, H., Rayport, S. G., & Kandel, E. R. (1981). Emergence of posttetanic potentiation as a distinct phase in the differentiation of an identified synapse in *Aplysia*. *Science, 213*, 1016–1018.

Rankin, C. H., & Carew, T. J. (1987). Development of learning and memory in *Aplysia*. II. Habituation and dishabituation. *Journal of Neuroscience, 7*, 133–143.

Rankin, C. H., & Carew, T. J. (1988). Dishabituation and sensitization emerge as separate processes during development in *Aplysia*. *Journal of Neuroscience, 8*, 197–211.

Rankin, C. H., & Carew, T. J. (1989). Developmental analysis in *Aplysia* reveals inhibitory as well as facilitatory effects of tail shock. *Behavioral Neuroscience, 103*(2), 334–344.

Rayport, S. G. (1981). Development of the functional and plastic capabilities of neurons mediating a defensive behavior in *Aplysia*. Unpublished doctoral dissertation, Columbia University, New York.

Rayport, S. G., & Camardo, J. S. (1984). Differential emergence of cellular mechanisms mediating habituation and sensitization in the developing *Aplysia* nervous system. *Journal of Neuroscience, 4*, 2528–2532.

Schacher, S., Kandel, E. R., & Woolley, R. (1979a). Development of neurons in the abdominal ganglion of *Aplysia californica*. I. Axosomatic synaptic contacts. *Developmental Biology, 71*, 163–175.

Schacher, S., Kandel, E. R., & Woolley, R. (1979b). Development of neurons in the abdominal ganglion of *Aplysia californica*. II. Non-neural support cells. *Developmental Biology, 71*, 176–190.

6

Tests of the Ontogeny of Learning and Memory: Issues, Methods, and Results

NORMAN E. SPEAR AND JERRY W. RUDY

Experimental psychologists and neuroscientists have often been criticized for relying on the rat as the primary animal for the study of learning and memory. This animal, nevertheless, remains the subject of choice in the vast majority of such studies, in part because more is known about the learning and memory capacities of the rat than of other nonhuman mammals. For research in development, the rat is especially attractive because a great deal of its development is postnatal and thus relatively accessible for study.

As a subject in developmental studies of learning, the rat has a long history. John B. Watson in his doctoral dissertation attempted to relate changes in the rat's learning capacity during ontogeny to brain growth (Watson, 1903), and several other studies reported prior to World War II used the rat in attempts to relate age and learning ability. These studies, however, were often irredeemably flawed. In what now must be considered a watershed paper, Campbell (1967) soundly criticized the early work and pointed to a number of problems, including the failure of these studies to (1) equate motivation level and reward magnitude across ages, (2) recognize age-related differences in the unlearned behaviors the subjects brought to the task, (3) control the effects of previous experiments, and (4) employ adequate measures of learning. Campbell (1967) followed his critique with a description of a series of experiments from his laboratory illustrating how some of these methodological pitfalls can be avoided and, in the process, provided an empirical foundation for the study of age-related differences in learning and retention for the next decade.

The studies of Campbell (1967) also provided an important technical development that has allowed more effective tests of learning and memory in the developing rat. These studies established a type of mildly punishing foot shock that is equally detectable and aversive for the rat from weaning through adulthood. This source of negative reinforcement, already useful methodologically for establishing learning and memory without otherwise intruding in the animal's daily regimen (as, for example, by deprivation of food), could henceforth be included in a wide age range of tests to assess learning without fear of confounding by age-related differences in motivation or reinforcement. The methodological points Campbell made remain as fundamental today as they were 20 years ago, and we highly recommend this paper to the uninitiated reader of our chapter.

Considerable progress was made in studying age-related changes in learning and retention in the 1960s and early 1970s, but with few exceptions the youngest rats included in developmental studies of learning and memory were well into their third week postnatal. During the mid-1970s, however, new techniques emerged to permit tests of the learning and memory capacities of perinatal rats less than a week old. For instance, an ingeniously simple device has permitted controlled infusion of liquids into the mouth of an infant rat long before the animal is capable otherwise of drinking controlled amounts of liquid on its own (Hall & Rosenblatt, 1977); other techniques have exploited the olfactory abilities of the infant rat, the infant's proclivity to return to its "home," and other infant-particular dispositions (Gemberling, Domjan, & Amsel, 1980; Johanson & Hall, 1979; Johanson & Teicher, 1980; Kenny & Blass, 1977; Rudy & Cheatle, 1977, 1979; Smotherman, 1982; for reviews, see Spear & Campbell, 1979).

Recent developments in the field have provided the contemporary researcher with a rich assortment of methods that can be used not only in identifying age-related changes in learning and memory throughout the perinatal period and beyond (e.g., Johanson, Hall, & Palefone, 1984; Kucharski & Hall, 1987; Castro, Paylor, & Rudy, 1987) but also in studying how learning and memory processes contribute to the general adaptation of the infant animal to its own special needs (e.g., Alberts & May, 1984; Galef & Clark, 1971; Terry & Johanson, 1987; Sullivan, Brake, Hofer, & Williams, 1980; Pederson, Williams, & Blass, 1982). We judge our task to be that of making a representative sampling of these advances accessible to the nonspecialists.

Whether learning proceeds more rapidly during early development than later is no longer a particularly useful question. Even if we believed that the basic mechanism of the learning we observe is subject to change during development, it is unclear that such a mechanism would be unitary in applying to all types of learning. And it would in any case be difficult to separate the development of such a mechanism from age-related changes in organizational features such as stimulus selection or encoding. Despite implicit assumptions in the developmental literature that younger animals possess a basic mechanism of learning that is relatively ineffective, or, alternatively, possess a special disposition for plasticity that makes them especially good learners, most studies of learning development are no longer phrased in those terms.

The conceptual basis for studying the development of learning and memory has instead been diverse. The impetus for most studies has been rather functionalistic in character, with additional orientation provided by the special problems of adaptation during development or general principles and theories of learning. During recent years the major studies of learning development have been concerned with the following issues: Does the capacity for learning within a particular sensory system develop independently of sensory-perceptual and motivational factors (e.g., Rudy, Vogt, & Hyson, 1984)? What features of the development of learning might help us to understand the phenomenon of infantile amnesia (Campbell & Spear, 1972; Campbell & Coulter, 1976)? What is the role of early learning as an adaptive mechanism in the development of regulatory mechanisms (Alberts, 1984; Gubernick & Alberts, 1984; Blass & Kehoe, 1987; Galef, 1985; Spear, Kucharski, & Hoffmann, 1985)? How is early learning affected by early brain insult or enrichment (e.g., Riley, 1990)? Do the mechanisms and content of learning change ontogentically (Spear, 1979a, 1984; Spear & Molina, 1987)? What can learning during early development tell us about the generality of learning theories based on adults (Amsel, 1986)? and what peculiarities in the developing brain might lead to peculiarities in the learning of developing animals (Amsel, 1986; Kucharski & Hall, 1987; Myslivecek & Hassmannova, 1983)?

For organizational purposes, we classify the methods and procedures we describe into somewhat arbitrary groupings. The first includes *nonassociative learning*. Here we review paradigms and procedures which reveal that the animal's behavior can be modified simply as a consequence of being exposed to some isolated feature of the environment. Next, we describe procedures that fall under the heading *associative learning*. In this case, reference is to those instances of experience-induced behavior change that can be shown to depend at least in part on the animal experiencing a relationship between two or more events. The arbitrariness of this categorization is obvious. For example, although the operations of a nonassociative procedure may involve the experimental variation of only a single event, the behavior change observed may in fact be mediated by associative learning initiated by that event. Likewise, nonassociative learning processes may contribute to the behavior changes produced by the experimental operations that define an associative learning procedure. A third section of the paper attempts to go beyond the basic associative paradigms. It introduces the procedures that enable the researcher to study the development of some of the interesting cognitive processes the rat can employ to learn about the consequences of its behaviors and the spatial layout of its environment. The final section addresses *infantile amnesia* and some methodological and procedural features in studies of ontogenetic change in memory—retention and forgetting.

The specific methdologies we describe are selected in part because they allow reasonably clear conclusions about the *ontogeny* of learning and memory. The methodological difficulties in arriving at conclusions about individual differences attributable to age have been illustrated throughout the present book. For interpretation of learning and memory, the difficulties center on the need to control potential differences in basic psychological processes that are likely to differ ontogentically and also are likely to influence learning and memory. It is the most fundamental processes—sensory detection, perception, motivation, reinforcement—that must be controlled.

The present review, thus, is necessarily selective. We focus on tests that have been devised in the context of some larger theoretical issue of development. These issues can be discussed only briefly in each case, but their resolution is in our view the ultimate value of the memory tests we describe; and, without them, the tests would be little more than an exercise in engineering.

NONASSOCIATIVE LEARNING ASSESSING THE INFLUENCE OF MERE EXPOSURE TO A STIMULUS

Perhaps the simplest question one can ask about learned influences on behavior is, How does an organism's behavior change as a consequence of being exposed to a particular environmental event? Although the most elementary of learning tests, the effects of mere exposure to a stimulus can lead to (1) increased preference for that stimulus (sometimes referred to as *reduced neophobia*), (2) diminished responding to that stimulus (habituation), (3) enhanced responding to that stimulus or some other stimulus (sensitization), or (4) reduction in the ability of that stimulus to acquire or express conditioned properties (latent inhibition). (Evidence is clear, incidentally, that (1) and (3) are different effects, and so are (2) and (4), despite their superficial similarity.) Examples of each outcome have been observed in the developing rat.

Exposure-Induced Preference Change

Several laboratories have reported that mere exposure to an odor results in the infant rat subsequently preferring that odor to a second novel odor (Leon, Galef, & Behse, 1977; Alberts & May, 1984; Caza & Spear, 1984). The work of Alberts and May (1980) is of interest because it illustrates not only how one can study the impact of stimulus exposure on preference but also how the general method can be exploited to reveal how learning processes contribute to the development of social behavior. Alberts and May's tests were to identify the contribution experience makes to filial huddling, the disposition to remain in physical contact with conspecifics that begins to emerge when the rat is about 2 weeks old. Olfactory stimulation was known to play a role in determining filial huddling (Brunjes & Alberts, 1979). Alberts and May developed a method for identifying the experiential determinants of olfactory stimulus control of huddling.

From birth through postnatal day 14 pups were removed from the mother for 4 hours and placed individually into a small plastic box that contained a cotton swab adulterated with a novel odorant. The effects of this exposure were tested on postnatal day 15 with a preference test that simulated huddling. The pups were placed into a test arena that contained identical furry surfaces in different locations, one scented with the odorant presented during training and the other with a novel odorant. Olfactory preference was assessed by measuring the total time a pup was in contact with each target. Pups displayed a strong preference for the target scented with the previously experienced odor.

Duration of mere exposure to an odor need not be long in order to affect olfactory preference; Caza and Spear (1984) found that only 3 min of mere exposure is sufficient to produce a substantial preference for an odor.

Habituation and Sensitization

With repeated exposure, the reflex-eliciting power of a stimulus diminishes. Reflex habituation has been well studied in adult animals (see Thompson & Spencer, 1966; Groves & Thompson, 1970). The most extensive study of ontogenetic changes in reflex habituation in the rat is provided by Stehouwer and Campbell (1978; see also Campbell & Stehouwer, 1979). They developed a preparation for studying the habituation of the leg-flexion response evoked by the application of brief 35 ms of electrical stimulation to the forepaw. They studied rats 3–15 days old and were able to observe serveral of the characteristics that define habituation (Thompson & Spencer, 1966): (1) the response declined with repeated application of the stimulus, (2) the magnitude of the decline was inversely related to stimulus intensity, (3) spontaneous recovery from habituation occurred over time, and (4) dishabituation, or the restoration of the habituated response that is produced by a second intense stimulus, was observed.

An important feature of the Stehouwer and Campbell research was their finding that the threshold shock intensity necessary to elicit the leg-flexion response was the same value for pups 3–15 days old. This finding suggests that there are no ontogenetic differences over this age range in detection of the eliciting stimulus. There were, however, striking age-related differences in how the pups responded. Pups 3–6 days old responded in a relatively diffuse manner: the leg-flexion response occurred repetitively, even after the eliciting stimulus had terminated, and was accompanied by gross whole body movement. In contrast, the 10–15-day-old animals' responses were more selective and phasically locked to the stimulus.

Pups at all ages habituated, but there were age-related differences. For instance, as the intensity of the shock increased, the older animals' responses decreased much less over training than did those of the youngest animals. Also, the effect of the dishabituation stimulus in restoring the habituated response was much greater for the 3-and 6-day-olds than for the 15-day-olds.

That the older rats' responses decreased relatively little with an increase in the intensity of the eliciting stimulus was interpreted to reflect the contribution of sensitization to performance. On this basis, and from complementary evidence, Stehouwer and Campbell inferred that the process underlying sensitization develops later than the habituation process (similar timing has been shown in the developing *Aplysia*; see chap. 5, this volume).

One final point regarding the habituation paradigm is of interest. In addition to providing a means of investigating relatively simple forms of learning, this paradigm may also be exploited as a way of studying the developing rat's sensory capacities. The general strategy was described by Rubel and Rosenthal (1975). It involves combining the habituation procedure with a generalization test. For example, one might habituate the subject to a tone of particular frequency and then test responding to stimuli of different frequencies. The degree to which the training and test stimuli are judged as similar can be indexed by comparing a subject's habituated response to the training stimulus with its response to the test stimuli. If the training and test stimuli are perceived as similar, habituation produced by the training stimulus should generalize to the test stimuli, but if the two are sensed as different, the test stimulus should evoke a stronger response. Both Campbell and Haroutunian (1983) and Rudy and Hyson (1984) have used this method with some success to study auditory frequency discriminations in infant rats.

Habituation of Exploration

Habituation of the animal's tendency to explore its environment also has been tested. A distinction between exploratory behavior and reflexive behavior is significant because the two classes appear to differ in their ontogeny and in their susceptibility to pharmacological influence (Williams, Hamilton, & Carlton, 1974). Techniques for observing fine-grain exploratory responses emitted by perinatal rats have not been developed because of their limited mobility. It is a relatively simple matter, however, to measure exploration in rats at least 2 weeks old.

Typically, the animal is placed in an open field area, and exploration is measured by recording the number of times the animal rears, grooms, sniffs, and so forth (Bronstein, Neiman, Wolkoff, & Levine, 1974; Shaywitz, Gordon, Klopper, Zelterman, & Irvine, 1979). The frequency of these responses declines with exposure to the eliciting environment. For instance, Feigley, Parsons, Hamilton, and Spear (1972; also see Parsons, Fagen, & Spear, 1973) have studied exploration by measuring the rats' tendency to poke their heads into a hole in a wall. This response is made with high frequency by rats at least 15 days old. Regardless of how exploration is measured, it habituates much less rapidly for 15-day-olds than for older rats (see Figures 6.1 to 6.3 in this chap.; Figures 3 and 4 in Bronstein et al., 1974; and Figure 4 in Feigley et al.,

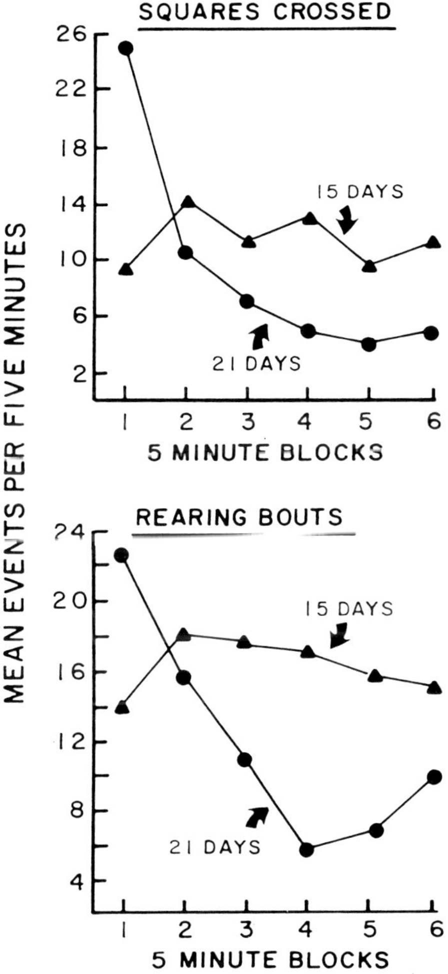

Figure 6.1. (top) Habituation of explorations in terms of number of squares crossed (area covered in an open field) as a function of age and time in the open field. (Adapted from "The Development of Habituation in the Rat" by P. M. Bronstein, H. Neiman, F. D. Wolkoff, and M. J. Levine, 1974, *Animal Learning & Behavior, 2*, Figure 3. Copyright by The Psychonomic Society, Inc. Reprinted by permission of the publisher.)

Figure 6.2. (bottom) Habituation of exploration in terms of mean number of rearing bouts in an open field as a function of age and time in the open field. (Adapted from "The Development of Habituation in the Rat" by P. M. Bronstein, H. Neiman, F. D. Wolkoff, and M. J. Levine, 1974, *Animal Learning and Behavior, 2*, Figure 4. Copyright by The Psychonomic Society, Inc. Reprinted by permission of the publisher.)

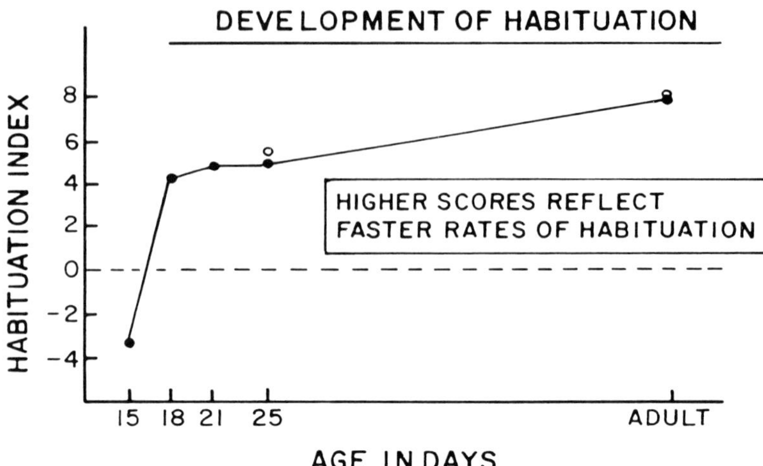

Figure 6.3. Habituation to headpoking as a function of age in the albino rat. The index of habituation is based on the decline in response frequency from the first 5-min trial to the third 5-min trial. (Adapted from "Development of Habituation to Novel Environments in the Rat" by D. A. Feigley, P. J. Parsons, L. W. Hamilton, and N. E. Spear, 1972, *Journal of Comparative and Physiological Psychology, 79,* Figure 4. Copyright by the American Psychological Reprinted by permission of the publisher.)

1972). Although the number of points tested in ontogeny are so far too few to be conclusive, this type of habituation seems on the basis of these experiments to change later in ontogeny than reflexive habituation.

Latent Inhibition

One of the most well-established phenomena in animal learning is the so-called latent inhibition or CS-preexposure effect (see Lubow, 1973, for a review). Exposure to a stimulus that later serves as a Pavlovian conditioned stimulus (CS) makes conditioning with that CS more difficult. This provides another way to assess learning as a consequence of mere stimulus exposure.

Rudy and Cheatle (1979) used this paradigm to investigate developmental changes in the rat's memory for olfactory stimulation. They exposed 2- and 8-day-old pups to a lemon scent for 30 min. Some time later the pups were exposed to a pairing of that odor and an illness-inducing drug. Prior exposure to the odor greatly attenuated conditioning of an aversion to the odor. The degree of the attenuation interacted with the pup's age and the interval separating preexposure and conditioning. The conditioned aversion was attenuated for the 8-day-olds whether the interval was 1, 4, or 8 hours. In contrast, the aversion of the 2-day-olds was attenuated only when the interval was 1 hour.

One implication of these age-related differences in latent inhibition may be that substantial maturation-induced change occurs between postnatal days 2 and 8 in the processing of the memory representing a novel odor. Other procedures of prior CS exposure have yielded that age-related differences support this general conclusion with respect to the conditioning phase of memory processing. After 6- and 12-day-old rats were allowed to drink chocolate milk infused through a cannula into their mouths, they subsequently were given pairings of chocolate milk and illness-inducing lithium chloride (LiCl). Although the prior exposure attentuated conditioning of an aversion to the chocolate milk among the 12-day-olds, there was no such attenuation among the 6-day-olds (Kraemer, Hoffmann, & Spear, 1988). With a different conditioning procedure involving pairings of a novel odor and a mild foot shock, moderate durations of preexposure to the odor impaired conditioning among 18-day-old rats but *faciliated* conditioning among 10-day-old rats, whereas longer durations of preexposure attenuated conditioning at both ages (Hoffmann & Spear, 1988).

Summary

Mere exposure to a stimulus can change the way in which a developing animal responds to it, and the degree of this change can depend on the animal's age. The first of four cases discussed was the increase in preference for an odor following the rat's exposure to the odor. This effect may have a functional role in promoting filial huddling and, consequently, regulation of body temperature during infancy. Recent evidence has indicated that an increase in preference for an odor can be induced further in perinatal rats if the odor is accompanied by any of a variety of arousing events—such as a tail pinch, brushing of the fur, or injection of amphetamine—which raises the question of whether these agents somehow enhance the mere exposure effect or act as more conventional unconditioned stimuli in Pavlovian conditioning (for review of these effects, see Leon et al., 1987, and chap. 17 this volume.)

Habituation and sensitization represent the second set of "mere exposure" phenomenon. We noted that all basic phenomena associated with habituation in adults have been observed in rat pups 3–15 days of age. Within this age range both sensitization and habituation increase with ontogeny, although the development of sensitization occurs somewhat later than that of habituation. This evidence was gathered primarily with what has been termed *respondent* or *exploratory* habituation. Older preweanlings have been tested primarily for *operant* habituation, represented by a decrease in exploration or in responding that is instrumental in the occurrence of a particular event. These tests suggest quite tentatively at this point that operant habituation may develop later than respondent habituation.

The third effect is the decrease in conditioning to a stimulus to which the animal has been preexposed, conventionally termed *latent inhibition*. Tests of this effect have also indicated an increase during preweanling development, with at least one result indicating that preexposure may facilitate rather than impair subsequent conditioning in relatively young infants.

ASSESSING ASSOCIATIVE PROCESSES

The methods discussed in this section were developed to assess associative processes in very young animals. The section explores how these processes contribute to behavioral adaptation and how they change during ontogenesis.

The term *associative learning* is a theoretical abstraction. One does not observe associative learning directly; one infers its occurrence. To conclude that associative learning has occurred is to say that the organism's behavior is a product of its experience with a relationship between two or more environmental events. Since the critical experimental variable is the relationship among the events, any assertion that the behavior being measured reflects a contribution of associative learning processes must be evaluated against a set of control conditions that rule out other interpretations of the data. The reader is referred to Rescorla (1967) for an illuminating discussion of the control issue.

Pavlovian (Classical) Conditioning

It is fair to say that most of the procedures developmental psychobiologists have employed recently to assess simple associative processes have been a variation of the Pavlovian or classical conditioning paradigm familiar to any psychology student. The animal is exposed to two events, the so-called conditioned stimulus (CS) and unconditioned stimulus (US), and the experimenter measures the effect of this experience on the so-called conditioned response (CR). The Pavlovian paradigm has two important virtues for investigators studying immature animals. First, it permits the experimenter to specify and control the characteristics of the to-be-associated events and to ensure that the animal is exposed to them. Second, the limited response repertoire of the perinatal rat does not preclude the expression of learning, as it might with instrumental learning procedures. The experimenter has more flexibility in choosing a behavior that is influenced by the animal's experience with the CS-US relationship and can serve as the index to the "association."

It is important to appreciate the sequence in which the rat's sensory systems emerge when discussing the ontogeny of associative processes. This issue was covered in some detail by Gottlieb (1971) and more recently by Alberts (1984). Here we simply note that the rat is altricial at birth and that during the first two weeks of life its behavior is controlled primarily by thermal-tactile, olfactory, and gustatory stimulation. Its auditory system does not begin to function until it is 10–12 days old, and its eyes do not open until it is about 15 days old. This sequence of sensory system development has obvious implications for the questions one asks about the associative processes of the infant, the development of these processes, and the methods used to answer the questions. We will use the order of sensory system development to organize the initial part of this section.

During the first 10 days after birth the rat can neither hear nor see. It is thus no surprise that most investigators have probed the rat's olfactory and gustatory systems during this period to gain insight into its learning and memory capacity.

Olfactory Conditioning

Studies of associative learning mediated by the olfactory system have revealed a surprising degree of competence in the neonate's associative learning process (Johanson & Teicher, 1980; Johanson & Hall, 1982, 1984). The research methods can be distinguished by whether the unconditioned stimulus was appetitive or aversive and whether the learning was indexed as a change in preference for the odor or in the odor's acquired control over one or more of the conditioned responses elicited by the US. For instance, if the US were milk, we might expect that the mouthing response elicited by milk would be elicited by odor.

Johanson & Teicher (1980) introduced an appetitive conditioning procedure that revealed associative processes in rat pups only 3 days old and that can also be applied effectively to study older animals (Johanson & Hall, 1982; Johanson, Hall, & Palefone, 1984). The general procedure consists of placing the pup into a warm, clear container and exposing it to paired presentations of an odor (CS) and an oral infusion of milk. Associative learning induced by this training has been inferred by measuring the effect that pairing of the CS and US has on (1) the pup's preference for the odor and (2) the extent to which the CS evokes components of the unconditioned response to the milk infusion. Typically, the conditioning procedure increases the pup's preference for the odor CS, as measured by the amount of time the pup spends in the vicinity of that odor CS relative to an alternative odor. In some circumstances the preference measure seems more sensitive to learning by 3-and 6-day-old pups than to that by pups either 1 or 9 days old (see Figure 6.4 this volume; Figure 1, Johanson & Hall, 1982).

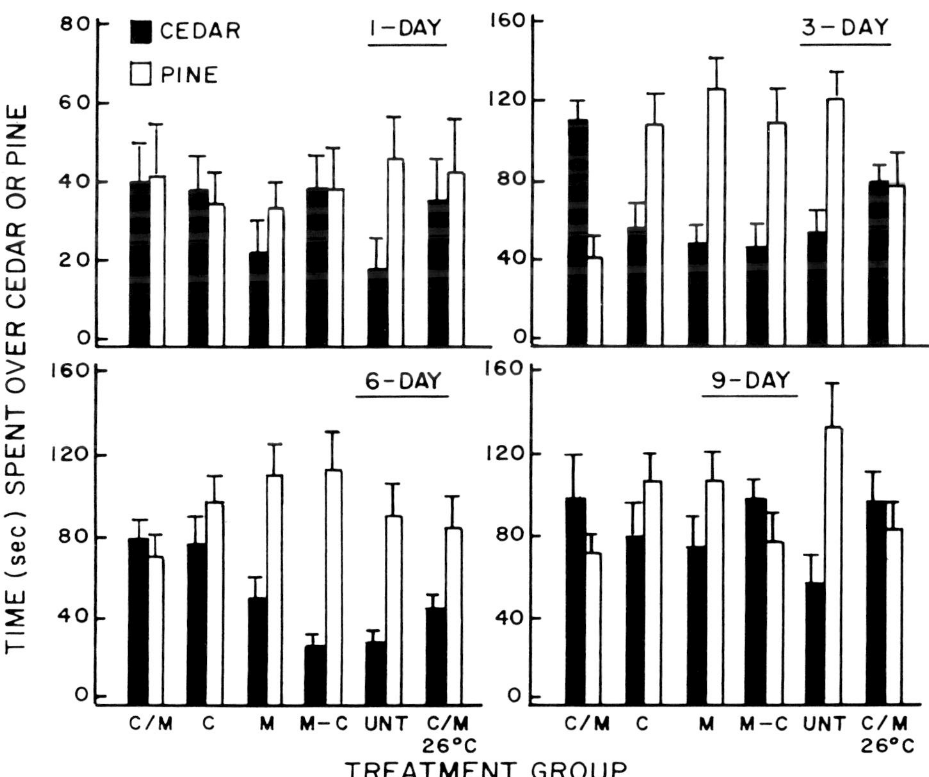

Figure 6.4. Mean time spent over cedar or pine for 1–9-day-old pups. Experimental pups (C/M) were given 10 cedar/milk pairings. Controls received cedar exposure alone (C); milk infusions alone (M); milk infusions followed by cedar exposure (M-C); no treatment (UNT); or cedar/milk (C/M) pairings at 26°C. The error bars indicate the standard error of the mean. (Adapted from "Appetitive Conditioning in Neonatal Rats: Conditioned Orientation to a Novel Odor" by I. B. Johanson & W. G. Hall, 1982, *Developmental Psychobiology, 15*, Figure 1. Copyright by John Wiley & Sons. Reprinted by permission of the publisher.)

Hall (1979) has provided a detailed description of the unconditioned responses elicited by milk infused into the pup's mouth (e.g., general activity, mouthing, and probing). Johanson et al. (1984) reported that odor CSs paired with milk infusions acquire the capacity to elicit these behaviors as conditioned responses. Space does not permit a detailed description of these behavioral categories. However, both the mouthing and probing responses were sensitive indicators of conditioning in pups 3 to 12 days of age, whereas the activity measure was sensitive only for pups 6 to 9 days old.

A number of researchers have studied how aversive unconditioned stimuli can condition odor-preference changes. The acquisition of conditioned odor aversions was first described by Rudy and Cheatle (1977, 1979). They exposed 2-day-old pups to a single paired presentation of lemon-scented pine shavings and illness induced by lithium chloride (LiCl). When these pups were tested 6 days later they displayed a reduced preference for the lemon scent in comparison to pups in a number of control conditions.

These experiments indicate a substantial capacity for the conditioning of aversions to odors during the ealry stages of postnatal life. That even the fetal rat is capable of rapid acquisition of olfactory aversions is suggested by the data of Smotherman (e.g., 1982), who paired a novel flavor (apple juice) introduced into the amniotic fluid with injections of LiCl into the fetus 2 or 3 days prior to the normal end of gestation. When these pups attained weaning age of 22 days postnatal, they had a significant aversion to the odor of apple juice. The immature nature of the prenatal gustatory system suggests that a major portion of the detection of the apple juice flavor was through olfaction (Smotherman & Robinson, 1987).

Although illness-inducing drugs consistently produced conditioned odor aversions in pups as young as 1 day old (e.g., Cheatle & Rudy, 1978) a somewhat mixed set of outcomes has been observed when other presumably aversive events have served as the unconditioned stimulus. For example, Haroutunian and Campbell (1979) studied changes in odor preferences in pups that ranged in age from 2 to 15 days when the US was either LiCl, intraperitoneal electric shock, or peripheral foot shock. For pups of all ages, significant conditioned odor aversions were produced by both the LiCl and intraperitoneal shock USs. In contrast, foot shock was not effective until pups were 14 days old.

Kucharski and Spear (1984), however, found that with foot shock as the US, conditioned odor aversion occurred readily in pups only 6 days old (see also Bryan, 1979; Kessler & Spear, 1980). A number of procedural differences between their study and Haroutunian and Campbell's can account for the differences in results. Perhaps the most important procedural difference was that in addition to exposing the pup to paired presentations of one odor (S +) and shock, Kucharski and Spear also exposed the pup to nonreinforced presentations of another odor (CS −). The inclusion of these CS − presentations was essential for conditioned aversions to CS + to be established in pups 6 days old, but its importance for conditioning diminished when older pups were tested (also see Miller, Jagielo, & Spear, 1989).

The use of peripheral aversive USs such as electrical shock with pups less than 7 days old has in fact proved to be paradoxical. For example, Camp and Rudy (1988) have reported circumstances in which, without a CS − , pups 6 days old acquire a conditioned preference for an odor repeatedly paired with foot shock, whereas pups

12 days old acquire a conditioned aversion to that odor when trained under the same conditions. Similarly, Sullivan, Hofer, and Brake (1986) have reported that pups 3 to 6 days old acquire a preference for an odor paired with pinching the pup's tail. Such results suggest that we do not yet understand the maturational changes during the second week after birth that occur in processes mediating the effectiveness of these stimuli.

Gustatory Conditioning

It is well known that adult rats rapidly acquire aversions to gustatory stimuli experienced prior to gastrointestinal distress (Garcia & Koelling, 1966; Revusky & Garcia, 1970). It is therefore not surprising that a similar paradigm for inducing conditioned taste aversion has been adapted to study ontogenetic changes in learning mediated by the gustatory system. Preweanling rats as young as 18 days postnatal can be conditioned and tested readily with ordinary drinking tubes, using essentially the same procedures used for adults (e.g., Baker, Baker, & Kesner, 1977; Steinert, Infurna, Jardula, & Spear, 1979). For younger rats developmental psychobiologists have taken advantage of the oral-infusion procedure (described earlier) to study the acquisition of taste aversions (Gemberling, Domjan, & Amsel, 1980; Hoffmann, Molina, Kucharski, & Spear, 1987; Vogt & Rudy, 1984). Following infusion of a sapid solution through a cannula into the rear of the pup's mouth, the pup is made ill by an intraperitoneal injection of LiCl solution. To assess the acquired aversion to the taste, the pup is given a second infusion of the solution. To measure how much of this solution the pup actually ingests, the animal is weighed just prior to the second infusion and again immediately afterward. Subtracting the pup's preinfusion weight from its postinfusion weight provides an index of how much fluid the pup consumed. A conditioned association between the taste and illness can be inferred if pups given the pairing of the two events ingest less of the solution than those in the proper control conditions that do not include such a pairing.

Using this general procedure some researchers have found that pups only 5 days old are capable of acquiring taste aversions (Gemberling et al, 1980; Hoffmann et al., 1987; Molina, Hoffmann, & Spear, 1986). Other researchers, however, have found no evidence of taste-aversion learning until the pup is 12 days old (Vogt & Rudy, 1984). Hoffmann et al. (1987) have discussed some of the procedural differences that may be important in determining when pups less than a week old will acquire the aversion. Thus, anyone wishing to study taste-aversion learning in the very young pup is well advised to pay careful attention to methodological detail.

Finally, we note that Hoffmann and Spear (1988) have used foot shock to establish a taste aversion 5- and 10-day-old rat pups. This observation is surprising given the well-known fact that it is difficult to use foot shock to establish taste aversion in adult rats (Garcia & Koelling, 1966; Pelchat, Grill, Rozin, & Jacobs, 1983). More consistent with the adult literature, however, is the observation by Hoffmann and Spear (1988) that the same number of taste-shock pairings do not establish a taste aversion in 15-day-old pups.

Auditory and Visual Conditioning

The rat's auditory and visual systems begin to function in the latter part of the second week following birth. The external meatus opens the ear canal to sound at about 12–13 days postnatal, and the eyes open when the pup is about 15 days old, although there is some evidence that the animal can respond to sound and light before these landmark events (Crozier & Pincus, 1927; Infurna, 1981). Based on results using both appetitive and aversive Pavlovian conditioning, however, not until about two days after the external meatus and eyelids open do the auditory and visual systems mediate associative learning.

Hyson and Rudy (1984) developed an appetitive conditioning procedure to study associative learning mediated by the auditory system. In their procedure, the US was a 0.1 ml 10% sucrose solution infused directly into the pup's mouth through a cannula implanted into the pup's cheek. This stimulus elicits a vigorous mouthing response in a maternal- and food-deprived pup that can be easily monitored by videotaping with a camera mounted directly below the clear Plexiglas floor of the conditioning chamber (Rudy & Hyson, 1984). Although rat pups 10 and 12 days old were able to detect the 2,000 Hz tone that served as the CS, Rudy and Hyson report that it was not until the pups were 14 days old that they conditioned to that tone.

Moye and Rudy (1985) developed an aversive fear-conditioning procedure to study associative learning in both the auditory and visual systems. Rats typically react to stimuli signaling the occurrence of a rapid-onset aversive event by remaining relatively motionless during the occurrence of these stimuli. Moye and Rudy (1985) employed an ultrasonic motion detector to monitor the behavioral activity of different-aged pups to a visual CS (a flashing light) that was repeatedly paired with the delivery of scrambled shock to the pups' feet. They found that even though the pups could detect the flashing light CS when 15 days old, the light paired with shock did not elicit a conditioned freezing response until the pups were 17 days old. Like Hyson and Rudy (1984), however, Moye and Rudy (1985) found that rats only 14 days old could condition to an auditory CS.

The use of Pavlovian conditioning procedures, coupled with increased sensitivity on the part of researchers to some of the contextual variables that are important to the normal functioning of the infant rat, has had a dramatic influence on our knowledge and understanding of the associative learning capacities of the perinatal rat and how they change during development. New methodologies have established that the newborn rat certainly possesses associative learning processes, and these processes may even be present in the fetal rat.

That the newborn or fetal animal may possess associative learning capabilities, however, should not obscure the fact that important maturational changes in the rat's associative learning abilities occur during the initial 3 weeks after birth. In fact, during ontogenesis the rat's associative learning abilities at any particular age appear to be intimately tied to the sensory events the animal is asked to associate. For instance, a newborn rat can associate odors with a variety of consequences, and by 8 days of age it can associate an odor experience with an illness episode that occurs an hour later (Rudy & Cheatle, 1979); yet the 15-day-old rat, eyes open and quite capable of detecting a visual stimulus such as a flashing light in an otherwise dark chamber, apparently cannot associate that light with an aversive US such as electrical foot shock (Moye & Rudy, 1985). This result cannot be due to the ineffectiveness of the shock to reinforce conditioning because 15-day-old rats will condition to an auditory or olfactory stimulus paired with the same shock US. It is not until the pup is 17 days old that it displays any ability to associate the light and shock. Thus, at an age when its associate learning capacities are adultlike when it is asked to learn about its olfactory world, the 15-day-old rat would be judged as incapable of associative learning involving a readily detectable visual stimulus and aversive foot shock.

Two implications of the preceding discussion are especially interesting. First, during the development of a particular sensory system there may be a period in which the system can detect and respond reflexively to a relevant source of stimulation although be unable to mediate associative learning involving that stimulus (see Rudy & Hyson, 1984, and Moye and Rudy, 1985, for a more detailed discussion of this point). Second, these developmental studies may be telling us that there is no unitary, localizable associative learning process, but instead there are multiple sites of associative processing that reside in each sensory system.

Independent of the above speculations, however, the basic sensory, associative, motivational, and motor processes necessary for the acquisition and expression of simple associations seem in place for the rat by 17 days postnatal and the stage set for more advanced cognitive functioning.

Instrumental (Operant) Learning: Simple to Complex

The Pavlovian paradigm has much to commend it when studying learning in very young animals. A developmental analysis of the learning and memory that centers on just the use of the Pavlovian paradigm, however, is unnecessarily restrictive and may miss the development of some of the more interesting cognitive processes involved in learning the consequences of the animal's own actions.

The ontogeny of instrumental learning has been studied in the preweanling most frequently with either of three sources of reward: access to an anesthetized adult rat, typically a lactating dam; escape (or avoidance) from mild electrical foot shock; or escape from forced swimming in water.

Learning to Approach an Anesthetized Dam

Rats as young as 7 days postnatal can learn a position discrimination in a Y-maze when the "correct" alternative includes access to an anesthetized dam and the "incorrect" alternative does not permit such access (Kenny & Blass, 1977). This learning paradigm also has been used to address questions about ontogenetic change in motivational determinants in the developing rat (e.g., Stoloff & Blass, 1983; Smith & Bogomolny, 1984).

Perhaps the most extensive use of an anesthetized dam as a reinforcer for instrumental learning, however, has been applied by Amsel and his colleagues in their systematic study of ontogenetic change in the influence of conditions of reinforcement (for a review, see Amsel, 1986). For these studies the pup must move from one to another end of a straight alley to obtain access to the dam. The index of learning is the progressive increase in speed of traversing the alley that accompanies practice. Although space does not permit a sufficient description of these extensive and illuminating studies, the general result has been to show dramatic and regular change during the first four postnatal weeks in the effects of nonreward on both immediate responding and subsequent persistence in the absence of reward.

Learning to Escape or Avoid Foot Shock

Learning of a simple escape response from mild foot shock has been studied during the first two postnatal weeks in terms of three behaviors required for escape: crawling onto a surrounding ledge, moving from one end of an alley to another; and turning in a particular direction. Nondirectional escape by climbing can be learned at 5 days postnatal, whereas directional escape by moving from one end of an alley to another apparently cannot be learned before 7 days postnatal (Misanin, Nagy, Keiser, & Bowen, 1971; Misanin, Chubb, Quinn, & Schweikert, 1974). The position-discrimination task could not be learned in conventional circumstances before 10 or 11 days of age (Misanin, Haigh, Hinderliter, & Nagy, 1973). However, in the context of the odor of the pup's home nest such learning has been found to occur in rat pups as young as 7 days postnatal (Smith & Spear, 1981).

Tests of ontogenetic differences in learning to *avoid* foot shock have included the widely used passive avoidance, in which withholding of a response, such as movement from one distinctive compartment to another, is necessary to avoid foot shock. Relative to the learning of an active, escape-like movement to avoid foot shock—readily acquired by 15-day-olds and with adultlike efficacy by 21-day-old weanlings—preweanling rats have special difficulty learning passive avoidance of the conventional variety (Feigley & Spear, 1970; Klein & Spear, 1969; Schulenburg, Riccio, & Stikes, 1971; Riccio, Rohrbaugh, & Hodges, 1968), with some improvement observed in the presence of home nest odors (Smith & Spear, 1978). Although passive avoidance learning had seemed unlikely before 7–10 days of age, it was difficult to reach a decision because so little forward movement of any kind occurs in the rat pup during the first postnatal week. By providing a steady stream of air in the direction of the to-be-shocked compartment, however, Myslivecek and his colleagues induced sufficient movement in perinatal rats to permit the observation that passive avoidance learning could be learned during the first postnatal day (Myslivecek & Hassmannova, 1983).

Although learning active avoidance in a particular direction once seemed beyond the capacity of rats during the first two postnatal weeks or so, earlier learning was observed, with a modification of the nondirectional escape procedures of Misanin et al. (1974). Rats 9 days of age learned to avoid foot shock by climbing in any direction from a well prior to the cessation of a 5-s vibrotactile warning signal and exhibited retention of that conditioned avoidance 24 h later if given a reactivation treatment (Misanin et al., 1987; Spear & Smith, 1978).

Complex Learning to Escape Water

By about 16 days of age, the rat is an excellent swimmer and adjusts well to water, but appears to be highly motivated to escape. These properties make water escape a desired alternative to the use of food deprivation or electrical shock as the motivation for instrumental learning. Recently, water-escape tasks have been used with considerable success to study the ontogeny of (1) visual discrimination learning, (2) place learning, and (3) conditional spatial discrimination learning.

1. To study visual discrimination learning, Rudy and Castro (1987) developed an aquatic version of the Lashley (1930) jumpstand that could be used with rats from 16 days old through adulthood. The rat is placed in a water-filled tank on a platform facing the discriminanda, which, in the case of a brightness discrimination problem, would be a black card and a white card. Behind one of the discriminanda is a platform onto which the rat may escape from the water. On each trial the rat is placed into the water and required to swim under the card that leads to the escape platform. Because younger preweanlings lose body heat more rapidly than older animals, the water

temperature ranges from 30–24°C, depending on the age of the pup. Using daily training sessions of 25 trials each, Rudy and Castro (1987) found that rats that begin training at 28 days postnatal can solve this problem in less than 15 trials, whereas rats that begin training when only 16–18 days old require about 60 trials to reach criterion. Castro and Rudy (in press) also have used this apparatus to study visual pattern discrimination learning (horizontal versus vertical black and white stripes) in normal and previously undernourished rats.

2. Morris (1981) developed a water-escape task that has provided a simple and reliable means of studying spatial learning that depends on the use of what O'Keefe and Nadel (1978) have termed *locale strategies*. Such problems require the animal to move directly to a goal from several directions, using the relationship among stimuli that may be very far from the goal to guide its behavior. In the Morris task, this property is realized by placing an escape platform hidden beneath the surface in a circular pool of water. The normal adult rat rapidly learns to locate the platform and to swim directly to it from different starting points. Rudy, Stadler-Morris, and Alberts (1987) found no evidence that rats less than 20 days old are capable of such location learning, although rats only 16–17 days are capable of swimming to a visible platform (see also Rudy & Paylor, 1988). Space does not permit a detailed discussion of the variety of measures of such behavior (heading direction, path length, quadrant search time, platform crossings) needed to ensure that the animal is using a locale strategy. It is thus strongly recommended that the potential user consult the original Morris (1981) paper, as well as papers by Sutherland and Dyck (1984), Rudy et al. (1987), and Castro and Rudy (1987) for details about this task and its proper interpretation.

3. The test for conditional spatial discrimination learning is based on a training procedure used to study short-term memory processes in the rat (e.g., Feldman & Gordon, 1979; Roberts, 1974; Grant, 1981). A trial consists of two episodes, a forced run in which the rat is forced to one side of a T-maze and a choice run in which the animal is allowed a free choice between the two alternative goals of the T-maze. The "correct" location of the goal on the choice trial is conditional upon the outcome of the forced run. A correct choice depends on remembering which arm was visited on the forced run and on using this information to discriminate the correct arm on the choice trial. Once the rat has mastered this discrimination problem, its capacity for short-term retention can be tested by varying the interval between the forced run and the choice run.

Castro, Paylor, and Rudy (1987) have developed a version of this test requiring that the rat solve the discrimination by escaping from water. They trained rats that were either 21, 28, or 38 days old on a win/shift version of this problem. The pups were given daily training sessions of 30 trials with a 10-s interval separating the forced run and choice run (see Castro et al., 1987, for details). The youngest rats required about 80 trials to reach a 9/10 correct criterion, whereas the oldest required only about 40 trials. After the animals had mastered the basic problem, their short-term retention was further assessed by increasing the interval between the forced run and the choice run. Animals at each age were able to choose correctly when the interval was 30 s, but when it was increased to 60 s, the youngest pups' behavior dropped to chance; and when the interval was increased to 180, the behavior of the 28-day-olds fell to chance. The oldest animals effectively chose the correct run with either the 60-s or the 180-s intervals.

These studies have revealed that the learning and memory capacities of the rat are by no means completely functional by 17 days of age, the point in development when simple associations seem readily acquired in all sensory systems. The relational processes needed for place learning and conditional discrimination may not be available to the 17-day-old but are available to the 21-day-old and improve thereafter.

Summary

The ontogeny of associative learning has been studied with both Pavlovian (classical) and instrumental (operant) conditioning. The ontogenesis of conditioning and subsequent age-related change depend upon the sensory system involved and the relative complexity of what is to be learned. Associative learning can be observed at younger ages in those sensory systems that are the earliest to emerge. Olfactory conditioning has been observed prenatally and shortly after birth, with continued increase in efficacy thereafter, depending on the unconditioned stimulus that is to be associated with the olfactory event. Conditioning with gustatory stimuli seems to emerge somewhat later in ontogeny, although still during the first postnatal week, and it too depends on the nature of the unconditioned stimulus. In both cases, early conditioning is more likely with unconditioned stimuli that are either gustatory or visceral in nature rather than those that are relatively peripheral (such as foot shock). Simple conditioning involving the auditory or visual systems occurs within a few days after these systems become functional in other behavioral respects. It is notable, however, that ontogenesis of the detection of a visual or auditory event precedes by a day or two efficacy in conditioning of an association involving that event.

Instrumental approach and avoidance learning occur during the first postnatal week. Escape learning is observed at earlier ages than is avoidance learning. Despite previous indications that the ontogenesis of passive ("inhibitory") avoidance learning might occur later than that of active avoidance learning, recent techniques have permitted the observance of passive avoidance learning a few hours after birth. To date, however, there are no reports of active avoidance learning among rats younger than 9 days postnatal. Tests of relatively complex instrumental learning have included visual discrimination learning, spatial location learning, and conditional spatial discrimination learning. In each of these tasks, increases in rate of learning continue throughout the fourth postnatal week or later in ontogeny.

RETENTION AND FORGETTING DURING ONTOGENY

The most common form of this aspect of memory processing is one that we all experience personally—the "infantile amnesia" noted by Freud that is seen in our general inability to remember the events of our infancy. That a 20-year-old human cannot remember any events that took place when he or she was 1 year old becomes interesting in contrast to the 40-year-old who can remember many of the events that occurred at age 21. Experimental tests with a 19-year retention interval are obviously impractical, however. For this and more subtle (yet more important), analytical reasons, infantile amnesia is best studied with an animal model.

An ontogenetic increase in retention has seemed inevitably to occur: the younger the preweanling rat the more rapid the forgetting. Among the many altricial animals that have been shown to exhibit infantile amnesia, the rat is the most widely studied. This is in part due to the rapid growth rate of the rat's brain. More immature than the newborn human brain at birth, the rat brain becomes essentially adultlike within 30 days. This allows convenient tests of theories that attribute infantile amnesia to brain growth.

An important feature of laboratory experiments investigating infantile amnesia with the rat is the control exerted over degree of original learning. This typically is accomplished operationally, by comparing retention for events that take place during infancy and adulthood only in circumstances in which immediate (very short-term) retention for these events is equivalent. It would be of little interest to determine that the events of adulthood are remembered better than those of infancy merely because they were learned better in the first place. The interest is in the differential retention that occurs despite equivalent learning.

A variety of training procedures have been used to reveal ontogenetic differences in retention of conditioning (Campbell & Coulter, 1976; Campbell & Spear, 1972). The breadth of these tests and some of their methodological features can be illustrated briefly. Markiewicz, Kucharski, and Spear (1986) compared the retention of Pavlovian conditioning established when rats were 12, 18, or 60 (adult) days postnatal and tested several days or several weeks later. At issue was the strength of the aversion exhibited after retention intervals of various lengths had intervened since the rat was given a few (no more than 6) pairings of a mild but annoying foot shock with either a particular odor, movement (vibration of floor), visual event (black or white), or ambient temperature. Procedural circumstances were arranged to ensure that conditioning assessed on immediate tests was equivalent across ages. One example of the more rapid forgetting exhibited by younger animals in all conditions is illustrated in Figure 6.5 (modified from Markiewicz et al., 1986).

Infantile amnesia for instrumental learning has been assessed with a variety of procedures, including passive avoidance learning, in which the animal must remember that to avoid a foot shock it must not enter a particular compartment (see Figure 6.6), and discriminated escape learning, in which the rat must remember that in order to escape an annoying foot shock it must turn left (or for some animals, right) at the choice point (see Figure 6.7).

Variation in procedural parameters yields corresponding variation in the developing rat's rate of forgetting. There is no doubt that infants can be made more resistant to forgetting of either Pavlovian or instrumental conditioning. This can be accomplished in several ways, such as increasing the number of training episodes or the temporal distribution of these episodes (e.g., Coulter, 1979; Spear, 1979; Sussmann & Ferguson, 1980). Yet despite procedural variations that alter dramatically the amount or rate of forgetting at any particular age, forgetting by younger preweanlings has nearly always been greater and never less than that by older animals. For instance, Pavlovian conditioning established by pairing a taste with injection of LiCl has seemed unusually resistant to forgetting even for rats 18–20 days postnatal; yet infantile amnesia in this case is nevertheless evident after a month's interval for pups conditioned at 18 days of age, and younger preweanlings show still more rapid forgetting (Campbell & Alberts, 1979; Steinert, Infurna, & Spear, 1980; Schweitzer & Green, 1982).

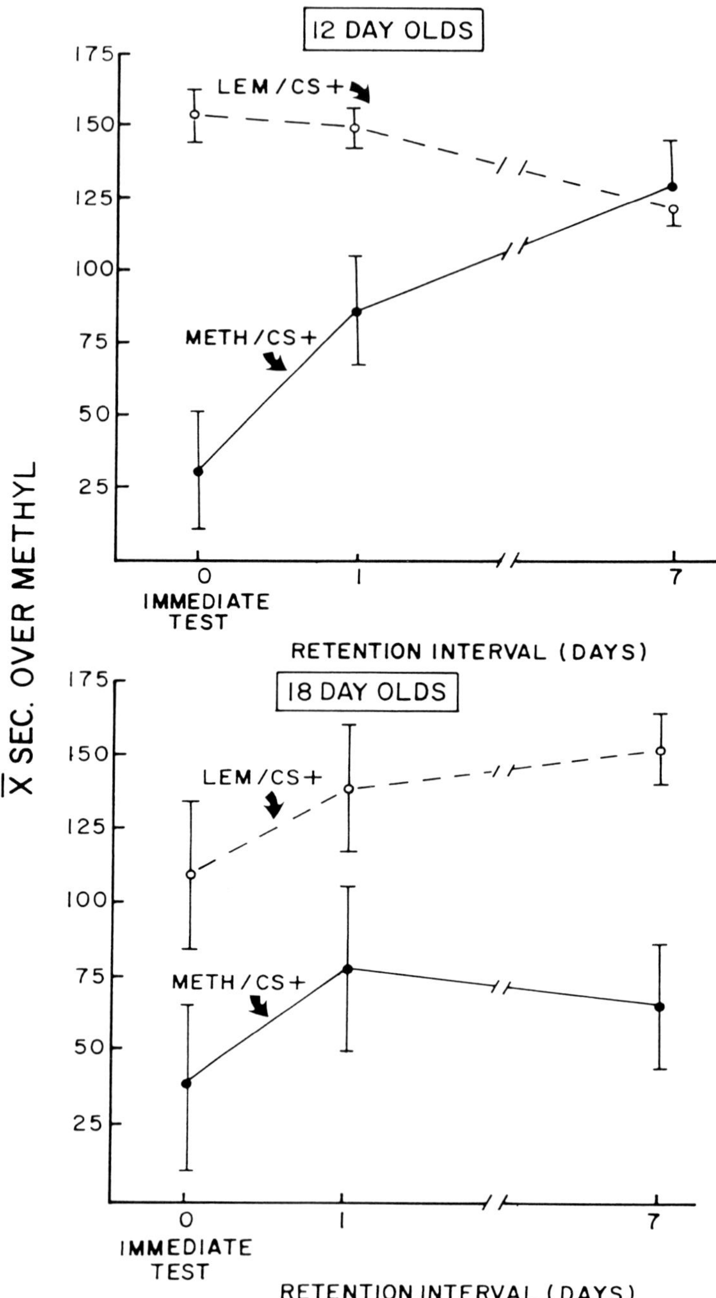

Figure 6.5. Mean number of seconds spent over the side of the test chamber odorized with methyl salicylate (the alternative side was odorized with lemon) as a function of whether the animals previously had lemon odor paired with foot shock (LEM/CS) or methyl salicylate paired with foot shock (METH/CS) and retention interval (0, 1, or 7 days). Top panel shows the forgetting functions for 12-day-olds and the lower panel for 18-day-olds. (Adapted from "Ontogenetic Comparison of Memory for Pavlovian Conditioned Aversions in Temperature, Vibration, Odor or Brightness" by B. Markiewicz, D. Kucharski, and N. E. Spear, 1986, *Developmental Psychobiology, 19*. Copyright by John Wiley & Sons. Reprinted by permission of the publisher.)

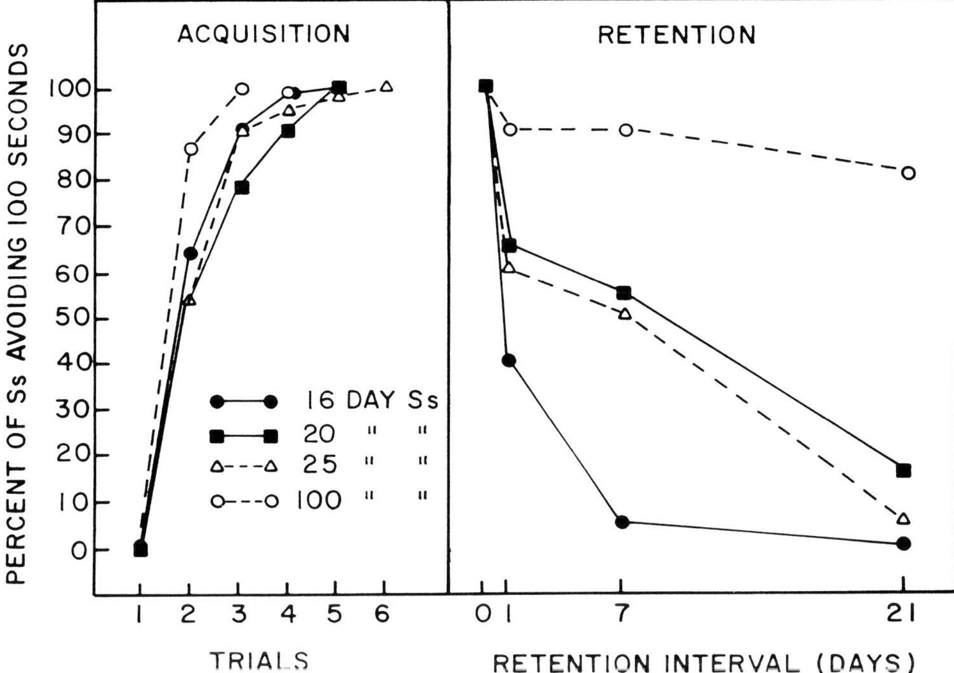

Figure 6.6. Retention of a passive avoidance lowered to the same degree (left panel) is shown in terms of the percentage of rats that avoided for 100 s entering into a previously shocked compartment, as a function of age and retention interval. (Adapted from "Species Differences in Ontogeny of Memory: Indirect Support or Neural Maturation as a Determinant of Forgetting" by B. A. Campbell, J. R. Misanin, B. C. White, and L. D. Lytle, 1974, *Journal of Comparative and Physiological Psychology, 87,* Figure 3. Copyright by the American Psychological Association. Reprinted by permission of the publisher.)

It has become evident that retention after very short intervals also is less effective for infants than for older rats. There have been three primary methods leading to this inference.

1. A simple approach is to use minimal conditioning procedures and merely test after short intervals. Procedures for both olfactory and visual conditioning have been applied, but they are directly analogous, and we need consider only the olfactory case, which has the advantage of being usable for rat pups too young to have opened their eyes. Conditioning is accomplished with a 20-s exposure to an odor, termed CS+ because it is accompanied by two foot shocks (during seconds 8–10 and 18–20 of the exposure), preceded by a 20-s exposure to a different odor, termed CS − because it is not accompanied by a foot shock. Greater forgetting following this minimal conditioning procedure is more rapid among the younger preweanlings (Miller, Jagielo, & Spear, 1989).

2. A more common procedure is a trace conditioning variation of the standard Pavlovian conditioning procedure. Whereas the standard procedure involves presenting the CS and US in close temporal proximity, with presentation of the US coincidental with or overlapping termination of the CS, in the trace conditioning paradigm a

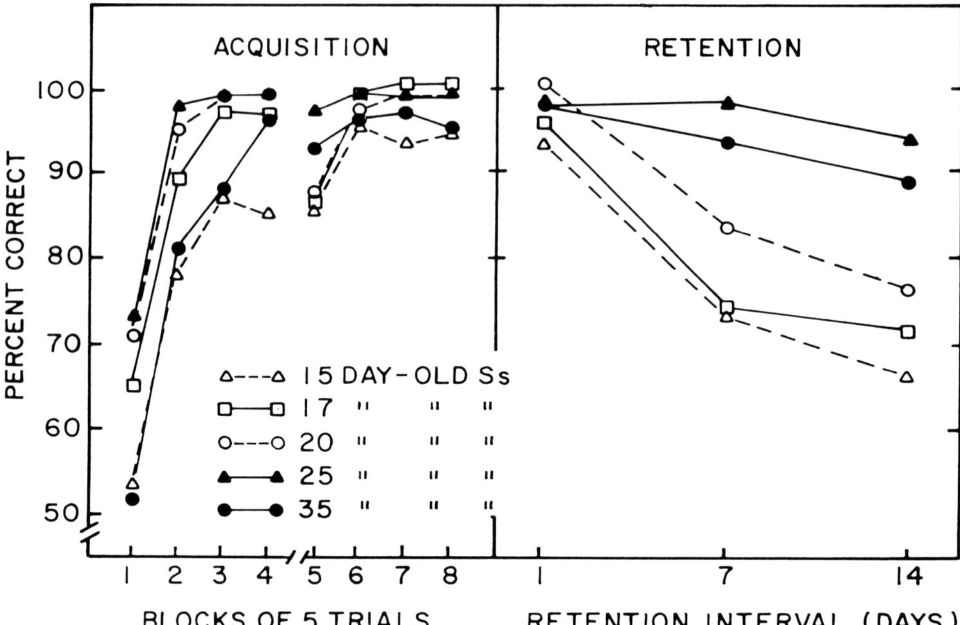

Figure 6.7. Retention of a position discrimination motivated by escape from shock in a T-maze following roughly equivalent acquisition (left panel). Retention (right panel) is shown as a function of retention interval and age. (Adapted from "Species Differences in Ontogeny of Memory: Indirect Support or Neural Maturation as a Determinant of Forgetting" by B. A. Campbell, J. R. Misanin, B. C. White, & L. D. Lytle, 1974, *Journal of Comparative and Physiological Psychology, 87*, Figure 1. Copyright by the American Psychological Association. Reprinted by permission of the publisher.

temporal gap is inserted between CS termination and US onset. The temporal gap makes acquisiton of the CS-US association theoretically dependent on the availability of a memory representation of the CS when the US occurs. To the extent that the trace conditioning paradigm does reflect memory for the CS at the time of the US, it provides a simple and powerful means for studying the rat's short-term memory for phasic events. It has been extensively exploited by developmental psychobiologists to produce a remarkably similar pattern of results. Regardless of the sensory system, there has been found an early period in development when the animal can condition to the particular CS when it is temporally contiguous with the US, but fails to condition as effectively when there is a temporal gap separating the CS and US. The important point is that the temporal gap has less effect on older rats. This observation was made by Rudy and Cheatle (1979) for the developing olfactory system and has been obtained for the gustatory system (Steinert, Infurna, Jardula, & Spear, 1979; Gregg, Kittrel, Domjam, & Amsel, 1978; Vogt & Rudy, 1984) and the auditory and visual systems (Moye & Rudy, 1987).

   3. Our earlier discussion of the relatively complex, conditional discrimination learning indicated that short-term retention of the signaling event at the time of the discrimination decision may continue to improve after 4 weeks of age.

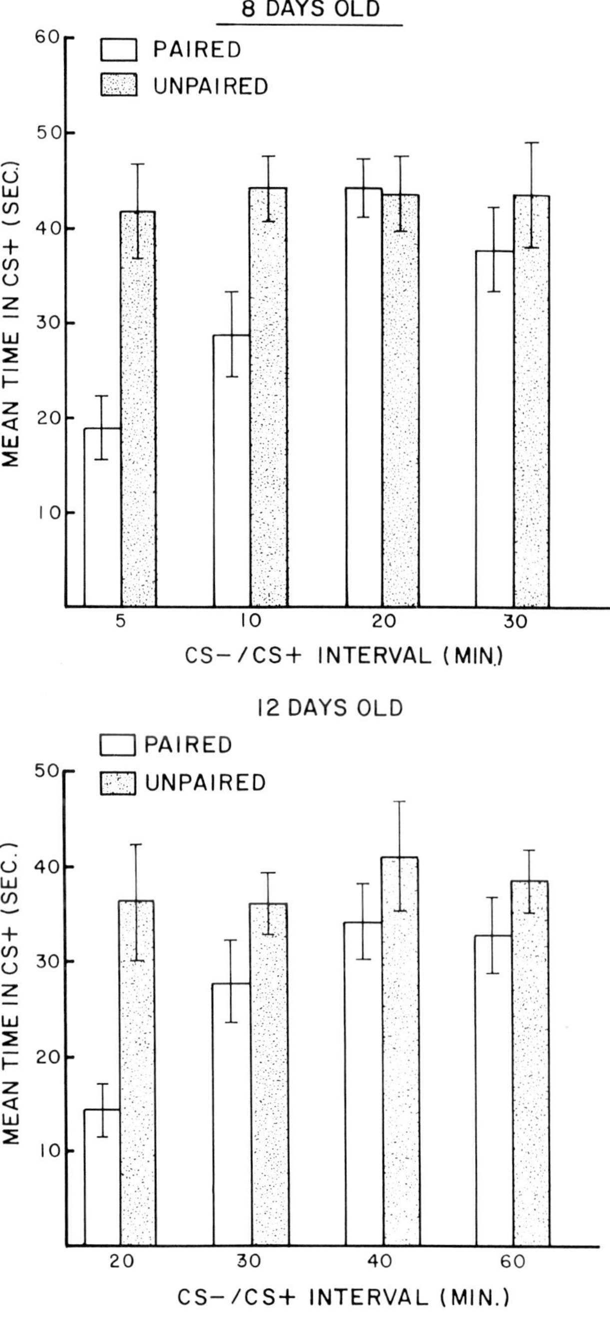

Figure 6.8. Mean seconds spent in the side of the chamber odorized with methyl salicytate (CS+, previously paired with shock) as a function of the delay during conditioning between presentation of the CS− (the presence of which is necessary for conditioning a CS+) and the CS+. This provides an estimate of the rat's forgetting the occurrence of the CS. The upper panel represents 8-day-olds and the lower panel 12-day-olds. (Modified from "Age Related Differences in Short-term Retention of Separable Elements of an Odor Aversion" by J. S. Miller, J. A. Jagielo, and N. E. Spear, 1989, *Journal of Experimental Psychology: Animal Behavior Processes, 15,* Figures 5 and 6. Copyright by the Society for the Experimental Analysis of Behavior, Inc. Reprinted by permission of the publisher.)

4. A final procedure permits assessment of short-term retention of an isolated olfactory or visual event. This test emerged through the observation that young preweanlings that otherwise show no conditioned aversion to a CS+ (in this case, an odor paired with foot shock) do so given a preceding CS− (a different odor, but no foot shock). This is termed the "cf effect" (Kucharski & Spear, 1984). We may therefore assess retention of the CS− in terms of the extent to which an aversion is formed to a subsequently presented CS+. In a specific test, an olfactory CS− was presented to 7-, 12-, and 20-day-old rats at intervals of 0–60 minutes prior to presentation of the olfactory CS+. Retention of the CS− was assessed by an immediate test of aversion to the CS+. Because no aversion to CS+ would be formed if the CS+ is experienced in isolation from the CS−, some retention of CS− can be inferred to the extent that a CS+ aversion is expressed. Despite surprisingly effective retention of this single olfactory event after several minutes, loss of this retention also was clearly more rapid the younger the preweanling (see Figure 6.8).

Alleviation of Infantile Amnesia

It has been suggested that if the forgetting of infancy were alleviated and the events of that period revealed to adults, a number of clinical psychologists—particularly those using therapies aimed at recovering the events of childhood—would suddenly be underemployed. A number of experiments have addressed the possibility of such alleviation with some success. It has been determined that a variety of instances of learning during infancy may be spared from the typically rapid forgetting thereafter by introducing cuing treatments either periodically throughout the retention interval or just prior to the retention test. The cuing treatments employed initially consisted of an entire learning trial, including all reinforcement contingencies (Campbell & Jaynes, 1966). Subsequently, however, it was determined that individual elements of the conditioning trial were sufficient (Riccio, Rohrbaugh, & Hodges, 1968; Spear & Parsons, 1976).

Did these treatments alleviate infantile amnesia? Given that the interest in infantile amnesia is as an instance of forgetting unlike that for events encountered at an older age, true alleviation of infantile amnesia can be established only when it is shown that the alleviating treatment affects the forgetting of infantile experiences more than the forgetting of adulthood experiences. The few appropriate tests have shown, however, that treatments found to alleviate the forgetting of infantile experiences accomplish the same or still greater reduction in forgetting by older animals (Spear & Parsons, 1976).

Two other treatments have seemed potentially to have the capacity for reducing infantile amnesia. One of these, more in terms of a preventative rather than an alleviating treatment, is to accomplish learning with a wide distribution of practice. Known to reduce forgetting in a variety of circumstances, there is some suggestion that a wide distribution of learning experiences may slow the forgetting of infants more than older animals, although the evidence is limited (Spear, 1979a).

A second treatment that may have the capacity for alleviating infantile amnesia is the imposition of a distinctive context during both conditioning and testing. Following a study showing that retention by 16-day-old rats was facilitated if they were injected with pentobarbitol both prior to conditioning and prior to testing (Concannon, Smith, Spear, & Scobie, 1979), a series of experiments by Richardson, Riccio, and their colleagues reported similar effects for preweanlings in separate experiments with a variety of distinctive contexts. These included a barbiturate-induced state, illness, and odors from the animals' home nest (Richardson, Riccio, & Devine, 1984; Richardson, Riccio, & Axiotis, 1986). Although there was some indication in the experiments by Concannon et al. (1979) that this distinctive context effect might be less apparent in 23-day-old rats than in 16-day-olds, there is no other evidence that this source of alleviation is any different for infantile experiences than for adulthood experiences. It is notable also that a distinctive context does not inevitably facilitate retention; a strong novel odor at both conditioning and testing can impair rather than facilitate retention of a brightness aversion (Lariviere, Chen, & Spear, 1990).

The conclusion is that although some reduction in the forgetting of the events of infancy can be accomplished, this reduction is not clearly different from that possible with regard to the forgetting of events that might occur later in life. It remains unclear whether infantile amnesia per se can be alleviated.

Summary and Comment on Theory

Rate of forgetting during the first three postnatal weeks is more rapid than that in adulthood, and the younger the animal during this early period, the more rapid the forgetting. These relationships hold for short and long retention intervals, for all sensory systems assessed in Pavlovian conditioning, and for all instrumental conditioning tests that have been applied. To our knowledge no reversals of this effect have been observed in methodologically sound experiments, in contrast to the several examples in which weanlings have learned more rapidly than adults, or young preweanlings have learned more rapidly than older preweanlings (Spear & Molina, 1987; Spear, Kraemer, Molina, & Smoller, 1988).

Why rate of forgetting is inversely related to age is a question about which we have learned only a little since Freud introduced infantile amnesia as a central part of psychoanalytic theory. We have learned that Freud was wrong about the notion that this effect reflects primarily repression of the events of one's infancy upon the realization that proper social behaviors pertaining to sex had not been achieved during that period. This notion is dismissable simply because the effect is observed in altricial animals such as the rat, whose social rules are quite different from those of the human.

We have also learned that the events of infancy are forgotten over short periods by infants more rapidly than the events of adulthood are forgotten over the same short periods by adults. This indicates that transition from infancy to adulthood is not a necessary condition for the especially rapid forgetting of the events of infancy.

Further understanding of infantile amnesia will likely arise from the two broad theoretical distinctions that have been drawn so far. The first is in terms of the locus of infantile amnesia: To explain infantile amnesia, one set of theories emphasizes subsequent growth of any of several kinds, including physical growth as well as the psychological growth responsible for age-specific differences in what is encoded for storage in memory and in how that storage takes place. The second distinction is in how permanent infantile amnesia is presumed to be. One set of theories expects that memories for the events of infancy ultimately are retrievable given the appropriate circumstances, which may be relatively innocuous; but the other set expects that infantile amnesia can be neither prevented nor alleviated without drastic physiological intervention (for elaboration on these distinctions, see Spear, 1978, 1979a, 1979b).

Freud addressed both issues in terms of his selective reconstruction model of infantile amnesia (Pillemer & White, 1989), which dealt indirectly with the issue of the effectiveness of recall, by means of contemporary encoding, of an event from a previous (infantile) encoding. Despite his medical background, Freud did not address how growth of the brain might influence infantile amnesia. Contemporary scientists have also largely ignored this issue, undoubtedly because it presents a difficult problem of experimental design and procedure (compare Campbell & Spear, 1972; Campbell, Misanin, White, & Lytle, 1974).

Freud thought the permanence of infantile amnesia was a relatively flexible matter in that apparently forgotten memories of infancy might potentially be revealed in adulthood in the right circumstances (such as psychoanalysis). Tests with rats have confirmed that forgetting of the events of infancy can be alleviated by a series of cuing treatments presented throughout the retention interval or just prior to the retention test and perhaps also by having a distinctive context present during both learning and testing. There is no evidence so far, however, that infantile amnesia is an incidence of forgetting that is unique to memories acquired during infancy. There is, instead, considerable evidence that forgetting of the events of infancy can be alleviated in the same way and to the same extent as forgetting of the events of adulthood. Despite this alleviation of forgetting by infants, alleviation of infantile amnesia as a unique source of forgetting has yet to be established.

GENERAL SUMMARY AND COMMENT

We have reviewed a wide variety of techniques that can be applied toward assessing learning and memory in the developing rat from the perinatal period through adulthood. The review is far from complete, however. Space limitations did not permit us to discuss, for example, the powerful procedures Campbell and his associates have developed to study ontogenetic change in conditioned autonomic responding (Sananes, Gaddy, & Campbell, 1988). We nevertheless hope that the sample we have provided illustrates the substantial opportunity these methods afford for a comprehensive understanding of the ontogeny of learning and memory or, secondarily, the study of the ontogeny of brain function through tests of learning, memory, and cognition in the rat.

ACKNOWLEDGMENTS

This research was supported by Grant 1 RO1 MH35219 from the National Institute of Mental Health and Grant 5 RO1 AA06634 from the National Institute of Alcohol Abuse and Alcoholism. The authors wish to express their appreciation to Norman G. Richter for technical assistance and to Teri Tanenhaus for preparation of the manuscript.

REFERENCES

Alberts, J. R. (1984). Sensory-perceptual development in the Norway rat: A view toward comparative studies. In R. Kail & N. E. Spear (Eds.), *Comparative perspectives on the development of memory* (pp. 65–102). Hillsdale, NJ: Erlbaum.

Alberts, J. R., & May, B. (1980). Ontogeny of olfaction: Development of the rat's sensitivity to urine and amyl acetate. *Physiology and Behavior, 24*, 965–970.

Amsel, A. (1986). Developmental psychobiology and behavior theory: Reciprocating influences. *Canadian Journal of Psychology, 40*, 311–342.

Baker, L. J., Baker, T. B., & Kesner, R. P. (1977). Taste aversion learning in young and adult rats. *Journal of Comparative and Physiological Psychology, 91*, 1168–1178.

Blass, E. M., & Kehoe, P. (1987). Behavioral characteristics of emerging opioid systems in newborn rats. In N. A. Krasnegor, E. M. Blass, M. A. Hofer, M. A. Smotherman, & W. P. Smotherman (Eds.) *Perinatal development: A psychobiological perspective* (pp. 61–84). New York: Academic Press.

Bronstein, P. M., Neiman, H., Wolkoff, F. D., & Levine, M. J. (1974). The development of habituation in the rat. *Animal Learning & Behavior, 2*, 92–96.

Brunjes, P. C., & Alberts, J. R. (1979). Olfactory stimulation induces filial huddling preferences in rat pups. *Journal of Comparative and Physiological Psychology, 93*, 548–555.

Bryan, R. G. (1979). Retention of odor-shock conditioning in neonatal rats: Effects of distribution of practice. Unpublished doctoral thesis, Rutgers University, New Brunswick, NJ.

Camp, L. L., & Rudy, J. W. (1988). Changes in the categorization of appetitive and aversive events during postnatal development of the rat. *Developmental Psychobiology, 21*, 25–42.

Campbell, B. A. (1967). Developmental studies of learning and motivation in infraprimate mammals. In H. W. Stevenson, E. H. Hess, & H. L. Rheingold (Eds.), *Early behavior: Comparative and developmental approaches* (pp. 43–72). New York: Wiley.

Campbell, B. A., & Alberts, J. R. (1979). Ontogeny of long-term memory for learned taste aversions. *Behavioral & Neural Biology, 25*: 139–156.

Campbell, B. A., & Coulter, X. (1976). The ontogenesis of learning and memory. In M. R. Rosenzweig & E. L. Bennett (Eds.), *Natural mechanisms of learning and memory* (pp. 209–235). Cambridge, MA: MIT Press.

Campbell, B. A., & Haroutunian, V. (1983). Perceptual sharpening in the developing rat. *Journal of Comparative Psychology, 1*, 3–11.

Campbell, B. A., & Jaynes, J. (1966). Reinstatement. *Psychological Review, 73*, 478–480.

Campbell, B. A., Misanin, J. R., White, B. C., & Lytle, L. D. (1974). Species differences in ontogeny of memory: Indirect support or neutral maturation as a determinant of forgetting. *Journal of Comparative and Physiological Psychology, 87*, 193–202.

Campbell, B. A., & Spear, N. E. (1972). Ontogeny of memory. *Psychological Review, 79*, 215–236.

Campbell, B. A., & Stehouwer, D. J. (1979). Ontogeny of habituation and sensitization in the rat. In N. E. Spear & B. A. Campbell (Eds.), *Ontogeny of learning and memory* (pp. 67–100). Hillsdale, NJ: Erlbaum.

Castro, C. A., Paylor, R. & Rudy, J. W. (1987). A developmental analysis of the learning and short-term memory processes mediating performance in conditional spatial discrimination problems. *Psychobiology, 15*, 308–316.

Castro, C. A., & Rudy, J. W. (1987). Early-life malnutrition selectively retards the development of distal- but not proximal-cue navigation. *Developmental Psychobiology, 20*, 531–537.

Caza, P. A., & Spear, N. E. (1984). Short-term exposure to an odor increases its subsequent preference in preweanling rats: A descriptive profile of the phenomenon. *Developmental Psychobiology, 17*, 407–422.

Cheatle, M. D., & Rudy, J. W. (1978). Analysis of second-order odor-aversion conditioning in neonatal rats: Implications for Kamin's blocking effect. *Journal of Experimental Psychology: Animal Behavior Processes, 4*, 237–249.

Concannon, J. T., Smith, G. J., Spear, N. E., Scobie, S. R. (1979). Drug cues, drug states, and infantile amnesia. In F. C. Colpaert & J. A. Rosencrans (Eds.), *First international symposium on drugs as discriminative stimuli* (pp. 353–396). Amsterdam: Elsevier.

Coulter, X. (1979). The determinants of infantile amnesia. In N. E. Spear & B. A. Campbell (Eds.), *The ontogeny of learning and memory*. Hillsdale, NJ: Erlbaum.

Crozier, W. J., & Pincus, G. (1927). Photic stimulation of young rats. *Journal of Genetic Psychology, 17*, 105–111.

Feigley, D. A., Parsons, P. J., Hamilton, L. W., & Spear, N. E. (1972). Development of habituation to novel environments in the rat. *Journal of Comparative and Physiological Psychology, 79*, 443–452.

Feigley, D. A., & Spear, N. E. (1970). Effect of age and punishment condition on long-term retention by the rat of active and passive avoidance learning. *Journal of Comparative and Physiological Psychology, 73*, 515–526.

Feldman, D. T., & Gordon, W. C. (1979). The alleviation of short-term retention decrements with reactivation. *Learning and Motivation, 10*, 198–210.

Galef, B. G., Jr., & Clark, M. M. (1971). Social factors in the poison avoidance and feeding behavior of wild and domesticated rats. *Journal of Comparative and Physiological Psychology, 75*, 341–347.

Garcia. J., & Koelling, R. A. (1966). Relation of cue to consequence in avoidance learning. *Psychonomic Science, 4*, 123–124.

Gemberling, G. A., Domjan, M., & Amsel, A. (1980). Aversive learning in 5-day old rats: Taste-toxicosis and texture-shock aversive learning. *Journal of Comparative and Physiological Psychology, 94*, 734–745.

Gottleib, G. (1971). Ontogenesis of sensory function in birds and mammals. In E. Tobach, L. R. Aronson, & E. Shaw (Eds.), *The Biopsychology of Development* (pp. 67–128). New York: Academic Press.

Grant, D. G. (1981). Intertrial interference in rat short-term memory. *Journal of Experimental Psychology: Animal Behavior Processes, 7*, 217–227.

Gregg. B., Kittrel, E. M. W., Domjan, M., Amsel, A. (1978). Ingestional aversion learning in preweanling rats. *Journal of Comparative and Physiological Psychology, 92*, 785–795.

Groves, P. M., & Thompson, R. F. (1970). Habituation: A dual-process theory. *Psychological Review, 77*, 419–450.

Gubernick, D. J., & Alberts, J. R. (1984). A specialization of taste aversion learning during suckling and its weaning-associated transformation. *Developmental Psychobiology, 17*, 613–628.

Hall, W. G. (1979). Feeding and behavioral activation in infant rats. *Science, 205*, 206–209.

Hall, W. G., & Rosenblatt, J. S. (1977). Suckling behavior and intake control in the developing rat pup. *Journal of Comparative and Physiological Psychology, 91*, 1232–1247.

Haroutunian, V., & Campbell, B. A. (1979). Emergence of interoceptive and exteroceptive control of behavior in rats. *Science, 205*, 927–928.

Hoffmann, H., Molina, J. C., Kucharski, D., & Spear, N. E. (1987). Further examination of ontogenetic limitations on conditioned taste aversions. *Developmental Psychobiology, 20:* 455–463.

Hoffmann, H., & Spear, N. E. (1988). Ontogenetic differences in conditioning of an aversion to a gustatory CS with a peripheral US. *Behavioral & Neural Biology, 50*, 16–23.

Hyson, R. L., & Rudy, J. W. (1984). Ontogeny of learning: II. Variation in the rat's unlearned and learned response to acoustic stimulation. *Developmental Psychobiology, 17*, 263–283.

Infurna. R. N. (1981). Daily biorhythmicity influences homing behavior, psychopharmacological responsiveness, learning, and retention of suckling rats. *Journal of Comparative and Physiological Psychology, 95*, 896–914.

Johanson, I. B., & Hall, W. G. (1979). Appetitive learning in 1-day-old rat pups. *Science, 205*, 419–421.

Johanson, I. B., & Hall, W. G. (1982). Appetitive conditioning in neonatal rats: Conditioned orientation to a novel odor. *Developmental Psychobiology, 15*, 379–397.

Johanson, I. B., & Hall, W. G. (1984). Ontogeny of appetitive learning: Independent ingestion as a model motivation system. In R. Kail & N. E. Spear (Eds.), *Comparative perspectives on the development of memory* (pp. 135–157). Hillsdale, NJ: Erlbaum.

Johanson, I. B., Hall, W. G., & Palefone, J. M. (1984). Appetitive conditioning in neonatal rats: Conditioned ingestive responding to stimuli paired with oral infusions of milk. *Developmental Psychology, 17*, 357–381.

Johanson, I. B., & Teicher, M. H. (1980). Appetitive conditioning in neonatal rats: Conditioned orientation to a novel odor. *Developmental Psychobiology, 15*, 379–397.

Kenny, J. T., & Blass, E. M. (1977). Suckling as incentive to instrumental learning in preweanling rats. *Science, 196*, 989–999.

Kessler, P., & Spear, N. E. (1980). Neonatal thyroxine treatment enhances classical conditioning in the rat. *Hormones and Behavior, 14*, 201–210.

Klein, S. B., & Spear, N. E. (1969). Influence of age on short-term retention of active avoidance learning. *Journal of Comparative and Physiological Psychology, 69*, 583–598.

Kraemer, P. J., Hoffmann, H. L., & Spear, N. E. (1988). Attenuation of the CS-preexposure effect after a retention interval in preweanling rats. *Animal Learning & Behavior, 16*, 185–190.

Kucharski, D., & Hall, W. G. (1987). New routes to early memories. *Science, 238*, 786–788.

Kucharski, D., & Spear, N. E. (1984). Conditioning of aversion to an odor paired with peripheral shock in the developing rat. *Developmental Psychobiology, 17*, 465–479.

Lashley, K. S. (1930). The mechanism of vision. *Journal of Genetic Psychology, 37*, 453–460.

Leon, M., Coopersmith, R., Lee, S., Sullivan, R. M., Wilson, D. A., & Woo, C. C. (1987). Neural and behavioral plasticity induced by early olfactory learning. In N. A. Krasnegor, E. M. Blass, M. A. Hofer, & W. P. Smotherman, *Perinatal Developmental: A Psychobiological Perspective* (pp. 145–168). New York: Academic Press.

Leon, M., Galef, B. G., & Behse, J. H. (1977). Establishment of pheromonal bonds in diet choice in young rats by odor preexposure. *Physiology & Behavior, 18*, 387–391.

Lubow, R. E. (1973). Latent inhibition. *Psychological Bulletin, 79*, 398–407.

Markiewicz, B., Kucharski, D., & Spear, N. E. (1986). Ontogenetic comparison of memory for Pavlovian conditioned aversions in temperature, vibration, odor or brightness. *Developmental Psychobiology, 19*, 139–154.

Miller, J. S., Jagielo, J. A., & Spear, N. E. (1989). Age-related differences in short-term retention of separable elements of an odor aversion. *Journal of Experimental Psychology: Animal Behavior Processes, 15*, 194–201.

Misanin, J. R., Chubb, L. D., Quinn, S. A., & Schweikert, G. E. (1974). An apparatus and procedure for effective instrumental training of neonatal and infant rats. *Bulletin of the Psychonomic Society, 4*, 171–173.

Misanin, J. R., Haigh, J. M., Hinderliter, C. F., & Nagy, Z. M. (1973). Analysis of response competition in nondiscriminated escape training of neonatal rats. *Journal of Comparative and Physiological Psychology, 85*, 570–580.

Misanin, J. R., Lariviere, N. A., & Hinderliter, C. F. (1987). The differential effects of reinstatement and reactivation on retention of active avoidance in pre-visual rats. Paper presented at the meeting of the Eastern Psychological Association.

Misanin, J. R., Nagy, Z. M., Keiser, E. F., & Bowen, W. (1971). Emergence of long-term memory in the neonatal rat. *Journal of Comparative and Physiological Psychology, 77*, 188–199.

Molina, J. C., Hoffmann, H., & Spear, N. E. (1986). Conditioning of aversion to alcohol orosensory cues in 5- and 10-day-old rats: Subsequent reduction in alcohol ingestion. *Developmental Psychobiology, 19*, 175–183.

Morris. R. G. M. (1981). Spatial localization does not require the presence of local cues. *Learning and Motivation, 12*, 239–260.

Moye, T. B., & Rudy, J. W. (1985). Ontogenesis of learning: VI. Learned and unlearned responses to visual stimulation in the infant hooded rat. *Developmental Psychobiology, 18*, 395–409.

Moye, T. B., & Rudy, J. W. (1987). Ontogenesis of trace conditioning in young rats: Dissociation of associative and memory processes. *Developmental Psychobiology, 20*, 405–414.

Myslivecek, J., & Hassmannova, J. (1983). The development of inhibitory learning and memory in hooded and albino rats. *Behavioral Brain Research, 8*, 151–166.

O'Keefe, J., & Nadel, L. (1978). *The hippocampus as a cognitive map*. Oxford, England: Clarendon Press.

Parsons, P. J., Fagen, T., & Spear, N. E. (1973). Short-term retention of habituation by infant, weanling and adult rats. *Journal of Comparative and Physiological Psychology, 84*, 543–545.

Pelchat, M. L., Grill, H. V., Rozin, P., & Jacobs, J. (1983). Quality of acquired responses to taste by *rattus norvegus* depends on type of associative discomfort. *Journal of Comparative Psychology, 97*, 140–153.

Pedersen, P. E., Williams, C. L., & Blass, E. M. (1982). Activation and odor conditioning of suckling behavior in 3-day-old albino rats. *Journal of Experimental Psychology: Animal Behavior Processes, 8*, 329–341.

Pillemer, D. B., & White, S. H. (1989). Childhood events recalled by children and adults. In H. W. Reese (Ed.), *Advances in child development and behavior* (Vol. 22). New York: Academic Press.

Rescorla, R. A. (1967). Pavlovian conditioning and its proper control procedures. *Psychological Review, 74*, 71–80.

Revusky, S. H., & Garcia, J. (1970). Learned associations over long delays. In G. H. Bower & J. T. Spence (Eds.), *The psychology of learning and motivation* (Vol. 4). New York: Academic Press.

Riccio, D. C., Rohrbaugh, M., & Hodges, L. A. (1968). Developmental aspects of passive and active avoidance learning in rats. *Developmental Psychobiology, 1*, 108–111.

Richardson, R., Riccio, D. C., & Axiotis, R. (1986). Alleviation of infantile amnesia in rats by internal and external cues. *Developmental Psychobiology, 19*, 453–462.

Richardson, R., Riccio, D. C., & Devine, L. (1984). ACTH-induced recovery of extinguished avoidance responding. *Physiological Psychology, 12*, 184–192.

Riley, E. P. (1990). The long-term behavioral effects of prenatal alcohol exposure in rats. *Alcoholism: Genetic and Experimental Research, 14*, 670–673.

Roberts, W. A. (1974). Spaced repetition facilitates short-term retention in the rat. *Journal of Comparative and Physiological Psychology, 86*, 164–171.

Rubel, E. W., & Rosenthal, M. (1975). The ontogeny of auditory frequency generalization in the chicken. *Journal of Experimental Psychology: Animal Behavior Processes, 1*, 287–297.

Rudy, J. W., & Castro, C. A. (1987). A developmental analysis of brightness discrimination learning in the rat: Evidence for an attentional deficit. *Psychobiology, 15*, 79–86.

Rudy, J. W., & Cheatle, M. D. (1977). Odor-aversion learning by neonatal rats. *Science, 198*, 845–846.

Rudy, J. W., & Cheatle, M. D. (1979). Ontogeny of associative learning: Acquisition of odor aversions by neonatal rats. In N. E. Spear & B. A. Campbell (Eds.), *Ontogeny of learning and memory* (pp. 157–188). Hillsdale, NJ: Erlbaum.

Rudy, J. W., & Hyson, R. L. (1984). Ontogenesis of learning: III. Variation in the rat's differential reflexive and learned responses to sound frequencies. *Developmental Psychobiology, 17*, 285–300.

Rudy, J. S., & Paylor, R. (1988). Reducing the temporal demands of the Morris place-learning task fails to ameliorate the place-learning impairment in preweanling rats. *Psychobiology, 16*, 152–156.

Rudy, J. W., Stadler-Morris, S., & Albert, P. (1987). Ontogeny of spatial navigation behaviors in the rat: Dissociation of "proximal"- and "distal"-cue based behaviors. *Behavioral Neuroscience, 101*, 62–73.

Rudy, J. W., Vogt, M. B., & Hyson, R. L. (1984). A developmental analysis of the rat's learned reaction to gustatory and auditory stimulation. In R. Kail & N. E. Spear (Eds.), *Comparative perspectives on the development of memory* (pp. 181–208). Hillsdale, NJ: Erlbaum.

Sananes, C. B., Gaddy, J. R., & Campbell, B. A. (1988). Ontogeny of conditioned heart rate to an olfactory stimulus. *Developmental Psychobiology, 21*, 117–134.

Schulenburg, C. J., Riccio, D. C., & Stikes, E. R. (1971). Acquisition and retention of a passive avoidance response as a function of age in rats. *Journal of Comparative and Physiological Psychology, 74*, 75–83.

Schweitzer, L., & Green, L. (1982). Acquisition and extended retention of a conditioned taste aversion in preweanling rats. *Journal of Comparative and Physiological Psychology, 96*, 791–806.

Shaywitz, B. A., Gordon, J. W., Klopper, J. H., Zelterman, D. A., & Irvine, J. (1979). Ontogenesis of spontaneous activity and habituation of activity in the rat pup. *Developmental Psychobiology, 12*, 359–367.

Smith, G. J., & Bogomolny, A. (1983). Appetitive instrumental training in preweanling rats: I. Motivational determinants. *Developmental Psychobiology, 16*, 119–128.

Smith. G. J., & Spear, N. E. (1978). Home environmental effects on withholding behaviors in conditioning in infant and neonatal rats. *Science, 202*, 327–329.

Smith, G. J., & Spear, N. E. (1981). Home environmental stimuli facilitate learning of shock escape, spatial discrimination in rats 7–11 days of age. *Behavioral & Neural Biology, 31*, 360–365.

Smotherman, W. P. (1982). Odor aversion learning by the rat fetus. *Physiology and Behavior, 29*, 769–771.

Smotherman, W. P., & Robinson, S. R. (1987). Psychobiology of fetal experience in the rat. In N. A. Krasnegor, E. M. Blass, M. A. Hofer, & W. P Smotherman (Eds.), *Perinatal development: A psychobiological perspective* (pp. 39–60). New York: Academic Press.

Spear, N. E. (1978). *Processing memories: Forgetting and retention.* Hillsdale, NJ: Erlbaum.

Spear, N. E. (1979a). Memory storage factors in infantile amnesia. In G. Bower (Ed.), *The psychology of learning and motivation* (Vol. 13, pp. 91–154). New York: Academic Press.

Spear, N. E. (1979b). Memory storage factors in infantile amnesia. In G. Bower (Ed.), *The psychology of learning and motivation* (Vol. 13, pp. 91–154). New York: Academic Press.

Spear, N. E., & Campbell, B. A. (Eds.). (1979). *Ontogeny of learning and memory.* Hillsdale, NJ: Erlbaum.

Spear, N. E., Kraemer, P. J., Molina, J. C., & Smoller, D. E. (1988). Developmental change in learning and memory: Infantile disposition for "unitization." In J. Delacour & J. C. S. Levy (Eds.) *Systems with learning and memory abilities: Proceedings of the workshop held in Paris, June 15–19, 1987* (pp. 27–52). Amsterdam: Elsevier/North Holland Press.

Spear, N. E., Kucharski, D., & Hoffmann, H. L. (1985). Contextual influences on conditioned taste aversions in the developing rat. In N. S. Braveman & P. Bronstein (Eds.), *Experimental assessments in clinical applications of conditioned food aversions* (pp. 42–53). New York: Annals of the New York Academy of Sciences.

Spear, N. E., & Molina, J. C. (1987). The role of sensory modality in the ontogeny of stimulus selection. In N. Krasnegor, E. M. Blass, M. A. Hofer, & W. P. Smotherman (Eds.), *Perinatal development: A psychobiological perspective* (pp. 85–110). New York: Academic Press.

Spear, N. E., & Parsons, P. (1976). Alleviation of forgetting by reactivation treatment: A preliminary analysis of the ontogeny of memory processing. In D. Medin, W. Roberts, & R. Davis (Eds.), *Processes in animal memory* (pp. 135–166). Hillsdale, NJ: Erlbaum.

Spear, N. E., & Smith, G. J. (1978). Alleviation of forgetting in neonatal rats. *Developmental Psychobiology, 11,* 513–530.

Stehouwer, D. J., & Campbell, B. A. (1978). Habituation of the forelimb-withdrawal response in neonatal rats. *Journal of Experimental Psychology: Animal Behavior Processes, 4,* 104–119.

Steinert, P. A., Inturna, R. N., Jardula, M. F., & Spear, N. E. (1979). Effects of CS concentration on long-delay taste aversion learning in preweanling and adult rats. *Behavioral & Neural Biology, 27,* 487–502.

Steinert, P. A., Infurna, R. N., & Spear, N. E. (1980). Long-term retention of a conditioned taste aversion in preweaning and adult rats. *Animal Learning & Behavior, 8,* 375–381.

Stoloff, M. L, & Blass, E. M. (1983). Changes in appetitive behavior in weanling aged rats: Transitions from suckling to feeding behavior. *Developmental Psychobiology, 16,* 349–453.

Sullivan, R. M., Brake, S. C., Hofer, M. A., & Williams, C. L. (1986). Huddling and independent feeding of neonatal rats can be facilitated by a conditional change in behavioral state. *Developmental Psychobiology, 19,* 625–635.

Sutherland, R. J., & Dyck, R. H. (1984). Place navigation by rats in a swimming pool. *Canadian Journal of Psychology, 38,* 322–347.

Sullivan, R. M., Hofer, M. A., & Brake, S. C. (1986). Olfactory-guided orientation in neonatal rats is enhanced by a conditioned change in behavioral state. *Developmental Psychobiology, 19,* 615–623.

Sussman, P. S., & Ferguson, H. B. (1980). Retained elements of early avoidance training and relearning of forgotten operants. *Developmental Psychobiology, 13,* 545–562.

Terry. L. M., & Johanson, I. B. (1987). Olfactory influences on the ingestive behavior of infant rats. *Developmental Psychobiology, 20,* 313–332.

Thompson, R. F., & Spencer, W. A. (1966). Habituation: A model phenomenon for the study of neuronal substrates of behavior. *Psychological Review, 73,* 16–43.

Vogt, M. B., & Rudy, J. W. (1984). Ontogenesis of learning: IV. Dissociation of memory and perceptual-alerting processes mediating taste neophobia in the rat. *Developmental Psychobiology, 17,* 601–611.

Watson, J. B. (1903). Animal education. *Contributions of a Psychological Laboratory at University of Chicago, 4,* 5–122.

Williams, J. M., Hamilton, L. W., & Carlton, P. L. (1974). Pharmacological and anatomical dissociation of two types of habituation. *Journal of Comparative and Physiological Psychology, 87,* 724–732.

7

Innovation in the Study of Social Learning in Animals: Developmental and Biological Perspectives

BENNETT G. GALEF, JR.

It is somewhat surprising that I should have been invited to contribute a chapter to a volume concerned with methodological innovations and their use in resolution of theoretical issues in behavioral development. I have always taken a perverse pride in the fact that, despite a near total lack of technical competencies, I have been able to make empirical contributions to a field as methodologically sophisticated as developmental psychobiology has become. I can't program a computer, extirpate a gland, wire the simplest of electrical circuits, implant an electrode, assay a hormone, stain a tissue sample, or carry out any of the myriad other technical chores that today's graduate student can perform with ease. Consequently, if one defines methodological innovation in a narrow sense, this chapter is complete in a single paragraph, even in a single sentence. My methodological contribution to developmental psychobiology has been nil.

Yet it seems to me (and, I suspect, to the editors of the present volume, since they asked me to contribute to it) that methodological innovations can be conceptual as well as material. And conceptual innovations are often as important as methodological ones in advancing our capacity to study developmental problems of theoretical importance.

A scientist's most important tools are the intellectual frameworks or perspectives within which he or she works. Changes in perspective can lead to improvements in methodology as useful as those that accompany increased ability either to measure dependent variables or control the application of independent variables. Though novel perspectives on a problem are less concrete than innovations in technique, novel perspectives may lead directly to improvements in method and are, therefore, worth considering in a volume such as the present one.

Animal social learning, the research area in which I have been working for more than 20 years, is a field with a long and complex history (Galef, 1988a, 1988b). It is an area that might be characterized, not entirely unfairly, as having produced remarkably little progress in advancing understanding of behavior, despite more than a century of experimentation and discussion.

In the late 1960s, when I first became interested in problems in social learning, the area was a subspecialty of animal learning. Most experimental work in the area—at least most such work conducted by psychologists—was concerned with demonstrating, in standard laboratory apparatus (Skinner box or T-maze), the existence of general behavioral capacities in animals (learning by imitation, social facilitation, or the use of conspecifics as discriminative stimuli) that might result in social learning. For example, in a well-known study, typical of psychological research in social learning during the 1960s, Chesler (1969) demonstrated that kittens that had observed their respective mothers press a lever to obtain food subsequently acquired the lever-press response more rapidly than did kittens that had observed strange female cats press a lever for food. In Chesler's view, this finding demonstrated that kittens could learn by imitation.

Zajonc (1969) was making much of the observations that (1) two human cyclists in competition rode faster than either did when alone (Triplett, 1897) and (2) that ants each dug more vigorously when in pairs than when in isolation (Chen, 1937). Such observations were interpreted as exemplifying social facilitation, the energizing of dominant behaviors by the presence of others.

Russell Church (1957a, 1957b), Richard Solomon (Solomon & Coles, 1954), and Vaughn Stimbert (1970a, 1970b) had conducted (or were conducting) studies demonstrating that the activities of one rat could serve as discriminative stimuli for another rat, indicating to the latter individual those occasions when a particular response would be reinforced. For example, Church (1957a) had shown that a rat could be taught to turn left or right in a T-maze by reinforcing the animal whenever it entered the same arm of the maze that had been entered by a trained, leader rat.

During the 1950s and 1960s, when general process theories of learning dominated experimental, animal psychology, advance in the study of social learning seemed to require identification of social learning analogs of such individual learning processes as operant and classical conditioning. If one could elucidate critical features of paradigmatic cases of social learning (e.g., learning by imitation, matched dependent behavior, social facilitation), then social learning could be studied in the same way and at the same level of abstraction as individual learning had been studied for decades (Jenkins, 1984).

The problem I had with such an approach was that the categories of social learning available to explain relevant phenomena seemed both arbitrary and restrictive. For example, a cyclist might ride faster when on the track with a competitor than when alone because of an increase in aerodynamic efficiency gained by using the competitor as a windbreak. Alternatively, a cyclist might ride faster in the former situation than in the latter because of increased effort in a competitive situation. If ants do dig more efficiently when in pairs than when alone, it is probably because of some form of pheromonal communication between them. Nothing appeared to me to be gained by studying an abstraction called *coaction* or *social facilitation*. The proposed explanatory terms seemed to interfere with, rather than encourage, analyses of the effects of social interaction on behavior. Furthermore, even if the existence of social facilitation or imitation learning could be convincingly demonstrated, why should one believe that already-identified types of social learning are more important than other, as-yet-undiscovered processes?

I would attribute much of whatever progress my students and I have made in studying social learning to a recasting of the conceptual framework within which we have worked. We chose to treat social learning as an aspect of behavioral development rather than as a type of animal learning. In consequence, instead of asking whether animals have the ability to learn by imitation, to exhibit social facilitation, or to use conspecifics as discriminative stimuli, we asked whether social interactions might be important in development of adaptive patterns of behavior.

Changing the nature of the question opened up new realms of inquiry and provided new sources of both data and hypotheses. First, if one is interested in the role of social learning in the development of adaptive behavioral repertoires, the behaviors one might investigate extend far beyond the Skinner box and T-maze; foraging, predation, homing, predator avoidance, tool use, nest building—the list of behaviors of potential interest seems endless.

Second, one could make use of the field literature to identify promising behaviors to investigate. Instances in which many members of one population of a species exhibit a pattern of behavior absent in other populations of that species suggest that social influence might play a role in the development of the idiosyncratic behaviors (traditions) observed: Song dialect learning (Marler & Tamura, 1964), sweet potato washing (Kawamura, 1959), milk bottle opening (Fisher & Hinde, 1949), as well as myriad other less-well-known instances of animal tradition (Galef, 1976), are defined as phenomena that, if studied in the laboratory, might reveal behavioral processes involved in social learning.

Third, because we would be attempting to determine the causes of differences in behavior among populations, there would be no bias toward demonstrating that one process rather than another was involved in the development of the differences under investigation. Instead of undertaking demonstration experiments to prove, for example, that animals can learn by imitation, or that social facilitation is a general phenomenon, we needed only to determine the necessary and sufficient conditions for the emergence of behaviors observed in natural circumstances. Whatever the behavioral process involved in the development of behavioral differences between populations, whether social or nonsocial, new or familiar, such an approach promised unbiased, logically satisfying analyses.

Last, and perhaps most important, approaching the study of social learning as a developmental process permitted integration of psychological and biological approaches to the study of behavior. Tinbergen (1963) had indicated that four different questions could be asked about any behavior: What is its cause, function, development, and phylogeny? Looking at social learning as a factor in the development of behaviors that had functions outside the laboratory allowed us to examine behaviors of interest from both psychological and biological perspectives.

I should make clear that I can claim little originality in adopting the general perspective outlined above. I had the good fortune to complete my dissertation research under the supervision of Paul Rozin. He communicated most effectively the utility of combining biological, developmental, and psychological approaches to the study of behavior. Furthermore, by 1970, the impact of ethology was steadily growing and the dominance of the general process approach to the study of animal learning was in decline. Analyses of, among other things, imprinting in precocial birds (Hess, 1973) and song learning in passerines (Marler & Tamura, 1964), as well as texts such as Hinde's (1966) *Animal Behaviour*, had demonstrated the heuristic value of considering learning as a factor in the development of adaptive behavioral repertoires in animals. Many others (e.g., Johnston, 1981; Kamil & Yoerg, 1982; Shettleworth, 1984) came independently to the conclusion that consideration of animal learning within a biological framework, as a developmental process, might be a fruitful way to approach the study of learning by animals.

I cannot even claim originality in application of developmental and biological approaches to the study of social influences on behavior; Harlow's (Harlow & Harlow, 1965) studies of reproductive behavior in rhesus monkeys and Marler's (Marler & Tamura, 1964) investigations into bird-song learning in white-crowned sparrows are clear antecedents. All I can claim is the good sense to follow illuminating precedent.

INVESTIGATIONS OF SOCIAL LEARNING IN MAMMALS

It is easy enough to say that investigation of necessary and sufficient conditions for the development of traditional patterns of behavior observed in field settings might provide a fruitful approach to the study of animal social learning. It is a different and somewhat more difficult matter to design such investigations in a satisfying fashion. There are problems both in finding phenomena susceptible to investigation in controlled settings and in establishing reasonable laboratory analogs of traditions observed in the outside world. All this is especially true if, as I am, one is interested in mammalian behavior.

It is an unhappy fact that there are even today no mammals that are good subjects for both laboratory and field work, at least not if one is working on a modest budget. Outside the laboratory, rodents, such as rats, mice, gerbils, and hamsters, so admirably preadapted to laboratory existence, are either shy, nocturnal and subterranean, or indigenous to distant lands. Those mammals easily observed in the wild are too expensive for most researchers to acquire and maintain in captivity in numbers adequate for analytic research. Compromises have to be made. My own choice has been to work with Norway rats, members of that much-maligned, near-perfect species for behavioral research in the laboratory. The cost has been a need to depend, perhaps too heavily, on fragmentary, often anecdotal, field reports and the face validity of laboratory analogs of events presumed to occur in nature.

It is sometimes relatively easy to create a compelling analog of naturally occurring events. More frequently one must be satisfied with the knowledge that, in spite of inadequacies, one has done one's best to capture important features of the natural world while maintaining enough control over a situation to permit experiments to proceed. The important thing is to keep clearly in mind that the goal of a study is to cast light on processes that, in natural circumstances, lead to development of behaviors of interest. One hopes that even a dim light falling on an intriguing object is of more value than greater illumination of less interesting things.

In the present section, I briefly outline the methods and results of three research programs conducted in my laboratory, each concerned with the role of social learning in the development of behavior. All are, I believe, examples of the utility of studying social learning within a biological framework, as a problem in behavioral development. In each case, my strategy was similar: Begin with field observation of a possible case of animal tradition. Bring the phenomenon into the laboratory. Attempt to determine the ways in which social interaction influences the course of development of the behavior in question.

Case 1: Following Adults to Food

Some years ago, Fritz Steiniger, an ecologist, was working for the German government as a rodent control officer. His job was to exterminate populations of wild Norway rats by poisoning them. In the course of his work, Steiniger discovered that if he used a poison bait in an area for an extended period of time, despite initial success, with the rats eating lots of poison and dying in large numbers, later acceptance of the bait was very poor. Steiniger (1950) observed that young born to those rats that had, by chance, survived their initial ingestion of poison rejected the poison bait without ever even sampling it. These offspring of survivors fed exclusively on safe diets available in the territories of their respective colonies. Steiniger attributed such "traditional" poison avoidance by the young to the effect of urine and feces deposited on the poison bait by those surviving adults that had learned to avoid it.

My coworker, Mertice Clark, and I were fortunate in that the phenomenon is robust and readily observable in captive wild rats living in the laboratory in small (1- by 2-m) enclosures (Galef & Clark, 1971b). By introducing nonlethal but nauseating concentrations of poison into one of two foods (Diets A and B) presented for 3 hours a day to our wild rat subjects, we could easily train them to avoid eating the poisoned food. Soon the adults would not eat the poisoned food even if offered uncontaminated samples of it. Next, we had to wait till the adults in a colony produced a litter and the litter grew to weaning age. Finally, we could observe the pattern of diet selection exhibited by weanlings born to colonies trained to eat only Diet A or only Diet B.

As Steiniger would have predicted, as long as the young were left in contact with adults, those young wild rats raised by colonies of adult wild rats poisoned when eating Diet A ate only Diet B; those raised by colonies poisoned when eating Diet B ate only Diet A. Both groups of young rats continued for a week or more after removal from contact with their parent colony to prefer the food the parent colony had eaten.

Steiniger (1950) had proposed that urine and feces deposited by adults in or near a food dissuaded their young from eating that food. However, when we conducted an experiment in which we offered young rats samples of Diets A and B uncontaminated by adult droppings, the young continued to eat only the diet that the adults of their colony were eating (Galef & Clark, 1971b). Furthermore, we found that rat pups reared by adults fed only Diet B, when offered a choice between Diets A and B, were just as biased in their preference for Diet B as were rat pups reared by adults that had learned to avoid Diet A and eat only Diet B. Not only were pups not avoiding a diet because the adults had marked it, the pups were not even learning to avoid the diet the adults of their colony were avoiding. Instead, pups were learning to eat the food the adults of their colony were eating.

Further studies revealed that young wild rats moving from nest site to feeding site were strongly biased to approach adults or sibs and to begin eating in their immediate vicinity (Galef & Clark, 1971a). The simple physical presence of a conspecific at one feeding site rather than another biased the young to eat the food located there.

Wild Norway rats are known to exhibit great hesitancy to eat any food they have not previously eaten (Barnett, 1958; Galef, 1970). In consequence, biasing the young to start feeding on one diet greatly reduces the probability they will eat available alternatives.

In our experiments, young rats only gave the appearance of having learned to avoid the diet that the adults of their colony had learned to avoid. In fact, as a consequence of a tendency to affiliate with adults, pups had eaten and had become familiar with the food that the adults of their colony were eating. As a result of a hesitancy to eat unfamiliar foods, the pups were ignoring available alternative foods.

Shortly after I completed these studies, I was describing the results to a colleague at a conference. She responded, "So, baby rates follow adult rats. What's new?"—a remark I felt completely missed the point. What was new was that we could now describe in detail behavioral processes sufficient to support the development of an adaptive, traditional pattern of poison avoidance in rats. It is true that nothing like imitation may have been involved, but that matters only if you are trying to demonstrate the existence of imitation learning, not if you are interested in the role of social influence in the development of animal behavior. We had been able to show that development of an apparently complex tradition of poison avoidance rested on humble foundations. The tendency of young rats to approach adults could function to produce poison avoidance in the laboratory and possibly in the field.

Case 2: Diving for Food

In the early 1970s, a group of researchers working at the University of Parma reported that many members of some colonies of wild rats living along the banks of the Po River in northern Italy dove for and fed on freshwater clams inhabiting the river bottom. No members of nearby colonies, with equal access to mollusks within their home ranges, fed on mollusks (Gandolfi & Parisi, 1972, 1973; Parisi & Gandolfi, 1974).

These observations were interpreted as indicating that predation on submerged prey spreads through a wild rat colony as the result of naive rats observing and imitating their diving fellows. If discovery of mollusks on the riverbed by colony members is a rare event and if naive colony members learn to dive as the result of observing diving individuals, then one would expect the observed bimodality in frequency of individuals diving in various colonies. The central question, of course, is what explains the occurrence of diving in some populations of rats but not in others. Is social learning actually involved?

Once again, my coworkers and I brought the phenomenon into the laboratory in an attempt to determine whether a traditional pattern of behavior, observed in a free-living rat population, might develop as the result of social learning of some kind. Because we lacked access to either freshwater clams or the Po River we had to exercise our imaginations a bit in constructing a laboratory analog of the situation in Italy. We studied development of the behavior of diving for plastic-film-wrapped chocolates in

20-gallon aquaria (Galef, 1980). Despite possible inadequacies in our attempt to capture the real world in microcosm, I think we learned some interesting things about the development of diving behavior in rat populations.

First, we were unable to induce naive rats to dive in 15 cm of water for chocolates by letting them interact for a month with trained, diver rats recovering chocolates from beneath the water's surface. Neither adult nor juvenile, wild nor domesticated rats, that had interacted freely with a conspecific that dove 100 times or more learned to dive. Naive adult rats were extremely reluctant even to enter the water and swim. They never put their heads under the surface. About 20% of juveniles learned to dive whether they had interacted with a diving conspecific or not (Galef, 1980).

The question that arises following a negative outcome like that just described is what to do next. It is here that one's perspective on a problem is particularly important. If our goal had been to demonstrate the importance of imitation or of social facilitation in the acquisition of diving behavior in rats, we would have abandoned the project. However, because we were interested in the possible role of social influence in the development of adaptive behavioral repertoires, our negative result was almost as interesting as a positive one would have been. Regardless of the outcome of the first experiment, we still have a phenomenon to explain—Why do members of some colonies of rats and not of others dive for mollusks along the banks of the Po River?

Further studies (Galef, 1980) revealed that, if reared until 2 months of age with access to water, both wild and domesticated rats will spontaneously enter the water and swim before reaching maturity. We also found that swimming rats will, with high probability, spontaneously dive and retrieve food from beneath the water's surface. These were surprising findings in that they suggested that what needed to be explained was not the development of diving in some colonies along the Po but the absence of diving in other colonies found there. Results of additional studies suggested that although almost all rats reared with access to water will learn spontaneously both to swim as juveniles and to dive for food as adults, even animals trained to swim and dive for food would prefer not to engage in those activities. Both swimming and diving for food could be greatly inhibited by the presence of adequate rations ashore.

The results of our series of studies suggest a tentative solution to the problem posed by the tradition of diving exhibited by Po River rats. Differences among colonies in frequency of diving depended not on social learning but on differences in colony access to food on land. Although the relevant experiments have not, so far as I know, been performed, the results of our analysis suggest that delivering a load of food to the territory of a diving colony would end its diving. Removing food from the territory of a nondiving colony would induce colony members to start diving. What had initially appeared to be a developmental problem turned out to be a motivational one. Resolution of the issue depended on focus on the phenomenon rather than on attempts to demonstrate the action of particular processes (Galef, 1980).

Case 3: Information Centers in Rats

The field literature provides not only traditions to analyze but also hypotheses about the development of traditions that may prove worth investigating in controlled settings. In 1970, Ward and Zahavi published a widely cited paper in which they argued that aggregation sites of birds or mammals might function as "information centres" where unsuccessful (or marginally successful) foragers could acquire useful information from their more successful fellows about where food could be found. Individuals that were having trouble finding food might identify more successful individuals at a central site and then follow them to food. Thus, social interaction in aggregations might serve as a substrate for development of adaptive patterns of foraging.

For my purposes, the notion of an information center, of a location distant from a resource where animals might acquire information about the resource, was more interesting than the particular method of information transmission Ward and Zahavi had proposed. I had always assumed, from lack of imagination more than anything else, that mammals in general and rats in particular would be able to communicate about a resource only in the presence of that resource. Von Frisch's (1954) classic studies of dance-language communication in honeybees had clearly shown that it was possible for a successful forager bee to direct hive mates to the food source it was exploiting. Wenner (Wenner, Wells, & Johnson, 1969) had shown that odors clinging to the body of successful forager bees could serve the same function. Although I was aware of such findings, bees seemed so behaviorally specialized for communication about resource availability that it was not obvious to me that extrapolation to generalists, such as Norway rats, was warranted.

In a bit of serendipitous good fortune, Barbara Strupp, in 1980 a graduate student with David Levitsky at Cornell, came to visit my laboratory in Hamilton and told me about her thesis research on communication of diet preference among rats. Strupp had shown that a naive rat, living in a cage adjacent to a conspecific and offered a choice between two diets, one of which was the diet its neighbor was eating, would prefer its neighbor's diet (Strupp & Levitsky, 1984). The rats were communicating with one another concerning a resource while that resource was present in their shared environment. Strupp was generous in providing details of her procedures. She and Levitsky were not interested in pursuing the line of research that both interested me and seemed possible using variants of their procedure. There was lots to do.

Steven Wigmore, a graduate student in my laboratory, and I proceeded to ask whether rats could communicate information to one another concerning diets they had eaten at distant times and places. In our basic experiment (Galef & Wigmore, 1983), a demonstrator rat was fed a diet and, after finishing eating, the animal was allowed to interact briefly with a naive observer rat. Then the observer was offered a choice between the diet its demonstrator had eaten and a roughly equipalatable alternative. The results of such studies surpassed our wildest hopes. Following interaction with their respective demonstrators, observers exhibited greatly enhanced preference for whatever diet their demonstrators had eaten. The effects of demonstrators' diets on observers' subsequent diet preferences lasted for days. We now know that they can last for weeks (Galef, 1989).

Within a developmental, functionalist perspective we now had a whole series of questions to answer: How is information communicated from demonstrator to observer? How does the communicated information affect diet selection by observers? What are the implications of the ability of experienced individuals to affect the diet preference of naive conspecifics for our understanding of the development of adaptive patterns of dietary selection? How might such a capacity function to enhance foraging efficiency and utilization of resources in natural environments?

Five years of research with a number of collaborators and students is beginning to provide some insight into the ways in which rats communicate about distant foods, insights that I believe may have important implications for a more general understanding of behavioral development in vertebrates. Below, I briefly summarize our findings (see Galef, 1986a, 1988c, for more detailed accounts) and discuss their implications for the study of behavioral development.

First, all our data point to the conclusion that olfactory cues passing from a demonstrator rat to an observer rat carry information which allows an observer to identify the diet eaten by its demonstrator (Galef & Wigmore, 1983). Second, we have found that such diet-identifying, olfactory cues can arise both from the digestive tracts of demonstrators and from traces of food clinging to their fur and vibrissae (Galef & Stein, 1985). Thus, we have identified both the medium of communication from demonstrator to observer (olfaction) and the sources of information that allows observers to know what food their respective demonstrators have been eating.

The question of why observers prefer their respective demonstrators' diets is less fully answered. We have shown repeatedly that simple exposure of an observer to a diet is not effective in increasing an observer's preference for that diet, while exposure of an observer to the same diet in the presence of a demonstrator does increase an observer's preference for the diet. The question, of course, is why this is so. Our data suggest that, in addition to diet-identifying, olfactory cues, demonstrator rats emit contextual, olfactory cues that, when experienced by observers at the same time as diet-identifying cues, result in profound alterations in observers' diet preferences (Galef & Stein, 1985). I am currently working in collaboration with Russell Mason and George Preti at the Monell Chemical Senses Center to try to identify chemically this elusive contextual determinant of diet preferences (Russ and George look after the chemistry, I worry about the behavior).

The role of information acquired from conspecifics in the development of adaptive patterns of diet selection by new recruits to a population has been of particular interest to me. Social influences on diet selection of the type described above, can be profound. Rats that have learned an aversion to a diet will abandon that aversion following interaction with conspecifics that have eaten that diet (Galef, 1985, 1986b). Rats will greatly increase their intake of unpalatable diets following interaction with fellows that have fed on it (Galef, 1986b). Aversions are formed less readily to those novel diets that conspecifics are eating than to other novel diets (Galef, 1986c). In sum, development of feeding repertoires by rats is not just the result of individual palatability preferences and individual learning about the postingestional consequences of various foods. Each rat can and will make use of the feeding behavior of others in developing its own patterns of diet selection.

Last, rats are able to integrate the information they acquire from conspecifics with individual learning about the distribution and value of resources. A rat that knows, as the result of its individual feeding history, where a number of different foods are sometimes to be found, but does not know which food is available on a particular day, will go directly to the appropriate feeding site after interacting with a conspecific that has eaten one of the intermittently available foods (Galef, 1983; Galef & Wigmore, 1983). A rat will more readily follow a conspecific to food if that conspecific has eaten a food known by the follower to be safe than if the conspecific has eaten a food known by the follower to be toxic (Galef, Mischinger, & Malenfant, 1987). Taken together, our results suggest not only that naive rats can extract information from conspecifics as to the food those conspecifics have been eating, but also that such information can play an important role in the development of adaptive patterns of both diet selection and foraging-site exploitation.

Conclusions

There is every reason to believe that social interactions can provide stimuli that bias the development of a wide range of behaviors. Each member of a species need not learn for itself what foods to eat, what paths to take, what potential predators to avoid. A naive individual can "assume" that live conspecifics have successfully solved many of the problems posed by an area they share. By adopting the behavior of successful others, adequate behavioral repertoires can be developed without incurring risks that are an inevitable consequence of learning on one's own.

Consideration of social learning within a biological framework, as an aspect of behavioral development with adaptive consequences, has proven a fruitful perspective, attracting a growing number of students of social learning. Curio (Curio, Ernst, & Vieth, 1978) has examined the role of social learning in the development of predator avoidance and mobbing behavior in European blackbirds. Cook and Mineka (Cook, Mineka, Wolkenstein, & Laitsch, 1985) are studying the effects of social interaction in the acquisition of snake avoidance in rhesus monkeys. Lefebvre (Palametta & Lefebvre, 1985), Mason and Reidinger (1982), and Terkel and Eisner (J. Terkel, personal communication, August 1986) are examining social effects on the development of adaptive patterns of feeding and foraging in, respectively, common pigeons, red-wing blackbirds, and roof rats. Baptista and Petrinovich (1984) and West and King (West, King, & Harrocks, 1983) have increased understanding of the role of social interaction in the development of bird song.

What has been for decades a rather conservative and uninteresting subarea of animal learning is evolving rapidly into a field of both substance and intrinsic interest. Although such evolution has profited in no small measure from increased sophistication in experimental methods, progress would have been impossible without innovation in the conceptual tools, the frameworks, and perspectives used to study social learning phenomena.

ACKNOWLEDGMENTS

The research reviewed here was supported by grants from the Natural Sciences and Engineering Research Council of Canada and the McMaster University Research Board and was carried out in collaboration with Mertice Clark, David Sherry, Pat Henderson, Linda Heiber, Steven Wigmore, Deborah Kennett, Moni Stein, Anna Mischinger, and Sylvie Malenfant. I thank Mertice Clark and Harry Shair for helpful critiques of earlier drafts.

REFERENCES

Baptista, L. F., & Petrinovich, L. (1984). Social interaction, sensitive phases, and the song template hypothesis in the white-crowned sparrow. *Animal Behaviour, 32*, 172–181.

Barnett, S. A. (1958). Experiments on "neophobia" in wild and laboratory rats. *British Journal of Psychology, 49*, 195–201.

Chen, S. C. (1937). Social modification of the activity of ants in nest building. *Physiological Zoology, 10*, 420–436.

Chesler, P. (1969). Maternal influence in learning by observation in kittens. *Science, 166*, 901–903.

Church, R. M. (1957a). Transmission of learned behavior between rats. *Journal of Abnormal and Social Psychology, 54*, 163–165.

Church, R. M. (1957b). Two procedures for establishment of "imitative behavior." *Journal of Comparative and Physiological Psychology, 50*, 315–318.

Cook, M., Mineka, S., Wolkenstein, B., & Laitsch, K. (1985). Observational conditioning of snake fear in unrelated rhesus monkeys. *Journal of Abnormal Psychology, 93*, 355–372.

Curio, E., Ernst, U., & Vieth, W. (1978). Cultural transmission of enemy recognition: One function of mobbing. *Science, 202*, 899–901.

Fisher, J., & Hinde, R. A. (1949). The opening of milk bottles by birds. *British Birds, 42*, 347–357.

Galef, B. G., Jr. (1970). Aggression and timidity: Responses to novelty in feral Norway rats. *Journal of Comparative Physiological Psychology, 70*, 370–381.

Galef, B. G., Jr. (1976). Social transmission of acquired behavior: A discussion of tradition and social learning in vertebrates. In J. S. Rosenblatt, R. A. Hinde, E. Shaw, & C. Beer (Eds.), *Advances in the study of behavior* (Vol. 6, pp. 77–130). New York: Academic Press.

Galef, B. G., Jr. (1980). Diving for food: Analysis of a possible case of social learning in wild rats (*Rattus norvegicus*). *Journal of Comparative and Physiological Psychology, 94*, 416–425.

Galef, B. G., Jr. (1983). Utilization by Norway rats (*Rattus norvegicus*) of multiple messages concerning distant foods. *Journal of Comparative Psychology, 97*, 364–371.

Galef, B. G., Jr. (1985). Socially induced diet preference can partially reverse a LiCl-induced diet aversion. *Animal Learning and Behavior, 13*, 415–418.

Galef, B. G., Jr. (1986a). Olfactory communication among rats: Information concerning distant diets. In D. Duvall, D. Miller-Schwarze, & R. M. Silverstein (Eds.), *Chemical signals in vertebrates IV: Ecology, evolution, and comparative biology* (pp. 487–506). New York: Plenum.

Galef, B. G., Jr. (1986b). Social identification of toxic diets by Norway rats. *Journal of Comparative Psychology, 100*, 331–334.

Galef, B. G., Jr. (1986c). Social interaction modifies learned aversions, sodium appetite, and both palatability and handling-time induced dietary preference in rats (*Rattus norvegicus*). *Journal of Comparative Psychology, 100*, 432–439.

Galef, B. G., Jr. (1988a). Evolution and learning before Thorndike: A forgotten epoch in the history of behavioral research. In M. Beecher & R. C. Bolles (Eds.), *Evolution and learning* (pp. 39–52). Hillsdale, NJ: Erlbaum.

Galef, B. G., Jr. (1988b). Imitation in animals: History, definition, and interpretation of data from the psychological laboratory. In T. Zentall and B. G. Galef, Jr. (Eds.), *Social learning: Psychological and biological perspectives* (pp. 3–28). Hillsdale, NJ: Erlbaum.

Galef, B. G., Jr. (1988c). Communication of information concerning distant diets in a social, central-place foraging species (*Rattus norvegicus*). In T. Zentall & B. G. Galef, Jr. (Eds.), *Social learning: Psychological and biological perspectives* (pp. 119–140). Hillsdale, NJ: Erlbaum.

Galef, B. G., Jr. (1989). Enduring social enhancement of rats' preferences for the palatable and the piquant. *Appetite, 13*, 81–92.

Galef, B. G., Jr., & Clark, M. M. (1971a). Parent-offspring interactions determine time and place of first ingestion of solid food by wild rat pups. *Psychonomic Science, 25*, 15–16.

Galef, B. G., Jr., & Clark, M. M. (1971b). Social factors in the poison avoidance and feeding behavior of wild and domesticated rat pups. *Journal of Comparative and Physiological Psychology, 75*, 341–357.

Galef, B. G., Jr., Mischinger, A., & Malenfant, S. A. (1987). Hungry rats' following of conspecifics to food depends on the diets eaten by potential leaders. *Animal Behaviour, 35*, 1234–1239.

Galef, B. G., Jr., & Stein, M. (1985). Demonstrator influence on observer diet preference: Analysis of critical social interactions and olfactory signals. *Animal Learning and Behavior, 13*, 31–38.

Galef, B. G., Jr., & Wigmore, S. W. (1983). Transfer of information concerning distant foods in rats: A laboratory investigation of the 'information centre' hypothesis. *Animal Behaviour, 31,* 748–758.

Gandolfi, G., & Parisi, V. (1972). Predazione su *Unio pictorum* L. da parte del ratto, *Rattus norvegicus* (Berkenhout). *Acta Naturali, 8,* 1–27.

Gandolfi, G., & Parisi, V. (1973). Ethological aspects of predation by rats, *Rattus norvegicus,* (Berkenhout) on bivalves, *Unio pictorum* L. and *Cerastoderma lamarcki* (Reeve). *Bolletino di Zoologia, 40,* 69–74.

Harlow, H., & Harlow, M. K. (1965). The affectional systems. In A. M. Schrier & F. Stollnitz (Eds.), *Behavior of non-human primates* (Vol. 2, pp. 287–334). New York: Academic Press.

Hess, E. H. (1973). *Imprinting.* New York: Van Nostrand.

Hinde, R. A. (1966). *Animal behaviour.* New York: McGraw-Hill.

Jenkins, H. M. (1984). The study of animal learning in the tradition of Pavlov and Thorndike. In P. Marler & H. S. Terrace (Eds.), *The biology of learning* (pp. 89–114). Berlin: Springer Verlag.

Johnston, T. D. (1981). Contrasting approaches to a theory of learning. *Behavior and Brain Sciences, 4,* 125–139.

Kamil, A. C., & Yoerg, S. I. (1982). Learning and foraging behavior. In P. P. G. Bateson & P. H. Klopfer (Eds.), *Perspectives in ethology: Vol. V. Ontogeny* (pp. 325–356). New York: Plenum.

Kawamura, S. (1959). The process of sub-culture propagation among Japanese macaques. *Primates, 2,* 43–60.

Marler, P., & Tamura, M. (1964). Culturally transmitted patterns of vocal behavior in sparrows. *Science, 146,* 1483–1486.

Mason, J. R., & Reidinger, R. F. (1982). Observational learning of food aversions in red-winged blackbirds (*Agelains phoeniceus*). *Auk, 99,* 548–554.

Palametta, B., & Lefebvre, L. (1985). The social transmission of a food-finding technique in pigeons: What is learned? *Animal Behaviour, 33,* 892–896.

Parisi, V., & Gandolfi, G. (1974). Further aspects of the predation by rats on various mollusk species. *Bolletino di Zoologia, 41,* 87–106.

Shettleworth, S. J. (1984). Learning and behavioral ecology. In J. R. Krebs & N. B. Davies (Eds.), *Behavioural ecology* (2nd ed., pp. 170–194). Sunderland, MA: Sinauer.

Solomon, R. L., & Coles, M. R. (1954). A case of failure of generalization of imitation across drives and across situations. *Journal of Abnormal and Social Psychology, 49,* 7–13.

Steiniger, von F. (1950). Beitrage zur Soziologie und sonstigen Biologie der Wanderratte. *Zeitschrift für Tierpsychologie, 7,* 356–379.

Stimbert, V. E. (1970a). A comparison of learning based on social or nonsocial discriminative stimuli. *Psychonomic Science, 20,* 185–186.

Stimbert, V. E. (1970b). Partial reinforcement of social behavior in rats. *Psychological Reports, 26,* 723–726.

Strupp, B. J., & Levitsky, D. A. (1984). Social transmission of food preference in adult hood rats (*Rattus norvegicus*). *Journal of Comparative Psychology, 98,* 257–266.

Tinbergen, N. (1963). On aims and methods of ethology. *Zeitschrift für Tierpsychologie, 20,* 410–433.

Triplett, N. (1897). The dynamogenic factors in pacemaking and competition. *American Journal of Psychology, 9,* 507–533.

von Frisch, K. (1954). *The dancing bees.* London: Methuen.

Ward, P., & Zahavi, E. (1970). The importance of certain assemblages of birds as "information-centres" for food-finding. *Ibis, 115,* 517–534.

Wenner, A. M., Wells, P. H., & Johnson, D. L. (1969). Honeybee recruitment to food sources: Olfaction or language? *Science, 164,* 84–86.

West, M. J., King, A. P., and Harrocks, T. J. (1983). Cultural transmission of cowbird song (*Molothrus ater*): Measuring its development and outcome. *Journal of Comparative Psychology, 97,* 327–337.

Zajonc, R. B. (1969). *Animal social psychology.* New York: Wiley.

C.

ONTOGENETIC ADAPTATION

The evolutionary pressures on the behavior and physiology of young organisms are twofold. Each structure and behavior should prepare for, or at least minimally interfere with, adult forms. Second, at every age the young organism must be adapted for its own particular environment. That second feature is the principle of ontogenetic adaptation. Although all organisms face these two pressures during ontogeny and must have balanced cost and benefits during evolutionary history, they are easiest to recognize in organisms whose environment and morphology change rapidly and dramatically.

In this section you will read about two very different approaches to this common problem. Stehouwer, in Chapter 8, has chosen to study the bullfrog, with what he describes as two adult stages—tadpole and frog. He examines the similarities and differences in the swimming behavior of tadpole and frog, looking especially at the transition between the two types of behaviors during metamorphosis. He is then able to examine neurophysiological and anatomical substrates upon which these related behaviors are based. For example, Stehouwer describes a technique for removal of the spinal cord of the frog and its maintenance for analytic studies.

We rarely think of the birth process as a kind of metamorphosis. Yet the changes in environment and morphology seen before and after birth make this a not too unlikely analogy. As with tadpole and frog, the young mammal goes from a fluid environment to one of air. The influence of gravity is quite changed. The method of feeding is dramatically different. In choosing the rat as an experimental subject, Smotherman and Robinson, in Chapter 9, have followed the lead of countless psychologists and psychobiologists. By examining this species during fetal life, however, the authors have in essence opened up an entirely new area for investigation. They describe techniques that allow examination and manipulation of the rat fetus, then allow its return to the uterus of the dam. These authors demonstrate that the fetus is not merely a passive recipient of input from its environment but an active participant in that environment.

In a sense, both chapters represent descriptions of comparative studies—not comparative in the traditional sense of choosing two different species but comparative studies within the same species. Both chapters emphasize the continuity of behavior and its controlling mechanisms throughout the entire life history of an organism. We hope that the reader will be stimulated by these two chapters to imagine further uses for these exciting new preparations.

Other chapters within this volume that address the issue of ontogenetic adaptation, at least tangentially, include those by Phifer (chap. 11), Williams (chap. 12), and Stopfer et al. (chap. 5).

8

Amphibian Metamorphosis: A Model of Vertebrate Motor Development

DONALD J. STEHOUWER

Many authors have observed that structures and functions of immature animals serve not only as precursors of adult structures and functions but also as adaptations to the environmental niche of the young (e.g., Hall & Oppenheim, 1987). This is particularly obvious in species that undergo complete metamorphosis, such as holometabolous insects and anuran amphibians, in which the larval forms do not even remotely resemble the adult forms. The tadpole is so different from the frog that it is easy, and not entirely inappropriate, to think of the two as different species. Indeed, the bullfrog tadpole is anatomically and physiologically more similar to tadpoles of other species than it is to the adult frog that it will become. In the bullfrog, metamorphic climax lasts approximately 2 weeks and separates a long larval period (2–3 years) from a postmetamorphic period (several years), both of which are periods of relative stasis. Consequently, one may view the frog as having two ''adult'' stages: one as a tadpole and one as a frog. Systems used exclusively by the tadpole are mature in that they do not change significantly during the larval period and clearly represent functional adaptations of the tadpole. Such systems are exemplified by the neural and muscular machinery that produce undulatory swimming. However, within this ''mature'' tadpole lurks an embryonic frog; the hindlimbs and their motoneurons, for example, develop from an embryonic state at the onset of larval life to a functionally mature state by the time of metamorphosis.

The major questions of interest to us ask how the ''embryonic'' systems of the frog are functionally integrated with the ''mature'' systems of the tadpole during ontogeny, with specific reference to the locomotor system. How is control of locomotion passed from one set of neural systems to the other? Which neural elements of the larva are retained and modified for subsequent use by the frog? Which are discarded? Which neural components of the frog arise de novo? How does the tadpole know when to ''turn off'' tadpole circuits and ''turn on'' frog circuits? What are the mechanisms by which this is accomplished? How do developing behaviors affect further neural development? What role does the changing environmental niche of the developing animal play in its neural and behavioral maturation? Although our current understanding is far from complete, we have begun to answer many of these questions. This chapter focuses primarily on the transition from larval undulatory swimming to adultlike pedal locomotion and how that transition relates to anatomical, physiological, ecological, and other behavioral changes of the transforming frog.

Although behavioral scientists have long been chastised for focusing too exclusively on the rat (e.g., Beach, 1950; Bitterman, 1960), the primary purpose of this chapter is not to exhort students of animal behavior to study a wider range of species. However, because metamorphosis of the frog offers unique opportunities for developmental psychobiologists that have not been realized, it is hoped that the chapter will serve to stimulate interest in the frog as a subject for developmental psychobiological study. Despite their apparent lack of behavioral flexibility (see Hodos & Campbell, 1969; McGill, 1960; Miller & Berk, 1979; van Bergeijk, 1967), frogs are particularly interesting from both phylogenetic and ontogenetic perspectives. Phylogenetically, amphibians pioneered the terrestrial habitat; ontogenetically, this role is replayed in the individual during metamorphosis, which in anurans (tailless amphibians) is a period of rapid morphological, physiological, and behavioral change. In addition, since tadpoles are independent animals that are not confined to an egg or a uterus, one can directly study development of behavior from their origin without fear of artifacts resultings from anoxia, anesthesia, or confinement. Frogs have long been attractive subjects for the study of neurophysiology and neuropharmacology because the spinal cord can be removed and viably maintained in Ringer's solution, allowing the use of sophisticated electrophysiological recording techniques and easy manipulation of constituents in the medium bathing the tissue. Thus, the frog offers the opportunity for a multidisciplinary attack on the development of behavior: Behavioral, neurophysiological, neuropharmacological, and anatomical data are readily acquired for correlative analysis.

The unique developmental history of the frog and the technical advantages of the frog have allowed experiments that have given us insight into some historical controversies and suggested new issues and reconceptualizations. In the remainder of this chapter I will describe how those technical advantages and the experiments they have made possible have influenced our thoughts about the ontogeny and neural control of behavior.

DEVELOPMENT AND METAMORPHOSIS

Anatomical and physiological specializations of the larval period are essentially mature from the onset of larval life and do not change appreciably until the onset of metamorphic climax, when the profound external morphological changes typically associated with metamorphosis (e.g., degeneration of the tail) also occur. Nevertheless, the body of the tadpole harbors an embryonic frog, which is inconspicuously maturing throughout the larval period.

The rate of larval maturation depends upon a variety of environmental variables, including population density (Cecil & Just, 1979; Richards, 1958; Rose, 1960), food density (D'Angelo, Gordon, & Charipper, 1941), temperature, and levels of trace elements, particularly iodine, in the water (see also Just, Kraus-Just, & Check, 1981). Consequently, the age and size of a tadpole are poor indexes of its maturity, and larval development of the frog is characterized by staging systems based on external morphological criteria, especially growth and differentiation of the hindlimbs (Taylor & Kollros, 1946; Gosner, 1960). Taylor & Kollros (1946) divided the larval period of development into 25 stages, which they denoted by Roman numerals to avoid confusion with the 25 embryonic stages previously described by Shumway (1940), which are denoted by Arabic numerals. The onset of larval stages is marked by independence from the yolk sac and the onset of free feeding and by the first appearance of the hindlimb primordia as barely detectable swellings. Before Stage VI these limb buds consist of undifferentiated mesenchyme, but the distal ends of the hindlimbs become flattened into foot paddles, and individual toes become distinguishable between Stages VI and X. From Stage XI to Stage XVII, the toes continue to differentiate, and the hindlimbs continue to increase in size. Finally, a surge in the levels of thyroid hormones greatly accelerates maturation and triggers the degenerative and synthetic changes of metamorphosis between stages XVIII and XV (metamorphic climax). Among the most pronounced of these are degeneration of the tail; eruption of the forelimbs, which have been developing internally in the gill chambers; migration of the eyes dorsomedially; replacement of porphyropsin by rhodopsin as the primary visual pigment; degeneration of the lateral line organs and the development of ears; degeneration of larval teeth; development of the adultlike jaw; development of the tongue; replacement of the long, coiled intestine of the primarily herbivorous tadpole by the stout, straight intestine befitting the carnivorous frog; degeneration of the gills; and septation of the lungs.

These changes are gradual and continuous throughout metamorphosis, but critical behavioral transitions occur between stages XXI and XXIII. During that span of approximately 3 days, the environmental preference of the transforming tadpole shifts from water to land; forelimb postural support emerges (Stehouwer, 1988); undulatory swimming ceases as frog kicks become the preferred mode of locomotion (Stehouwer & Farel, 1984); the level of gross locomotor activity declines to postmetamorphic levels (Stehouwer, 1986; 1988); and the primary burden of respiration shifts from the gills to the lungs and skin (Dupre, Taylor, & Frazier, 1985).

These are but a few of the nonneural changes that occur prior to and during metamorphic climax. Many of the hormonal and physiological changes necessary to support transformation and emergence from water are not mentioned because their direct implications for behavior are less apparent. However, the rapid and dramatic morphological and physiological transformations described above make metamorphic climax a period of precarious existence for frogs; they are poorly adapted behaviorally and physiologically for either aquatic or terrestrial life (Wassersug & Sperry, 1977). Furthermore, they are particularly vulnerable to predation at this time, so there are probably strong evolutionary selection pressures against prolonged metamorphosis (Arnold & Wassersug, 1978). Although the larval period of the bullfrog may range from 1 to 3 years (Dickerson, 1969), metamorphic climax spans only about 14 days in our laboratory.

ORGANIZATION AND DEVELOPMENT OF LOCOMOTION

Development of neural activity underlying different modes of locomotion in the bullfrog can be studied electrophysiologically in vitro using the isolated nervous system of the tadpole. The ventral roots of the lumbar enlargement of the bullfrog tadpole are composed of two discrete rootlets (Silver, 1942; Taylor, 1943; Stehouwer & Farel, 1980). Axons of primary motoneurons comprise one rootlet (medial fascicle), and those of hindlimb motoneurons comprise the other (lateral fascicle). These two populations differ with respect to their ontogenetic development, the morphology and location of their cell bodies within the spinal cord, and the peripheral distribution of their axons. Primary motoneurons, which innervate axial swimming muscles of tadpoles (Silver, 1942; Letinsky, 1974), are born during gastrulation of the embryo (Lamborghini, 1980) and are fully mature by the first larval stage. In contrast, hindlimb motoneurons do not begin to proliferate until the final embryonic stages and continue to proliferate, migrate, and differentiate through midlarval stages (Beaudoin, 1955; Farel & Bemelmans, 1980). The round cell bodies of primary motoneurons extend ventromedially throughout the spinal cord (Forehand & Farel, 1982), whereas the spindle-shaped hindlimb motoneurons are found ventrolaterally, and only in the lumbar enlargement (Silver, 1942; Taylor, 1943). Although fascicles containing axons of primary and hindlimb motoneurons exit the spinal cord within a common sheath, they are easily dissected free of each other for independent electrophysiological study (e.g., Stehouwer & Farel, 1980, 1985).

Method of Study

Surgery and Dissection

Because amphibians are ectothermic, it is possible to anesthetize them via hypothermia, which effectively blocks ongoing neural activity. Tadpoles are placed in a slurry of ice and water for 30–45 min prior to surgery, whereas frogs are simply packed in crushed ice for 30–45 min. Hypothermia is safe, easy to reverse, and depresses cardiovascular function. This last feature is particularly valuable, because reduced bleeding allows better visualization throughout surgery and enhances postoperative recovery. The ease of reversibility and the absence of residual chemical anesthetics or their metabolites allow for immediate electrophysiological experimentation under physiological conditions that are as normal as possible.

All surgeries and dissections are performed with the aid of a Unitron dissection microscope (Samuels Scientific Instruments) having a zoom lens, yielding a net range of magnification ranging from $7\times$ to $45\times$. The bullfrog is a particularly useful amphibian for surgical manipulation because the larvae are large, typically 8 cm from snout to tail and weighing 5 g. A second feature of the bullfrog that makes it especially advantageous for electrophysiological study is that, as an amphibian, the metabolic requirements of its nervous system are much lower than those of mammals. Consequently, the entire CNS, including the brain, can be removed and maintained in Ringer's solution for electrophysiological study in vitro (e.g., Stehouwer, 1986; Stehouwer & Farel, 1980). Furthermore, one can use partially reduced preparations that consist of the CNS and selected portions of afferent or efferent systems for both behavioral and electrophysiological studies (e.g., Stehouwer, 1986; Stehouwer & Farel, 1983, 1985). Through appropriate use of these preparations, we have been able to distinguish between the contributions of mechanisms endogenous to the CNS (e.g., central pattern generators) and those derived from sensory input (e.g., reflexes) and to determine the role of sensory inputs in the expression of different behaviors during ontogeny.

Behavior

We use primarily time-lapse and single-frame videotape analyses to study meta-morphic changes in behavior. With time-lapse videography, we can observe subjects continuously for long periods of time, enabling us to study changes in habitat preference (Stehouwer, 1988), gross locomotion (Stehouwer & Farel, 1984), and the onset of predatory behavior during metamorphosis. Conversely, single-frame analyses allow us to expand the time base and study details of locomotor coordination (e.g., Stehouwer, 1986; Stehouwer & Farel, 1984) and of predatory strikes. For time-lapse videography we use standard video cameras in conjunction with Panasonic VHS videocassette recorders (Model NV-8050, many vendors available, e.g., Tallahassee Camera Center) that were designed for use in surveillance and security systems. Those machines are able to record and play standard VHS videotapes at eight different speeds, yielding from 2 h of total recording time (60 frames/s) to 10 days of total recording time (1 frame/2 s), in addition to manual single-frame advance capability. To avoid blurred images during rapid movements, we use Panasonic Omnimovie VHS camcorders with high-speed shutters (1/1,000 s) to create videotapes for subsequent single-frame analyses.

Analyses of videotaped behavioral episodes are performed with the assistance of an Apple IIe computer equipped with a digitizing tablet. The amount of gross locomotion is measured as the number of quadrants entered in a circular container, typically during a 4-h period. The videotapes are scored at $12\times$ to $24\times$ real-time speed with the help of the computer. Coordination is analyzed after videotaping the subject through the bottom of a glass circular container at 60 frames/s. For each frame of the videotape, the X-Y coordinates of the snout, rump, and each joint of the hindlimbs of the animal are digitized by projecting the video image onto the surface of the digitizing tablet. These data are saved on floppy disk and, using software written by the author, the angles of the hip, knee, and ankle joints are calculated for each frame and then plotted as a function of time. Other software, also written by the author, reconstructs "stick figures" of the animal and displays them on the computer monitor, either as single frames or as animated sequences (see Stehouwer, 1986).

Electrophysiology

We have used several different kinds of preparations for electrophysiological study, ranging from the isolated nervous system to the intact animal. Between these two extremes we have used preparations of varying degrees of dissection, selectively leaving different afferents, efferents, and effectors intact. We describe here only the procedure used for experiments on the isolated CNS; procedures for other prepara-tions vary only in detail. The CNS is removed by exposing the brain and spinal cord dorsally and then severing the spinal roots, cranial nerves, and denticulate ligaments. The CNS is mounted in a Plexiglas recording chamber and bathed in oxygenated glucose-Ringer's solution at 14–18 °C. All recordings of spontaneous activity are obtained bilaterally from primary and hindlimb motoneurons via suction electrodes (A-M Systems) applied individually to the medial and lateral fascicles of the ninth ventral roots (VR 9/M and VR 9/L, respectively), 1 to 2 mm from their exit from the spinal cord. In most experiments, electrodes are capacity-coupled to the pre-amplifiers. Although the isolated CNS remains physiologically active for about 48 h, we take a conservative approach and use a fresh preparation for every experimental session, which typically lasts about 8 h.

Undulatory Swimming

Locomotion of the bullfrog tadpole, like that of other vertebrates and invertebrates (see Delcomyn, 1980; Grillner, 1985), is controlled cooperatively via central pattern generators and peripherally generated feedback (Stehouwer & Farel, 1980, 1981). Central pattern generators are defined as the central neural circuits that underlie performance of coordinated, meaningful behaviors. Demonstrating the existence of a pattern generator requires one to demonstrate that the efferent output in an animal lacking movement-related peripheral feedback (e.g., a paralyzed or deafferented animal) closely resembles that of an intact, behaving animal. However, activity patterns of central mechanisms isolated from peripheral influences virtually always differ in detail from their activity pattern during actual execution of a behavior, sometimes making it difficult to distinguish between pattern generators underlying similar behaviors (e.g., McClellan, 1982; Ayers, Carpenter, Currie, & Kinch, 1983), and causing some researchers to question the usefulness of the term (e.g., Bassler, 1986). Nevertheless, there seems to be little problem with the term as long as the relationship between the central pattern generator and the corresponding behavior is unambiguous, and the term will be used in the present discussion.

Behavior

Locomotion of the bullfrog tadpole is accomplished via alternate contractions of axial muscles on the left and right sides of the body. The resulting undulatory wave, which passes down the body in a cephalocaudal direction, is the product of a rostrocaudal delay in contraction of axial muscles (Stehouwer & Farel, 1980) and of passive biomechanical transmission, as when one cracks a whip (see Blight, 1977). Within any given segment, axial muscles of the left and right sides contract in alternation (Stehouwer & Farel, 1980). The frequency of tail beating is directly related to swimming speed (Stehouwer & Farel, 1984; Wassersug & Hoff, 1985) and ranges from 2 to 30 beats per second, although rates climb above about 10 per second only under unusual circumstances and are sustained for only a few seconds (Stehouwer & Farel, 1984).

Neural Control

There are at least three ways in which an undulatory wave could be propagated down the body of the tadpole: (1) biomechanically, as described above; (2) reflexively, in which contraction of one myotome elicits proprioceptive reflexes that elicit contraction of the next; and (3) through endogenous coordination of spinal cord activity. By comparing locomotor activity of different preparations, we have been able to dissociate the contributions made by these different mechanisms.

Central locomotor controls can be studied without peripheral contributions by using the isolated CNS, in which the activity underlying swimming (fictive swimming) is easily identified (see Stehouwer & Farel, 1980). Fictive swimming is spontaneously expressed as alternating bursts of action potentials in axons of the primary

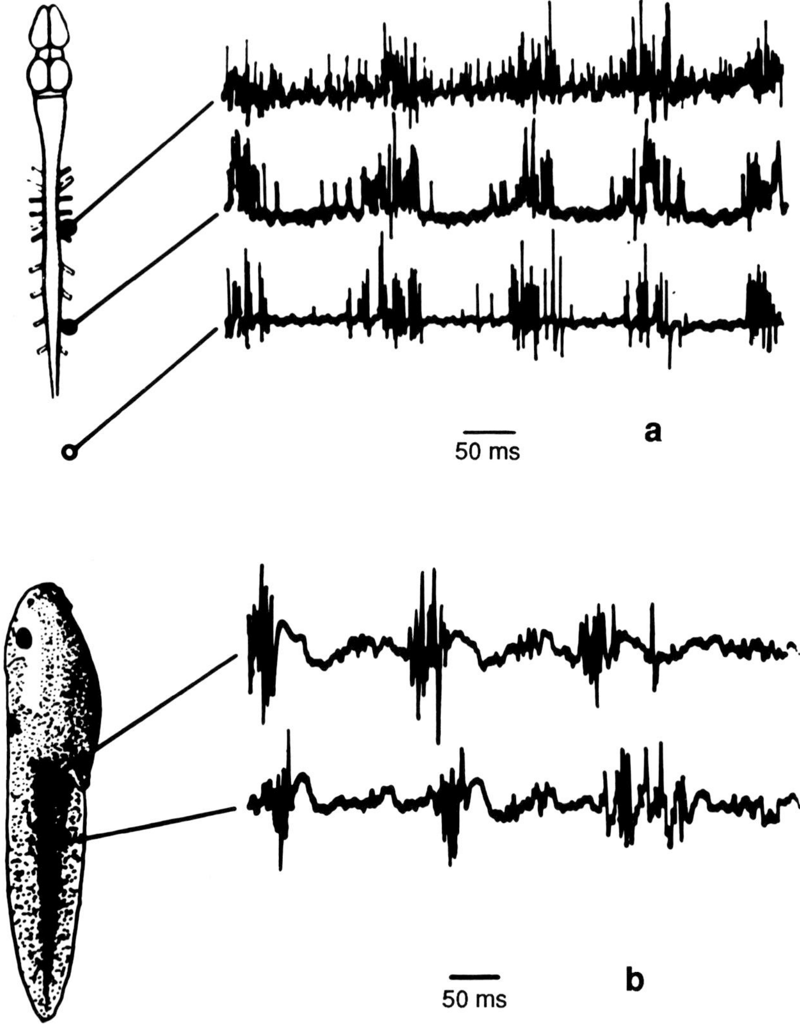

Figure 8.1. Homolateral motoneurons of the isolated nervous system burst in synchrony the entire length of the spinal cord, as shown in records obtained from homolateral fifth, ninth, and fourteenth ventral roots (top). In contrast, there is a rostrocaudal delay in muscle contractions of the freely swimming tadpole, as shown by electromyographic recordings (bottom). From "Central and Peripheral Controls of Swimming in Anuran Larvae" by D. J. Stehouwer and P. B. Farel, 1980, *Brain Research, 195.* Copyright by Elsevier Science Publishers. Reprinted by permission of the publisher.

motoneurons on opposite sides of the cord, which innervate the axial swimming muscles on the left and right sides of the body and tail of the tadpole. There is intersegmental coupling of activity as well, so that primary motoneurons on the same side burst in synchrony at all levels of the cord. A series of complete transverse sections at levels ranging from the cerebrum to the rostral end of the spinal cord showed that spontaneous fictive locomotion persists as long as the brain stem remains intact; more caudal transections abolish spontaneous locomotor activity. However, electrical stimulation of the rostral end of a transected spinal cord elicits episodes of fictive swimming that are indistinguishable from spontaneous episodes. Thus, the spinal cord generates the alternating contraction of heterolateral motoneurons and the intersegmental coupling of homolateral motoneurons, whereas the brainstem is necessary for activation of those spinal pattern generators.

In intact freely swimming tadpoles, activity of homolateral myotomes is coupled, but contraction of any given myotome lags behind that of more rostral ones. Intersegmental coupling persists following complete spinal deafferentation, but homolateral myotomes contract synchronously, like the primary motoneurons of the isolated CNS (Figure 8.1). Although deafferented tadpoles undulate, swimming is much less efficient than that of intact tadpoles (Stehouwer & Farel, 1980). These results suggest that the segments are coordinated centrally but that coordination is modified by spinal afferents. However, central and reflex generation of intersegmental coordination are not mutually exclusive, and activity of myotomes rostral and caudal to a complete spinal transection (Brenner & Stehouwer, 1988) remain coordinated in freely swimming tadpoles, showing that reflex mechanisms are able to coordinate activity even in the absence of central conduction. Thus, efficient swimming requires an interaction between properly timed muscle contractions, which result from the cooperation of central and peripheral mechanisms of coordination, and biomechanical factors (see Blight, 1977; and Wassersug & Hoff, 1985, for further discussion).

Development of Pedal Locomotion

Neural Control

Having satisfied ourselves that we had at least a rudimentary understanding of the fundamental factors involved in the control of undulatory swimming, we turned to the more intriguing question, How is the transition from undulatory swimming to pedal locomotion accomplished? Do limb movements become "individuated" from a total action pattern that initally includes undulations of the trunk, as Coghill (1929) suggested for Amblystoma? Or do limb movements originate independently from those of the trunk, as Windle (1944) suggested for mammals? Our use of the frog enabled us to examine the emergence of pedal locomotion without possible artifacts due to hypoxia and to separate central from peripheral contributions to coordination.

Development of interlimb coordination was studied by recording the spontaneous activity of hindlimb motoneurons in vitro and comparing it to concurrent recordings of activity of primary motoneurons (Stehouwer & Farel, 1985). Hindlimb motoneurons were found to receive rhythmic synaptic inputs as early as Stage III, an early larval stage in which the hindlimbs are small, undifferentiated buds. Hindlimb motoneurons on the left and right sides of the cord undergo alternating rhythmic depolarizations that occur at the same frequency as bursting in primary motoneurons, but are phase-shifted approximately 180° with respect to bursts in homolateral primary motoneurons. Hindlimb motoneurons at this stage of development are small and undifferentiated and are still in the early stages of proliferation and migration (Farel & Bemelmans, 1980). Between stages VIII and XIV, bursts of action potentials gradually emerge at the peaks of the depolarizations seen at younger stages (Figure 8.2). Bilaterally synchronous burst activity of hindlimb motoneurons was first seen at

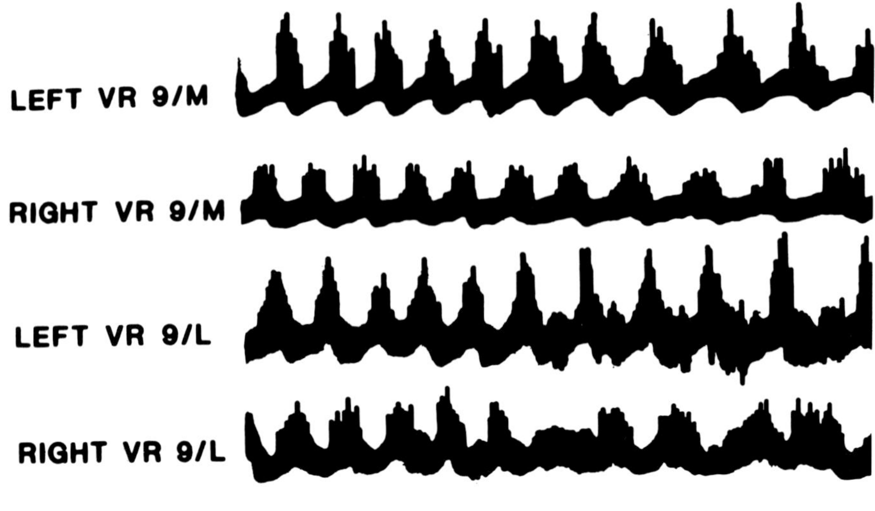

Figure 8.2. Beginning at Stage VIII, rhythmic depolarizations in hindlimb motoneurons (VR 9/L) generate bursts of action potentials at their peaks, as shown in this record obtained from a Stage X preparation. Bursting in hindlimb motoneurons alternates between the left and right sides, and is phase-shifted 180° with respect to bursting in homolateral primary motoneurons (VR 9/M). From "Development of Hindlimb Locomotor Mechanisms in the Frog" by D. J. Stehouwer and P. B. Farel, 1985, *Neurophysiology, 53.* Copyright by the American Physiological Society. Reprinted by permission of the publisher.

Stage XIV and became relatively more frequent until it was the dominant form of activity by the end of metamorphosis. Alternating and synchronous coordination of bursting described above corresponds to interlimb coordination during stepping and frog kicks, respectively. However, the absolute amount of all forms of spontaneous activity declines during metamorphic climax (Stehouwer & Farel, 1985).

Activity recorded from peripheral hindlimb nerves of the isolated nervous system was used to study development of locomotor coordination within a limb. Peripheral nerve fibers first enter the limb bud at Stage II (Letinsky, 1974) and are fasciculating until Stage VI (Taylor, 1943). The hindlimb consists of undifferentiated mesenchyme until Stage VI, and the process of muscle differentiation and cleavage in the thigh proceed through Stage XII (Dunlap, 1967). The activity of hindlimb peripheral nerves is discretely patterned at the earliest stage at which we could perform the dissection (Stage X), although the reliability with which this pattern was observed increased through Stage XIII, probably because of the associated greater ease of dissection. Bursting in motoneurons innervating hip flexors occurred in synchrony with bursts of homolateral primary motoneurons; bursting in motoneurons innervating hip extensors was phase-shifted approximately 180° (Stehouwer & Farel, 1985). The antiphasic bursts of flexor and extensor motoneurons of the hip, which were present at Stage X and remained constant throughout development, were appropriate for repeated limb flexion and extension of the hip and were coordinated with bursts of primary motoneurons so that flexion of the hip would coincide with contraction of homolateral axial muscles. This interpretation was confirmed by showing that the isolated nervous system, when connected to the hindquarters by only the ventral roots of limb-innervating segments, produced repetitive stepping in which hip flexion was coincident with contractions of homolateral axial muscles (Stehouwer & Farel, 1985).

Locomotor Behavior

The emergence of patterned activity in hindlimb motoneurons (fictive stepping) in early larval stages was quite a surprise because casual observation in the laboratory suggested that hindlimb locomotor behavior does not appear until the onset of metamorphosis. This delay between the onset of function in locomotor circuits and the onset of the corresponding locomotor behaviors suggested tonic inhibition of hindlimb locomotion and caused us to rephrase our experimental question. Rather than ask, How does CNS maturation result in the onset of behavior? we asked, What are the ontogenetic triggers that turn on, or disinhibit, the already capable hindlimb locomotor circuits?

Because the hindlimbs dramatically increase in size and because the tadpole preference changes from water to the shoreline during metamorphosis (Stehouwer, 1988), we hypothesized that the resulting increase in sensory stimulation of the limbs might trigger their use. In order to test this hypothesis, we examined hindlimb locomotor behavior on three different substrates chosen to mimic those that the tadpole would normally encounter as it matured. We tested tadpoles at different stages of development in deep water, the normal habitat prior to metamorphosis; in a shallow plastic pan containing 1 mm of water, to provide an intermediate amount of tactile stimulation of the limbs; and in a wooden box lined with screen mesh, chosen to mimic the postmetamorphic terrestrial environment. Results of this experiment confirmed this hypothesis; hindlimb locomotion began no earlier than Stage XVII in deep water, at Stage XIV on the slippery wet surface, and as early as Stage XII on the rough dry surface (Figure 8.3). Thus, environmental conditions determine, at least in part, when hindlimb locomotor behavior is first expressed. They also show that

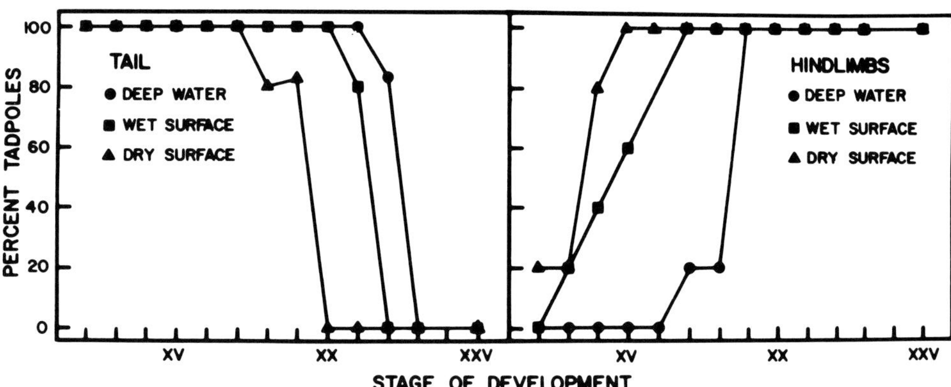

Figure 8.3. The stage at which larvae ceased using their tails (left panel) and began using their hindlimbs (right panel) was hightly dependent on the substrate upon which they were tested. Between stages XIV and XXII, depending on test condition, tadpoles frequently used their tails and hindlimbs in conjunction. From "Development of Hindlimb Motor Behavior in the Frog" by D. J. Stehouwer and P. B. Farel, 1984, *Developmental Psychobiology, 17.* Copyright by John Wiley & Sons. Reprinted by permission of the publisher.

hindlimb locomotor behavior, like activity of the underlying central pattern generators (CPGs), can be expressed by very immature animals long before its normal onset.

Despite the difference in timing, the *sequence* of development of hindlimb locomotor behavior in the intact tadpole is the same as that of the corresponding neural mechanisms as recorded in vitro; very young tadpoles use only axial muscles, which are joined by stepping movements of the hindlimbs during metamorphic climax. Finally, undulations of the tail cease, and sychronous kicks become the dominant mode of locomotor activity. Coordination between undulations and stepping is highly variable in the aquatic environment, which contrasts sharply with the highly stereotyped coordination between the fictive swimming and fictive stepping in vitro and suggests that coordination between the trunk and limbs is modified by afferent input. Therefore, the ontogenetic emergence and topography of behavior depend not only on the maturational status of the underlying neural circuits but also on the test conditions.

Hierarchical Organization

The fact that locomotor circuits are functionally competent long before their normal expression also suggests that those circuits are tonically inhibited. The experiments just described show how sensory input can disinhibit locomotor circuits; in the next experiments we attempted to disinhibit hindlimb locomotor behavior through surgical removal of the source of inhibition. Because locomotion in vertebrates is organized hierarchically, with spinal pattern generators under direct control of brain stem locomotor regions (Garcia-Rill, 1986; Grillner, 1981; Shik & Orlovsky, 1976), we spinally transected tadpoles and juvenile frogs in an attempt to disinhibit spinal controls of hindlimb locomotion (Stehouwer, 1986). Undulatory swimming of tadpoles was immediately abolished by the transection but recovered between postoperative days 8 and 32. The hindlimbs of tadpoles as young as Stage IX became very active

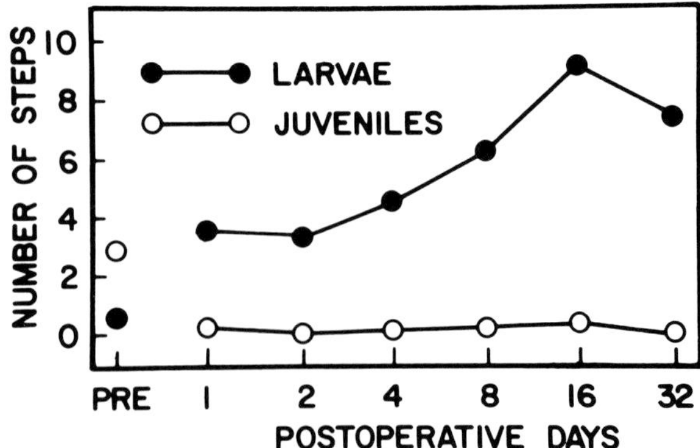

Figure 8.4. Prior to spinal transection (PRE), stepping in response to threshold-intensity cutaneous stimulation of the hindfoot was greater in juvenile frogs than in tadpoles. However, this relationship was reversed after medullospinal transection; spinal tadpoles showed an increase, and spinal juveniles a decrease, in the number of steps taken. From ''Behavior of Larval and Juvenile Bullfrogs (*Rana catesbeiana*) Following Chronic Spinal Transection'' by D. J. Stehouwer, 1986, *Behavioral and Neural Biology, 45*. Copyright by Academic Press, Inc. Reprinted by permission of the publisher.

immediately following the transection and displayed long sequences of coordinated stepping in response to cutaneous stimuli, whereas spinal frogs simply hopped once or twice (Figure 8.4). Deafferentation of the hindlimbs abolished stepping, suggesting that hindlimb reflexes, rather than CPGs, mediate stepping of spinal tadpoles. Hopping was the dominant mode of locomotion in froglets before and after the transection, whereas stepping was the only mode of hindlimb locomotor activity released by the transection in tadpoles. These results suggest that stepping reflexes are tonically inhibited by the brain prior to metamorphosis and that hopping emerges as a direct result of spinal cord maturation.

To further localize the source of inhibition of stepping reflexes, we transected the neuraxis just rostral to the cerebellum. Undulatory swimming persisted in tadpoles, which also failed to exhibit the stepping characteristic of spinally transected tadpoles. Thus, activation of spinal CPGs and inhibition of stepping reflexes both seem to arise from the hindbrain of the tadpole. There was little difference between hopping of intact froglets and that of tadpoles that had received brain stem transections, but many transected frogs displayed postural asymmetries that probably resulted from asymmetrical damage to vestibular pathways at the site of transection.

Morphological Change

The transition from undulatory swimming to pedal locomotion correlates well with the morphological changes associated with metamorphosis. Particularly important are degeneration of the tail and eruption of the forelimbs. Newly erupted forelimbs interfere with swimming of tadpoles by reducing streamlining and increasing drag (Wassersug & Sperry, 1977), and degeneration of the tail makes it a less effective organ of propulsion. Conversely, the forelimbs are essential for the posture essential to locomotion in postmetamorphic frogs (Stehouwer, 1988), and degeneration of the tail eliminates dead weight that hinders pedal locomotion (Wassersug & Sperry, 1977).

Not all of the effects of morphology on locomotion are mechanical, however. For example, I have already discussed the increased size of the hindlimbs and how the resulting increased tactile stimulation enhances hindlimb locomotor activity. This suggested to us that inducing artificial "metamorphosis" that consisted only of amputation of the tail might precociously induce hindlimb stepping in premetamorphic tadpoles. To test this hypothesis, we examined tadpoles at two stages of development in three treatment conditions. At Stage XIV or Stage XVII, the number of steps by threshold-intensity tactile stimulation of the foot was recorded for each subject. Subjects were then divided into three groups: (1) a maturational control group that were not manipulated until the final test (Control); (2) an experimental group whose tails were amputated immediately following the first behavioral test and were tested repeatedly (Amputated); and (3) a control group that were tested at each interval with subjects in group 2, but whose tails were amputated only 3 h prior to the final test (Amputated Control). Figure 8.5 shows that there were no differences between the less mature subjects in the three groups, but that tail amputation clearly altered the reponse of the late premetamorphic tadpoles. Amputated and Amputated Control subjects showed an immediate increase in the number of steps following amputation of their tails, whereas Control subjects and Amputated Control subjects prior to amputation of the tail showed no such increase, ruling out both maturation and repeated testing as causes for increased stepping. This result suggests that, as meta-

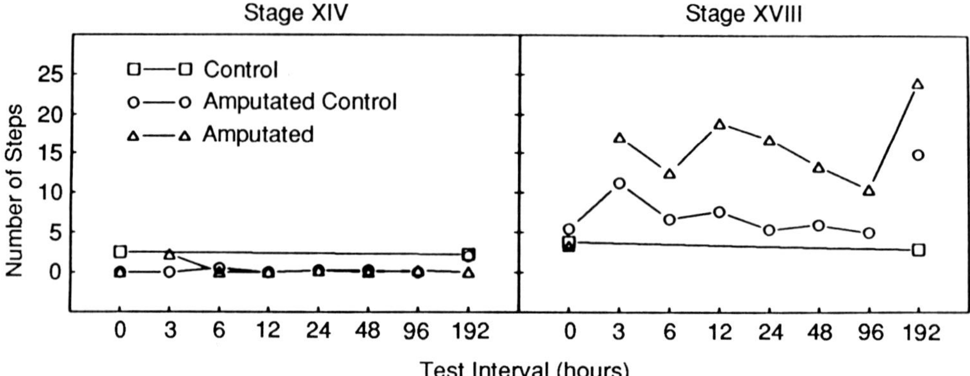

Figure 8.5. There is little hindlimb stepping in response to light cutaneous stimulation in either intact Stage XIV or intact Stage XVII tadpoles. Stepping increased markedly in Stage XVII but not in Stage XIV tadpoles following amputation of the tail. From "Behavior of Larval and Juvenile Bullfrogs (*Rana catesbeiana*) Following Chronic Spinal Transection" by D. J. Stehouwer, 1986, *Behavioral and Neural Biology, 45*. Copyright by Academic Press, Inc. Reprinted by permission of the publisher.

morphosis approaches, tadpoles become sensitive to loss of the tail and/or its associated neuronal circuitry, which triggers a central "switch" that activates hindlimb locomotor circuits.

CONCLUSIONS AND PERSPECTIVES

I began this chapter with the suggestion that some characteristics of the young serve to adapt the young to their own particular environment and some serve as necessary precursors of adult characteristics. In contrast to immature mammals, it is more difficult to think of the tadpole as a precursor than as an adaptation, given the radical differences in morphology and behavior in frogs before and after metamorphosis. Those differences are reflected in the natural history of the frog; it is a highly derived amphibian that has acquired many anatomical and functional specializations during the course of evolution. That is, modern frogs represent a "splinter group" that were not in the direct lineage that gave rise to reptiles, birds, and mammals. Thus, some features of frogs evolved prior to their evolutionary divergence (primitive characters), whereas others were acquired after the evolutionary path of frogs diverged from the mainstream (derived characters). In the former case studies of frogs are likely to offer insight into mammals and other vertebrates; in the latter case they may not. Therefore, one must consider carefully the character under study before generalizing to other species.

Ontogenetic Determinants of Behavioral Expression

The experiments described in this chapter show that undulatory swimming of the tadpole is produced in large part by CPGs. Patterned, locomotion-related activity in hindlimb motoneurons begsins early in larval development and is phase- and frequency-locked to activity in motoneurons innervating axial muscles. This suggests that the two forms of locomotion may depend on shared CPGs. It is important in the interpretation of results to distinguish between results of behaving animals and those obtained in vitro; in the latter, sensory modulation and other forms of central activity arising from sensory input are absent. In the intact frog, the development of locomotor behavior, particularly the transition from undulatory swimming to appendicular locomotion, is multiply determined. The necessary spinal circuitry develops during early larval life, but is apparently inhibited by descending inputs because spinal transection augments stepping behavior. As the tadpole proceeds through metamorphosis, the hindlimbs grow in size, and its ecological preference shifts from water to the shoreline, making contact between the hindlimbs and substrate more probable, an event that facilitates appendicular locomotion. At the same time, the tail is degenerating, a change in morphology that facilitates hindlimb locomotion, as shown by tail-amputation experiments. Thus, the emergence of appendicular locomotion is dependent on confluent developmental changes in intrinsic neuronal circuits, sensory inputs, supraspinal modulation dependent on those inputs, and gross morphology of the animal. Those same factors play a role in the development of other behaviors in the frog and other vertebrates.

Perhaps most obviously, development of any behavior depends upon maturation of the underlying neuronal circuits mediating that behavior. Experiments described in this chapter have shown that development of hindlimb locomotor circuits in the frog begins very early and that those circuits coexist with those underlying undulatory swimming for a considerable length of time before they are expressed (Stehouwer & Farel, 1983, 1985). Similarly, neural circuits underlying adultlike feeding behavior in rats seem to coexist with those for suckling for a long period of time prior to their normal onset of expression near weaning (Hall & Williams, 1983). Therefore, in both instances, existence of the circuitry is a necessary but not sufficient condition for the expression of the behavior in question. Inhibition and disinhibition of behavioral circuits by other circuits in the brain can also determine the ontogenetic pattern of behavioral expression. In the case of the tadpole, hindlimb stepping can be released by high spinal transection (Stehouwer, 1986). In the kitten, decerebration results in precocious onset of climbing and stalking behaviors (Bignall & Schramm, 1974). In early human development, coordinated behavior may appear, disappear, and then reappear (e.g., Hofer, 1981; McGraw, 1939), suggesting sequential inhibition and disinhibition of circuits mediating those behaviors.

Sensory input can often serve as a trigger for the onset of activity in circuits mediating specific behaviors, either by disinhibition or by providing a necessary and specific source of excitation. Manipulation of the environmental substrate alters the time of onset of appendicular locomotion in the frog (Stehouwer & Farel, 1984), and the ambient temperature determines whether adultlike feeding is observed in neonatal rats (Johanson & Hall, 1980). Hatching of chickens is triggered, at least in part, by the specific posture of the embryo in the egg; if adult chickens are tucked into glass "eggs" in the embryonic posture, they display hatching behaviors (Bekoff & Kauer, 1984; Bekoff & Sabichi, 1987). Hatching in quail can be triggered by the sounds of hatching made by others in the clutch (Vince, 1973), and a variety of motor behaviors can be elicited precociously in the infant rat by aversive stimuli (Gard, Hard, Larsson, & Petersson; 1967). In the frog, the onset of predatory behavior at the completion of metamorphosis coincides with the anatomical development of retinal ganglion cells believed to have "bug-detecting" properties (Pomeranz, 1972).

Morphological transformation of the frog is so profound that it cannot be ignored, and some of those changes can be mimicked experimentally. Amputation of the tail of premetamorphic tadpoles increases the level of hindlimb locomotor activity in premetamorphic tadpoles (Figure 8.5) and improves the leaping ability of transforming tadpoles (Wassersug & Sperry, 1977), suggesting that tail degeneration, per se, may contribute to the onset of appendicular locomotion. Morphological development in other animals is a more gradual and often overlooked variable that determines the ontogeny of behavior. However, a few studies attest to the pervasive importance of morphological constraints on behavior during development. For example, young lobsters flee when faced with a threat, whereas adult lobsters fight defensively. The ontogenetic transition between flight and defense depends on the allometric growth of the tail and claws. In young lobsters, the tail is bigger and the claws smaller, relative to its body, than the tail and claws of the adult. The large tail and small claws of the juvenile favor flight, whereas the small tail and large claws of the adult favor defense. Furthermore, adult lobsters immediately revert back to flight if their claws are lost (Lang, Govind, Costello & Green, 1977). Locomotion of 8-week-old kittens differs considerably from that of adults, largely because their limbs are longer, relative to their bodies, than those of adults. Consequently, homolateral limbs usually step in unison rather than in alternation, apparently to prevent the forelimbs and hindlimbs from colliding (Peters, 1983). Similarly, Thelen, Fisher, Ridley-Johnson, and Griffin (1982) have suggested that limb morphology plays an important role in the ontogeny of coordinated stepping in human infants. Phylogenetic (Lieberman, 1984) and ontogenetic (Buchwald, 1988) development of human speech depends not only on the presence of the requisite neural controls but also on development of the vocal apparatus. Specifically, the larynx of both newborns and ancestors is much higher in the pharynx than in modern adult humans, making formation of many speech sounds impossible. Thus, the infant's limited repertoire of crying reflects morphological constraints as much as it does CNS immaturity. Finally, a recent comparison of different species shows that small mammals (e.g., squirrels) locomote in a crouched posture, whereas large mammals (e.g., horses) locomote in a more upright posture. The upright posture of large animals is required in order to reduce stress on the bones of their limbs; the crouched posture of small mammals confers greater range and flexibility of motion (Biewener, 1989). These examples suggest that changes in size and allometric growth are pervasive biomechanical determinants of behavioral ontogeny that deserve further study.

From the multiple determination of behavioral ontogeny, two important conclusions must be drawn. First, one cannot deduce from the absence of behavior the absence of capacity for a behavior or of its underlying neural controls. As the examples above illustrate, there may be many reasons, other than neural immaturity, why a particular behavior is not expressed during a given phase of ontogeny. And although one can infer the existence of underlying neural mechanisms from the presence of a behavior, one cannot assume that they are necessarily the same mechanisms used for that same behavior or a very similar behavior later in life. For example, the jaw-closing refelx of the newborn rat is mediated by polysynaptic pathways, whereas that of the adult is mediated disynaptically (Thexton & Griffiths, 1979). Second, all of the classes of contributing variables described above interact to chart the course of development. In the case of the frog, the change in environmental preference and morphological changes of hindlimb growth enhance the probability that the hindlimbs will come into contact with the environmental substrate, thus augmenting the sensory input to the hindlimbs, enhancing hindlimb locomotor behavior. By the same token, degeneration of the tail results in a loss of afferents from the tail and death of motoneurons innervating the tail. The loss of sensory inputs from the tail may bias the animal toward use of the hindlimbs, and loss of tail motoneurons may, at the same time, induce reorganization of central motor controls (Finlay, Wikler, & Sengelaub, 1987), further facilitating use of the hindlimbs. Similar interactions are seen in the development of laterality of lobster claws; use of a claw during a critical period of development is thought to induce lateralization of the CNS, which in turn induces one claw to differentiate into a "crusher" while the other remains a "cutter" (Govind & Pearce, 1986). So, in this case, sensory input alters CNS organization, which in turn exerts centrifugal effects on peripheral morphology.

Development of Locomotion in Other Vertebrates

Although phyletic relationships between modern amphibia are not entirely clear (Parsons & Williams, 1963), Ranid frogs are highly derived amphibians (Kluge & Farris, 1969) that probably bear little resemblance to the stem amphibians that also gave rise to reptiles and, ultimately, birds and mammals (Young, 1962). Interesting and instructive analogies may be drawn from studies of frogs, as illustrated above, but one must be very careful in assuming homologies between frogs and other animals. For example, saltatory locomotion (hopping) of frogs likely evolved as a specialization in only some orders of amphibia (Gans & Parsons, 1966), so its neural controls may not be homologous to those of hopping and galloping in birds and mammals. However, alternate gaits are generalized and widespread, and the ontogenetic sequence of lateral undulations followed by the appearance of limb movements was probably present in stem amphibians and has been reported during the embryological development of extant representatives of all classes of terrestrial vertebrates (e.g., amphibians: Coghill, 1929; Abu Gideiri, 1971; reptiles: Decker, 1967; birds: Hamburger, Balaban, Oppenheim, & Wenger, 1965; mammals: Angulo y Gonzalez, 1932; Barcroft & Barron, 1939; Hooker, 1936; Humphrey, 1978; Narayanan, Fox, & Hamburger, 1971; Windle & Griffin, 1931). Therefore, it seems likely that the pervasive alternate gait evolved once and that the ontogenetic sequence of maturation has persisted throughout the vertebrate radiations.

The suggestion that interlimb coordination is derived from mechanisms underlying undulations is not a new one. Although differing in its specifics, Coghill (1929) argued that independent limb movements become individuated from mass action patterns that included undulatory movements and that this was a general principle of development. This view contrasts sharply with that of Windle (1944), who argued that local limb movements arose first and became integrated during further ontogenetic development. Space does not permit detailed comparisons of development of locomotion in frogs to recent studies of the development of locomotion in chickens (e.g., Bekoff, 1976; Bekoff, Stein, & Hamburger, 1975; Landmesser & O'Donovan, 1984; O'Donovan & Landmesser, 1987), rats (e.g., Bekoff & Lau, 1980; Bekoff & Trainer, 1979), and humans (e.g., Thelen, 1979; Thelen, Bradshaw, & Ward, 1981; Thelen, Ridley-Johnson, & Fisher, 1983), but in none of those species is there a single study examining the relationship between undulatory movements and patterned limb activity. However, the hypothesis that pattern generators producing embryonic lateral undulations are necessary precursors of those that later coordinate the limbs would explain why lateral trunk flexions are the first embryonic movements in the turtle (Decker, 1967). As adults, turtles have a hard carapace and total absence of somatic trunk muscles (Ariens-Kappers, Huber, & Crosby, 1960). The derivation of interlimb mechanisms of coordination from embryonic mechanisms underlying undulations would also explain why, in adult newts, alternating coordination of the hindlimbs requires the integrity of thoracic segments, even though those segments do not innervate the limbs and are unnecessary for coordination within a limb (Brandle & Szekely, 1973). Similarly, segments immediately rostral to limb-innervating segments are thought to be of primary importance for interlimb coordination in cats (Kato, 1988). It is worth noting that Humphrey (1944) described neurons in the spinal cord of embryonic humans that were believed to be homologous to primary neurons of amphibians. The question of the ontogenetic and phylogenetic development of coordination between limbs in mammals would be greatly enhanced by neuroanatomical and electrophysiological studies that attempted to trace those mechanisms back to their embryonic origins.

REFERENCES

Abu Gideiri, Y. B. (1971). The development of locomotory mechanisms in *Bufo regularis*. *Behavior*, *38*, 121–131.

Angulo y Gonzalez, A. W. (1932). The prenatal development of behavior in the albino rat. *Journal of Comparative Neurology*, *55*, 395–442.

Ariens-Kappers, C. U., Huber, G. C., & Crosby, E. C. (1960). *The comparative anatomy of the nervous system of vertebrates, including man* (Vol. 1). New York: Hafner Publishing Co.

Arnold, S. J., & Wassersug, R. J. (1978). Differential predation on metamorphic anurans by garter snakes (Thamnophis): Social behavior as a possible defense. *Ecology*, *59*, 1014–1022.

Ayers, J., Carpenter, G. A., Currie, S., & Kinch, J. (1983). Which behavior does the lamprey central motor program mediate? *Science*, *221*, 1312–1314.

Barcroft, J., & Barron, D. H. (1939). The development of behavior in the foetal sheep. *Journal of Comparative Neurology*, *70*, 477–502.

Bassler, U. (1986). On the definition of central pattern generator and its sensory control. *Biological Cybernetics*, *54*, 65–69.

Beach, F. A. (1950). The snark was a boojum. *American Psychologist*, *5*, 115–124.

Beaudoin, A. R. (1955). The development of lateral motor column cells in the lumbosacral cord in *Rana pipiens*. *Anatomical Record*, *121*, 81–96.

Bekoff, A. (1976). Ontogeny of leg motor output in the chick embryo: A neural analysis. *Brain Research*, *106*, 271–291.

Bekoff, A., & Kauer, J. A. (1984). Neural control of hatching: Fate of the pattern generator for the leg movements of hatching in post-hatching chicks. *Journal of Neuroscience*, *4*, 2659–2666.

Bekoff, A., and Lau, B. (1980). Interlimb coordination in 20-day-old rat fetuses. *Journal of Experimental Zoology*, *214*, 173–175.

Bekoff, A., & Sabichi, A. L. (1987). Sensory control of the initiation of hatching in chicks: Effects of a local anesthetic injected into the neck. *Developmental Psychobiology*, *20*, 489–495.

Bekoff, A., Stein, P. S. G., & Hamburger, V. (1975). Coordinated motor output in the hindlimb of the 7-day chick embryo. *Proceedings of the National Academy of Science*, *72*, 1245–1248.

Bekoff, A., & Trainer, W. (1979). The development of interlimb coordination during swimming in postnatal rats. *Journal of Experimental Biology*, *83*, 1–11.

Biewener, A. A. (1989). Scaling body support in mammals: Limb posture and muscle mechanics. *Science*, *245*, 45–48.

Bignall, K. E., & Schramm, L. (1974). Behavior of chronically decerebrated kittens. *Experimental Neurology, 42*, 519–531.

Bitterman, M. E. (1960). Toward a comparative psychology of learning. *American Psychologist, 15*, 704–712.

Blight, A. R. (1977). The muscular control of vertebrate swimming movements. *Biology Reviews, 52*, 181–218.

Brandle, K., & Szekely, G. (1973). The control of alternating coordination of limb pairs in the newt (*Triturus vulgaris*). *Brain, Behavior and Evolution, 8*, 366–385.

Buchwald, J. S. (1988). The development of language. In E. Meisami & P. S. Timiras (Eds.), *Handbook of human growth and developmental biology* (Vol. 1, Pt. B; pp. 285–294). Boca Raton, FL: CRC Press.

Cecil, S. G., & Just, J. J. (1979). Survival rate, population density and development of a naturally-occurring Anuran larvae (*Rana catesbeiana*). *Copeia, 1979*, 447–453.

Coghill, G. E. (1929). *Anatomy and the problem of behavior*. New York: Macmillan.

Comer, C., & Grobstein, P. (1981). Tactually elicited prey acquistion behavior in the frog, *Rana pipiens*, and a comparison with visually elicited behavior. *Journal of Comparative Physiology, 142*, 141–150.

D'Angelo, D., Gordon, A., & Charipper, H. (1941). The role of the thyroid and pituitary glands in the anomalous effect of inanition on amphibian metamorphosis. *Journal of Experimental Zoology, 87*, 259–277.

Decker, J. (1967). Motility of the turtle embryo, *Chelydra serpentina* (L.) *Science, 157*, 952–954.

Dickerson, M. C. (1969). *The frog book*. New York: Dover.

Dunlap, D. G. (1967). The development of the musculature of the hindlimb in the frog, *Rana pipiens*. *Journal of Morphology, 119*, 241–258.

Dupre, R. K., Taylor, R. F., & Frazier, D. T. (1985). Static lung compliance during the development of the bullfrog, *Rana catesbelana*. *Respiration Physiology, 59*, 331–338.

Farel, P. B., & Bemelmans, S. E. (1980). Retrograde labeling of migrating spinal motoneurons in bullfrog larvae. *Neuroscience Letters, 18*, 133–136.

Finlay, B. L., Wikler, K. C., & Sengelaub, D. R. (1987). Regressive events in brain development and scenarios for vertebrate brain evolution. *Brain, Behavior, and Evolution, 30*, 102–117.

Forehand, C., & Farel, P. B. (1982). Anatomical and behavioral recovery from the effects of spinal cord transection: Dependence on metamorphosis in anuran larval. *Journal of Neurosicence, 2*, 654–662.

Gans, C., & Parson, T. S. (1966). On the origin of the jumping mechanism in frogs. *Evolution, 20*, 92–99.

Garcia-Rill, E. (1986). The basal ganglia and the locomotor regions. *Brain Research Review, 11*, 47–63.

Gard, C., Hard, E., Larsson, K., & Petersson, V. (1967). The relationship between sensory stimulation and gross motor behavior during the postnatal development in the rat. *Animal Behavior, 15*, 563–567.

Gosner, K. L. (1960). A simplified table for staging anuran embryos and larvae with notes on identification. *Herpetologica, 16*, 183–190.

Govind, C. K., & Pearce, J. (1986). Differential reflex activity determines claw and closer muscle asymmetry in developing lobsters. *Science, 233*, 354–356.

Grillner, S. (1981). Control of locomotion in bipeds, tetrapods and fish. In *Handbook of physiology. The nervous system*, (Vol. II, Sect. 1, Pt. 2). Bethesda, MD: American Physiological Society.

Grillner, S. (1985). Neurobiological bases of rhythmic motor acts in vertebrates. *Science, 228*, 143–149.

Hall, W. G., & Oppenheim, R. W. (1987). Developmental psychobiology: Prenatal, perinatal and early postnatal aspects of behavioral development. *Annual Review of Psychology, 38*, 91–128.

Hall, W. G., & Williams, C. L. (1983). Suckling isn't feeding, or is it? A search for developmental continuities. In J. S. Rosenblatt (Ed.), *Advances in the study of behavior* (Vol. 13). New York: Academic Press.

Hamburger, V., Balaban, M., Oppenheim, R., & Wenger, E. (1965). Periodic motility of normal and spinal chick embryos between 8 and 17 days of incubation. *Journal of Experimental Zoology, 159*, 1–14.

Hodos, W., & Campbell, C. B. G. (1969). *Scala naturae.* Why there is no theory in comparative psychology. *Psychology Review, 76*, 337–350.

Hofer, M. A. (1981). *The roots of human behavior.* San Francisco: Freeman.

Hooker, D. (1936). Early fetal activities in mammals. *Yale Journal of Biology and Medicine, 8*, 579–602.

Humphrey, T. (1944). Primitve neurons in the embryonic human central nervous system. *Journal of Comparative Neurology, 81*, 1–45.

Humphrey, T. (1978). Function of the nervous system during prenatal life. In Uwe Stave (Ed.), *Perinatal physiology*, New York: Plenum.

Johanson, I. B., & Hall, W. G. (1980). The ontogeny of feeding in rats: III. Thermal determinants of early ingestive responding. *Journal of Comparative Physiology and Psychology, 94*, 977–992.

Just, J. J., Kraus-Just, J., & Check, D. A. (1981). Survey of chordata metamorphosis. In L. I. Gilbert & E. Frieden (Eds.), *Metamorphosis: A problem in developmental biology.* New York: Plenum.

Kato, M. (1988). Longitudinal myelotomy of lumbar spinal cord has little effect on coordinated locomotor activities of bilateral hindlimbs of the chronic cats. *Neuroscience Letter, 93*, 259–263.

Kluge, A. G., & Farris, J. S. (1969). Quantitative phyletics and the evolution of anurans. *Systematic Zoology, 18*, 1–32.

Lamborghini, J. E. (1980). Rohon-Beard cells and other large neurons in *Xenopus* embryos originate during gastrulation. *Journal of Comparative Neurology, 189*, 323–333.

Landmesser, L. T., & O'Donovan, M. J. (1984). Activation patterns of embryonic chick hindlimb muscles recorded in ovo and in an isolated spinal cord preparation. *Journal of Physiology, 347*, 189–204.

Lang, F., Govind, C. K., Costello, W. J., & Green, S. I. (1977). Developmental neuroethology: Changes in escape and defensive behavior during growth of the lobster. *Science, 197*, 682–685.

Letinsky, M. S. (1974). The development of nerve-muscle junctions in *Rana catesbeiana* tadpoles. *Developmental Biology, 40*, 129–153.

Lieberman. P. (1984). *The biology and evolution of language.* Cambridge, MA: Harvard University Press.

McClellan, A. D. (1982). Movements and motor patterns of the buccal mass of pleurobranchaea during feeding regurgitation and rejection. *Journal of Experimental Biology, 98*, 195–211.

McGill, T. E. (1960). Response of the leopard frog to shock in an escape-learning situation. *Journal of Comparative Physiology and Psychology, 53*, 443–445.

McGraw, M. (1939). Swimming behavior in the human infant. *Journal of Pediatrics, 15*, 485–490.

Miller, R. R., & Berk, A. M. (1979). Sources of infantile amnesia. In N. E. Spear & B. A. Campbell (Eds.), *Ontogeny of learning and memory.* Hillsdale, NJ: Erlbaum.

Narayanan, C. H., Fox, M. W., & Hamburger, V. (1971). Prenatal development of spontaneous and evoked activity in the rat (*Rattus norvegicus albinus*). *Behavior, 40*, 100–134.

O'Donovan, M. J., & Landmesser, L. (1987). The development of hindlimb motor activity studied in the isolated spinal cord of the chick embryo. *Journal of Neuroscience, 7*, 3256–3264.

Parsons, T. S., & Williams, E. E. (1963). The relationships of the modern Amphibia: A re-examination. *Quarterly Review of Biology, 38*, 26–53.

Peters, S. E. (1983). Postnatal development of gait behaviour and functional allometry in the domestic cat (*Felis catus*). *Journal of Zoology, London, 199*, 461–486.

Pomeranz, B. (1972). Metamorphosis of frog vision: Changes in ganglion cell physiology and anatomy. *Experimental Neurology, 34*, 187–199.

Richards, C. M. (1958). The inhibition of growth in crowded *Rana pipiens* tadpoles. *Physiological Zoology, 31*, 138–151.

Rose, S. M. (1960). A feedback mechanism of growth control in tadpoles. *Ecology, 41*, 188–199.

Shik, M. L., & Orlovsky, G. N. (1976). Neurophysiology of locomotor automatism. *Physiological Reviews, 56*, 465–501.

Shumway, W. (1940). Stages in the normal development of *Rana pipiens*. I. External form. *Anatomical Record, 78*, 139–144.

Silver, M. L. (1942). The motoneurons of the spinal cord of the frog. *Journal of Comparative Neurology, 77*, 1–40.

Stehouwer, D. J. (1986). Behavior of larval and juvenile bullfrogs (*Rana catesbeiana*) following chronic spinal transection. *Behavioral and Neural Biology, 45*, 120–134.

Stehouwer, D. J. (1988). Metamorphosis of behavior in the bullfrog (*Rana catesbeiana*). *Developmental Psychobiology*, *21*, 383–395.

Stehouwer, D. J., & Farel, P. B. (1980). Central and peripheral controls of swimming in anuran larvae. *Brain Research*, *195*, 323–335.

Stehouwer, D. J., & Farel, P. B. (1981). Sensory interactions with a central motor program in anuran larvae. *Brain Research*, *218*, 131–140.

Stehouwer, D. J., & Farel, P. B. (1983). Development of hindlimb locomotor activity in the bullfrog (*Rana catesbeiana*) studied in vitro. *Science*, *219*, 516–518.

Stehouwer, D. J., & Farel, P. B. (1984). Development of hindlimb locomotor behavior in the frog. *Developmental Psychobiology*, *17*, 217–232.

Stehouwer, D. J., & Farel, P. B. (1985). Development of hindlimb locomotor mechanisms in the frog. *Journal of Neurophysiology*, *53*, 1453–1466.

Taylor, A. C. (1943). Development of the innervation pattern in the limb bud of the frog. *Anatomical Record*, *87*, 379–412.

Taylor, A. C., & Kollros, J. J. (1946). Stages in the normal development of *Rana pipiens* larvae. *Anatomical Record*, *94*, 7–23.

Thelen, E. (1979). Rhythmical stereotypes in normal human infants. *Animal Behaviour*, *27*, 699–715.

Thelen, E., Bradshaw, G., & Ward, J. A. (1981). Spontaneous kicking in month-old infants: Manifestation of a human central locomotor program. *Behavioral and Neural Biology*, *32*, 45–53.

Thelen, E., Fisher, D. M., Ridley-Johnson, R., & Griffin, N. J. (1982). Effects of body build and arousal on newborn infant stepping. *Developmental Psychobiology*, *15*, 447–453.

Thelen, E., Ridley-Johnson, R., & Fisher, D. M. (1983). Shifting patterns of bilateral coordination and lateral dominance in the leg movements of young infants. *Developmental Psychobiology*, *16*, 29–46.

Thexton, A. J., & Griffiths, C. (1979). Relex oral activity in decerebrate rats of different age. *Brain Research*, *175*, 1–9.

van Bergeijk, W. (1967). Anticipatory feeding behavior in the bullfrog (*Rana catesbeiana*). *Animal Behaviour*, *15*, 231–238.

Vince, M. A. (1973). Some environmental effects on the activity and development of the avian embryo. In G. Gottlieb (Ed.), *Behavioral embryology*. New York: Academic Press.

Wassersug, R. J., & Hoff, K. (1985). The kinematics of swimming in anuran larvae. *Journal of Experimental Biology*, *119*, 1–30.

Wassersug, R. J., & Sperry, D. G. (1977). The relationship of locomotion to differential predation on *Pseudacris triseriata* (Anura: Hylidae). *Ecology*, *58*, 830–839.

Windle, W. F. (1944). Genesis of somatic motor function in mammalian embryos: A synthesizing article. *Physiological Zoology*, *17*, 247–260.

Windle, W. F., & Griffin, A. M. (1931). Observations on embryonic and fetal movements in the cat. *Journal of Comparative Neurology*, *52*, 149–188.

Young, J. Z. (1962). *The life of vertebrates*. New York: Oxford University Press.

9

Accessibility of the Rat Fetus for Psychobiological Investigation

WILLIAM P. SMOTHERMAN AND SCOTT R. ROBINSON

Behavioral development does not start at birth. Any woman who has been pregnant is aware that the fetus is active in utero. Until recently, in fact, maternal reports of fetal movement constituted the principal means of access to the fetus for the study of behavior. Innovative technologies, such as external fetal monitoring and real-time ultrasound, envisioned for monitoring fetal well-being and managing high-risk pregnancies, have recently been applied to the description of fetal movement (deVries, Visser, & Prechtl, 1982, 1985) and sensory development (Birnholz, 1984; Birnholz & Benacerraf, 1983). In the last five years, these methods have begun to offer an alternative to indirect inference about the behavioral abilities of the human fetus. The advent of an electronic window on fetal development is challenging long-held notions of the fetus as an intrauterine passenger. Clearly, the fetus is responsive and exhibits patterned motor activity. Understanding the relationship between prenatal activity and postnatal behavior, however, requires the application of detailed quantitative and experimental methods.

Aside from practical concerns, such as improving clinical care of premature infants, there are fundamental issues of theoretical importance that will benefit from study of the fetus. Is the fetus's behavior shaped by adaptation to its intrauterine environment or by the needs of preparation for future environments? On the one hand, it is apparent that the fetus is an active organism that responds to changes in the intrauterine environment. Some elements of the fetal repertoire, therefore, may serve as ontogenetic adaptations, improving the fit between the fetus and its surroundings. On the other hand, at no point during development does a young mammal experience a more radical change of environment than at the time of birth. The behavioral competence of the newborn argues that the neonate is prepared immediately after birth to interact with its postnatal environment. Preparation for postnatal life implies the existence of behavioral continuity: The roots of neonatal competence lie in the prenatal period. The issues of ontogenetic adaptation, preparedness, and continuity are especially relevant for the fetal period, but in broader perspective apply to the problem of behavioral development over the entire lifespan. Because the distinction between these issues is most clear at this point in development, understanding of prenatal behavior as well as processes of behavioral ontogeny in general should result from extension of a perspective of behavioral epigenesis to encompass events that precede birth.

In order to build an empirical foundation for addressing these issues, we have employed a comparative approach to the study of fetal behavior. Experimentation with human fetuses is an ethically sensitive and emotionally charged issue. Moreover, guidelines for the study of human fetuses place limitations on what can be learned from direct study of human fetuses. Therefore, comparative study of nonhuman animal fetuses is imperative. In this chapter, we describe methods currently employed in our laboratory to study the behavior of the rat fetus. Direct fetal observation necessitates (1) humanely desensitizing the pregnant rat without exposing fetuses to general anesthesia, (2) surgically preparing the uterus and fetuses to bring them into direct view, and (3) maintaining the viability of subject fetuses throughout the period of observation. In addition, behavioral study is greatly aided by (4) methods of manipulating the local physical and sensory environment of the fetus. The technical ability to view the fetus is incomplete, however, without (5) a correspondingly precise vocabulary of fetal movement and rigorous methods for describing and quantifying fetal behavior. These surgical and analytical tools make the rat fetus accessible for behavioral investigation.

PREPARATION OF THE PREGNANT RAT

The principal difficulty with devising ways to directly observe live fetuses is satisfying the conflicting needs of maintaining a normal physiological environment for the fetus while minimizing discomfort and trauma for the mother. The first attempts to view animal fetuses involved study of dying fetuses (Preyer, 1885). At the other extreme, researchers of the 1920s and 1930s often studied fetuses while mothers were under the effects of complete general anesthesia (Swenson, 1926). Under these conditions, fetuses show little behavior. Excluding classical methods, such as decerebration produced by ligature of the carotid arteries (Angulo y Gonzalez, 1932), two techniques were generally available prior to the 1980s that permitted direct observation of fetuses while circumventing the effects of general anesthesia. Physical transection of the spinal cord was developed early in fetal research (Windle & Becker, 1940) and has been employed in recent studies of fetal motor activity (Narayanan, Fox, & Hamburger, 1971; Kirby, 1979; Kodama & Sekiguchi, 1984). Physical transection of the spinal cord is effective in producing posterior paralysis and eliminating sensation in the lower limbs and abdomen. A number of variant techniques are available for transecting the spinal cord, but most involve some degree of surgical incision through the skin, dorsal musculature, and vertebral column to gain access to the spinal cord.

An alternative procedure developed in the 1960s—chemomyelotomy—achieves the same end as spinal transection with a simple intraspinal injection of ethanol (Basmajian & Ranney, 1961). Chemomyelotomy results in complete interruption of nervous transmission within the spinal cord at the site of injection and paralysis of the hindquarters and lower abdomen. This procedure has also been used effectively in the study of fetal behavior (Narayanan, Narayanan, & Browne, 1982). We currently use the following procedure to perform a chemomyelotomy on pregnant rats (Smotherman, Richards, & Robinson, 1984). The rat is placed in a jar suffused with anesthesia-grade ethyl ether until fully anesthetized. Ether is used for initial anesthesia because it takes effect quickly and is eliminated from the body rapidly after the rat is exposed to room air. An alternative short-acting anesthetic, such as halothane, may be preferred for species that are especially sensitive to ether.

Once anesthetized, the rat is removed from the jar and placed in a prone position on a surgical board. A small area of fur overlying the midback is removed with clippers. To facilitate visual guidance of the injection, a small midline incision (5–10 mm) is made in the skin, not extending into the musculature. The injection of 100% ethanol (100 μl for 300–400 g subjects, delivered at room temperature) is accomplished by inserting a 25-gauge needle, oriented perpendicular and dorsal to the spine, into the space between the first and second lumbar vertebrae. This site is identified by counting dorsal spinal processes posterior to the last pair of ribs. The tip of the needle is then directed caudally within the vertebral foramen and the ethanol injected in a 1–2 s pulse. The head and thorax of the rat should be slightly elevated immediately after injection to prevent rostral flow of ethanol within the vertebral canal. After the injection, prepared rats typically exhibit hyperventilation, extension of both hindlimbs, and nipple erection. When learning this technique, investigators may find it necessary to apply a nose cone with ether until the injection is delivered. But after training, the entire procedure requires only 1–2 min after removal of the rat from the ether jar, and no supplemental anesthetic is required.

Spinal transection and chemomyelotomy are outwardly similar in their effects on the pregnant rat. Both result in irreversible spinal blockade that permits surgical externalization of the uterus without the use of general maternal anesthesia. Under either of these preparations, fetuses can be observed without evidence of physiological impairment for periods exceeding one hour. However, spinal transection and chemomyelotomy are not equivalent in their effects on spontaneous fetal behavior. With mothers prepared by spinal transection, fetuses exhibit slightly more motor activity than fetuses from mothers prepared by chemomyelotomy (Smotherman, Richards, & Robinson, 1984). While differences in overall activity do not pose great problems for comparison between procedures, of greater concern are independent and unpredictable effects of the two procedures on specific patterns of fetal behavior. Differential behavioral effects, where present, seem to be most pronounced in older fetuses (day 20).

We believe there are several objective reasons for preferring chemomyelotomy over spinal transection as a method of preparing pregnant rats for fetal study: (1) Chemomyelotomy is simple in application and is readily taught to investigators or students unfamiliar with the procedure. (2) The effects of chemomyelotomy are consistent between subjects, providing high-quality fetal preparations with low rates of accidental mortality (about 1%). (3) Injection of ethanol into the spinal cord, in contrast to the surgical excision of dorsal musculature and vertebral segments employed in spinal transection, results in very little local trauma to the prepared female. This fact alone may contribute to the differential effects these two procedures have on fetal behavior. Therefore, while we would not wish to ignore the utility of spinal transection or eliminate its use in specific situations, we generally prefer the use of chemomyelotomy in behavioral applications, if for no other reason than to facilitate comparison of research findings obtained in different laboratories.

Spinal transection and chemomyelotomy are irreversible procedures; pregnant rats must be sacrificed after fetal observation. However, certain research questions, especially those concerned with issues of continuity between the prenatal and postnatal period, will benefit from an alternative preparation that is reversible. Lidocaine spinal anesthesia provides a means for direct observation of fetuses without terminating gestation (Smotherman, Robinson, & Miller, 1986). This procedure involves the same anesthesia and injection protocol as is used in performing a chemomyelotomy. A lidocaine solution containing epinephrine (2% lidocaine + 0.001% epinephrine) is injected into the spinal cord at the same level as chemomyelotomy (L1–L2). Intraspinal injection of 100 μl of the lidocaine solution is effective in satisfying the principal objectives of a reversible procedure: complete abdominal and hindlimb paralysis, consistently long periods of spinal anesthesia (in excess of 50 min), and complete recovery after anesthesia. After recovery, when fetuses are observed through the transparent wall of the uterus (see below), mothers can continue the pregnancy, deliver pups at term, and exhibit apparently normal maternal behavior. The potential for this reversible maternal preparation to facilitate longitudinal comparison of fetal and pup behavior is obvious.

Unlike spinal transection, there is no evidence that reversible lidocaine spinal anesthesia differs in its effects on fetal behavior from chemomyelotomy. Similar levels of overall fetal activity, synchronous movements, repertoire diversity, and patterning of individual fetal movements have been reported in the two preparations, suggesting that they are true alternatives (Smotherman, Robinson, & Miller, 1986). Lidocaine anesthesia should be preferred in experiments calling for longitudinal analysis of an individual subject through advancing stages of gestation and extending into the postnatal period. Chemomyelotomy offers the advantage of extended periods of observation, permitting manipulation of the intrauterine environment (Smotherman & Robinson, 1988a) and temporal analyses of fetal behavior (Smotherman, Robinson, & Robertson, 1988; Robinson & Smotherman, 1988). Both procedures result in prepared females that remain generally quiet during observation sessions and show no outward signs of discomfort. We view the development of these maternal preparations as a significant advance over the methods commonly used and ultimately discontinued during the early years (1920–1940) of fetal research.

PREPARATION OF SUBJECT FETUSES

Once the pregnant rat has been prepared with some form of spinal blockade, it is placed in a Plexiglas holding apparatus that elevates the head and body 45° above the horizontal (Smotherman, Richards, & Robinson, 1984). The rat is secured in the apparatus with a velcro jacket. The lower abdomen is shaved, and a low midventral incision (about 3 cm in length) is performed. The female and apparatus are then placed in a buffered isotonic saline solution (Locke's solution [Galigher & Kozloff, 1971]) with temperature regulated at 37.5°C ± 0.5°C. Water depth is adjusted to about the tip of the sternum, ensuring that the head is above and the incision below the water surface. Both horns of the uterus, and the constituent fetuses, are gently externalized through the abdominal incision into the saline bath. This manipulation is accomplished without placing strain on points of uterine attachment at the ovarian ends or at the cervix. The mother and fetuses remain undisturbed in this environment until a period of 15–20 min after the termination of general anesthesia has elapsed. This delay is fully adequate to ensure no residual effects of ether anesthesia on the mother or, more critically, on the behavior of fetuses (Kirby, 1979; Smotherman, Richards, & Robinson, 1984). The female should be monitored during this period to prevent uterine torsion from occluding blood flow within the uterus.

The fetus can be thought of as occupying a physical environment that is separated from the external world by successive envelopes or barriers. These envelopes, progressing from the outermost to the innermost, include the maternal abdomen, the wall of the uterus, the extraembryonic membranes (amnion and chorion), and amniotic fluid. Investigation of fetal behavior is greatly facilitated by stripping away successive envelopes, creating a series of different physical environments in which the spontaneous or stimulus-evoked behavior of fetuses can be observed. Each fetal environment offers different advantages and drawbacks for experimental study of the fetus.

Immersion of the pregnant rat and externalization of the uterus into the saline bath effectively strip away the outermost envelope surrounding the fetus. At this level of preparation individual fetuses can be viewed through the semitransparent wall of the uterus. Because the uterus is undisturbed throughout observation, observation of rat fetuses In Utero (as we have referred to this environment) most closely approximates normal intrauterine conditions during gestation. The wall of the uterus becomes thinner and more transparent as gestation proceeds. For this reason, observation In Utero is impractical earlier than about day 16 of gestation.

A second level of fetal preparation involves externalization of a single conceptus into the saline bath (In Amnion). The conceptus, which comprises a fetus, amniotic fluid, and surrounding membranes, is delivered through a small (10–20 mm) transverse incision in the uterine wall. Care should be taken in locating this incision, as it can influence the physiological condition of the fetus and the length of time that it can be observed. The incision should be placed on the side of the uterus facing away from uterine blood vessels and sites of placental attachment. If a fetus in a terminal position (closest to the ovary) is selected for study, a standard protocol we have often adopted in our research, a single incision can be made between the subject and other fetuses in the horn. An incision in this location minimizes the potential for adjacent fetuses to press against the placenta of the subject fetus and induce placental separation from the uterus. If a subject is selected from a miduterine position, a second incision isolating the subject from ovarian and caudal neighbors can be made. A second incision is especially useful when many fetuses occur in the same uterine horn or late in gestation, when intrauterine crowding is greatest. After externalization, the fetus In Amnion can be observed in considerable detail through the transparent amniotic and chorionic membranes. Throughout observation In Amnion the fetus remains attached to its placenta, which is not delivered through the incision and remains within the uterus. Because the extraembryonic membranes become very fragile and are prone to rupture near term, observation of rat fetuses In Amnion on day 21 is technically difficult.

The ultimate envelope around the fetus—the extraembryonic membranes—can be removed in a third-level preparation that provides the clearest view of the fetus and unobstructed access for experimental manipulation (Ex Utero). After an incision is made in the uterine wall (as in the In Amnion preparation), a small cut is made in the chorion near the head of the fetus. The incision in the membrane is made before delivery of the conceptus from the uterus. In this way the fetus is gently delivered headfirst into the bath while the chorion, amnion, and placenta are protected within the uterus. Because the fetus remains attached to the placenta by means of the umbilical cord, delivery into the saline bath does not impair fetal viability. Twisting of the umbilical cord at this point, or permitting a neighboring fetus to press against the cord, can lead to occlusion of umbilical circulation. One method of preventing this is to make a secondary incision in the uterus caudal to the neighboring fetus. The cord should be monitored throughout observation. We have routinely observed fetuses Ex Utero on days 17–21 of gestation for periods in excess of 30 min without evidence of fetal or placental compromise (Smotherman & Robinson, 1986a; Smotherman, Robinson, & Robertson, 1988).

We believe there is no "best" preparation in which to study fetal behavior. Each of the three fetal environments described above offers unique advantages and drawbacks for particular experimental needs (Smotherman & Robinson, 1988b). Fetuses observed In Utero remain in close proximity to contiguous siblings, permitting direct observation of prenatal behavioral interactions. If the pregnant rat is prepared by the reversible spinal anesthetic method, fetuses In Utero can be replaced within the maternal abdomen, the abdominal incision closed with sutures, and the litter allowed to continue gestation until normal parturition. The chemical constituents of the In Uterus or In Amnion environments can be altered experimentally by injecting various solutions into the amniotic fluid. Chemical manipulation of the amniotic fluid tends to persist in the fetal environment and thus provides a useful tool for investigating mid- to long-term effects of different chemical cues on spontaneous fetal behavior (Smotherman & Robinson, 1985), prenatal learning (Stickrod, Kimble, & Smotherman, 1982; Smotherman, 1982a, 1982b; Smotherman & Robinson, 1985), and morphological development (Moessinger, 1983). Short-term responsiveness to chemical stimuli is most conveniently studied with the fetus Ex Utero, where it is available for physical or surgical manipulation. The clearer view of the fetus afforded by observation Ex Utero also facilitates creation of real-time videotape or film records of fetal behavior. Playback of recorded behavior at reduced speed or frame by frame is necessary for close inspection and analysis of rapid, often subtle motor performances by the fetus (Bekoff & Lau, 1980).

In addition to serving particular experimental designs, observation of spontaneous fetal behavior in these three preparations has provided evidence that fetuses are responsive to changes in their physical environment. Especially late during gestation, when fetal body size and diminished amniotic fluid volume exacerbate physical crowding within the uterus, the spontaneous behavior of fetuses observed In Utero and Ex Utero is different in many respects (Smotherman & Robinson, 1986a). In view of this finding, one may wonder whether undisturbed fetuses that remain within the maternal abdomen, an environment that preserves all four envelopes surrounding the fetus, also exhibit different behavior. One approach to addressing this question is to use a minimally invasive optical device to visualize fetuses within the uterus inside the maternal abdomen (In Situ).

The endoscope we have employed to visualize rat fetuses In Situ is the same instrument widely used in arthroscopic surgery (Smotherman & Robinson, 1986b). It consists of a slender telescopic tube, a light source and fiber-optic cable, and a protective external sheath (Karl Storz Endoscopy-America). When connected to a microvideo camera, a wide-angle image can be displayed on a video monitor or recorded with standard equipment. The endoscope's small diameter (5 mm) permits it to be inserted through a very small incision in the pregnant rat's abdomen. The resultant views of the fetus vary with fetal size (and therefore gestational age) and the position of the abdominal incision. Endoscopic images are never as clear as afforded by direct observation of fetuses within an externalized uterus, but they are sufficient to measure rates of overall fetal activity and document gross patterns of motor behavior. In general, the repertoire and levels of activity exhibited by fetuses observed endoscopically In Situ are comparable to those expressed by fetuses observed In Utero.

Because each of the four environments in which fetuses may be observed (In Situ, In Utero, In Amnion, and Ex Utero) offer unique advantages for particular experimental needs, all are appropriate for the study of fetal behavior. However, a trade-off exists between the similarity of a particular observation environment to the conditions that exist during undisturbed pregnancy and the degree of control over stimulus conditions afforded the experimenter. In moving from endoscopic visualization In Situ to direct observation Ex Utero, fetuses experience less physical restraint. On the other hand, controlled presentation of chemical or tactile stimuli and examination of detailed motor responses is greatly facilitated by stripping away successive envelopes that surround the fetus. We have found that rather than limiting investigation, study of fetuses in different environments actually offers advantage in terms of generating hypotheses about normal behavioral development.

To illustrate, consider the different lines of research that originally led to independent investigation of behavioral and cardiac response to hypoxia produced by occlusion of the umbilical cord, chemosensory development and responsiveness of fetuses to chemical stimuli infused into the mouth, and the ability of fetuses to form conditioned associations in utero. By necessity, each of these areas of inquiry required study of fetuses in environments that permitted manipulation of the fetus and/or its sensory environment (In Amnion or Ex Utero). Briefly, these experiments led to discoveries that (1) umbilical occlusion results in behavioral hyperactivity and expression of stereotypic action patterns (Smotherman & Robinson, 1987b). (2) rat fetuses express different patterns of behavioral activation to some chemical stimuli (such as lemon or milk), but fail to respond at all to other stimuli (such as sucrose) (Smotherman & Robinson, 1988a), and (3) fetuses are capable of forming associations between neutral chemical stimuli paired with aversive events (such as the illness reaction produced by i.p. injection of lithium chloride) (Smotherman & Robinson, 1985). Only when this information became available to us was a scenario suggested in which fetal learning could occur under naturalistic conditions.

Olfactory stimuli can be transmitted to the fetus during normal pregnancy by transplacental transfer into the amniotic fluid (Hepper, 1987) or more directly by exposure of olfactory receptors to blood-borne odorants (Maruniak, Silver, & Moulton, 1983). Accidental occlusion of the umbilical cord, leading to transient episodes of fetal hypoxia, is a relatively common event even in unremarkable pregnancies (Mann, 1986). Chance association of exposure to olfactory cues in utero with a transient hypoxic episode could confer behavioral significance to these odors. We have recently begun to evaluate this naturalistic hypothesis in an experimental paradigm by pairing intraoral infusions of sucrose with a brief period of umbilical cord occlusion. Findings to date indicate that fetal conditioning does occur under these circumstances.

MANIPULATION OF FETUSES AND THEIR SENSORY ENVIRONMENT

In some respects, the fetus is a fragile organism. At term, only a small portion of the skeleton is ossified, and tissues offer little resistance to external force. While this fact underscores the need to manipulate fetuses carefully, it also can be made to serve the aims of the experimenter. Implantation of intraoral cannulas, installation of subcutaneous cardiac leads, and transection of the central nervous system can be accomplished with simple, rapid procedures that result in minimal trauma. Fetuses with cannulas or neural transections remain healthy and active and provide nearly ideal subjects for the study of early sensory, neural, and motor development.

Prior to 1980, much of what was widely believed about the sensory abilities of fetuses was based on histological examination of aborted fetuses (Humphrey, 1953). Direct behavioral assessment of fetal sensation was limited to experiments with a single sensory modality: touch. Evoking fetal movements by tactile stimulation is an experimental technique that was widely used in fetal reflexology during the 1920s and 1930s, and which has changed little in the past 50 years (Angulo y Gonzalez, 1932; Narayanan, Fox, & Hamburger, 1971). Punctate stimuli are delivered with a handheld filament or needle; stimulus intensity can be quantified by varying the flexibility of the filament, and thus the force applied, with a pressure aesthesiometer consisting of von Frey filaments of different diameter (Stoelting Co.). Punctate stimulation has been used to describe developmental changes in simple motor reflexes and to chart body areas of the fetus that exhibit different thresholds of tactile sensitivity.

An alternative form of tactile stimulation involves repeated stroking of the fetus with a soft brush (Smotherman & Robinson, 1988a). Because stroking is administered to an area rather than a point, it is less suitable for detailed mapping of tactile sensitivity over the surface of the fetus. Stroking can be effective, however, in producing distinct effects on behavior, such as manipulation of overall levels of fetal activity. Tactile stimulation, whether by punctate contact or stroking, can be administered in a general way through the amniotic sac, the wall of the uterus, or even the maternal abdomen. However, the uterus, extraembryonic membranes, and amniotic fluid cushion the fetus, attenuate tactile stimuli, and yield results that suggest elevated tactile thresholds. Because the efficiency of transmission of mechanical force through these envelopes varies during gestation, controlled application of tactile stimuli requires that the fetus be removed from the amniotic environment and tested Ex Utero.

It is perhaps not surprising that fetuses are sensitive to touch. During early postnatal life, however, pups also depend on other forms of sensory stimulation to interact with their environment. Pups utilize olfaction and gustation to locate and suckle from the nipple (Blass & Teicher, 1980), olfaction to recognize the mother (Leon, 1974) and discriminate kin from nonkin (Hepper, 1986), and vestibular and thermal sensitivity to maintain a huddle and regulate body temperature (Alberts & Cramer, 1988). Recent indirect evidence has suggested that rats possess a functional chemical sense before birth (Pedersen, Stewart, Greer, & Shepherd, 1983; Pedersen & Blass, 1982; Smotherman, 1982a; Stickrod, Kimble, & Smotherman, 1982). This conclusion is confirmed by direct behavioral assessment of fetal responsiveness to chemical stimuli (Smotherman & Robinson, 1985; 1988a), findings that have suggested a possible role for fetal olfaction in the development of chemical recognition (Hepper, 1987). Several techniques now exist for presenting chemical stimuli to the fetus in utero that allow for progressively greater control over the timing and intensity of stimulus exposure.

The most indirect method of exposing fetuses to chemical stimuli is perhaps the most relevant for fetal development. Specific substances in the diet of the pregnant rat can be transported by maternal circulation and diffuse across the placenta to the fetus. For example, garlic contains a sulfur-based compound (allyl sulfide) that is readily transmitted across the placenta. Pregnant rats that ingest garlic in their diet bear offspring that behave differently in the presence of garlic odor (Hepper, 1988). The effect of prenatal exposure may be due either to diffusion of allyl sulfide into the amniotic fluid, which is consumed by rat fetuses (Smotherman & Robinson 1988b), or to direct access of chemical receptors to blood-borne odors (Maruniak, Silver, & Moulton, 1983). While manipulation of maternal diet most closely approximates the conditions under which fetuses are normally exposed to external olfactorants, it does not permit measurement of fetal response at the time of exposure, assessment of age-related changes in chemosensation, or control over the moment of stimulus presentation or the intensity of stimulation.

Results paralleling the manipulation of maternal diet can be obtained by injecting chemical cues into the amniotic fluid that surrounds the fetus (Blass & Pedersen, 1980; Stickrod, 1981). Intra-amniotic injection is performed with the pregnant rat under general anesthesia. The uterus is externalized through a midline laparotomy and permitted to rest on the abdomen of the anesthetized, supine female. Chemical cues are introduced into the amniotic fluid of individual fetuses by injection of a small volume (20μl) of the test solution in an isotonic saline carrier. Solutions should be injected at the body temperature of the fetus. Use of a very fine needle (30 gauge) minimizes loss of amniotic fluid during injection. Further loss of amniotic fluid can be prevented by coating the site of injection with petroleum jelly or inserting the needle through the placenta. After injection, rotation of the conceptus within the uterus offsets the sites of injection through the uterine wall and the extraembryonic membranes and seals the amniotic sac from loss of fluid. After injection, the uterus is rinsed with isotonic saline and gently replaced within the maternal abdomen. The abdominal incision is closed with sutures and/or stainless steel wound clips and swabbed with a betadyne solution. Mothers recover quickly from surgery and exhibit no adverse effects.

Fetuses exposed to chemical stimuli by intra-amniotic injection are not harmed and can be allowed to complete normal gestation. By exposing the fetus to a chemical stimulus and replacing it within the maternal abdomen, the chemosensory experience of the fetus can be manipulated several days before testing. Alternatively, intra-amniotic injection can be used to manipulate the local environment of the fetus minutes or seconds before testing when the female is immersed in the water bath. In this case, testing can take place with the fetus In Utero or In Amnion, although the latter preparation probably increases the rate of diffusion of the chemical cue from the amniotic fluid. Intraamniotic injection has been performed with rat fetuses between days 17 and 21 of gestation. Comparison of fetuses manipulated by this procedure at different ages should take into account differences in stimulus concentration brought about by changes in amniotic fluid volume that normally occur during gestation (Smotherman & Robinson, 1988b).

Greater control over stimulus exposure (timing and concentration) is provided by direct infusion of test solutions into the mouth of the fetus (Figure 9.1). Intraoral

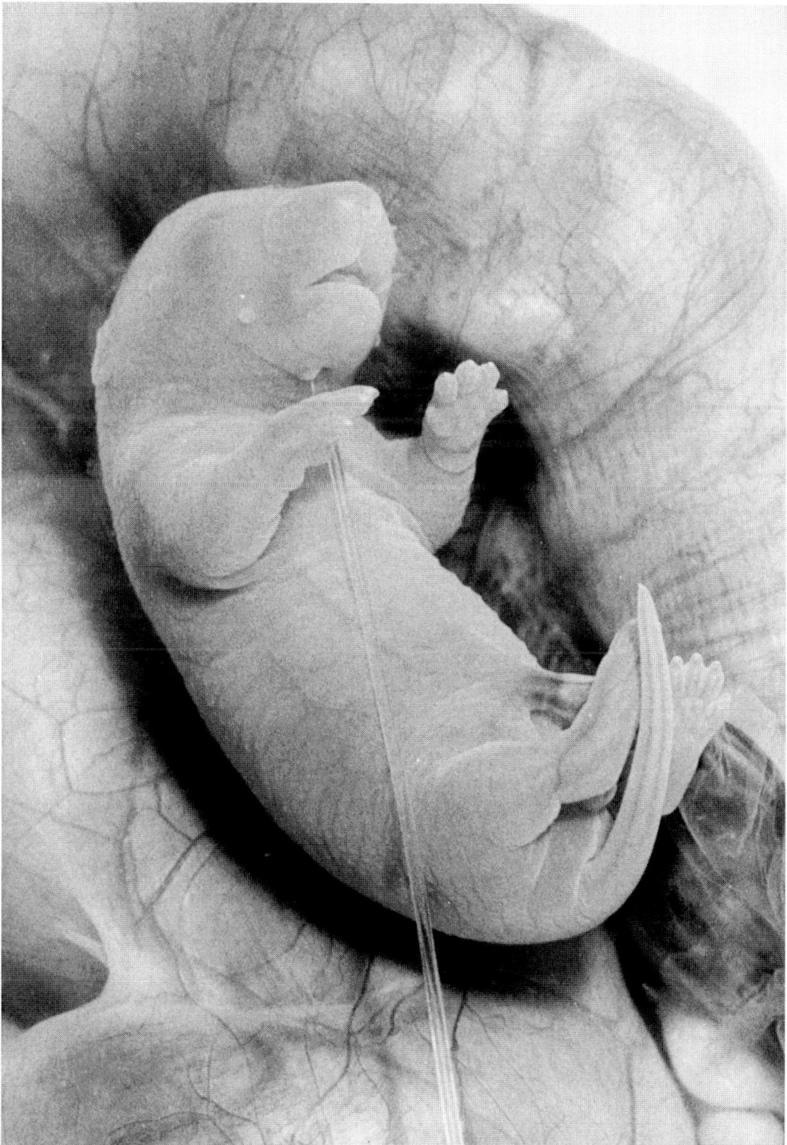

Figure 9.1. Rat fetus prepared for observation Ex Utero on day 20 of gestation. The fetus has been fitted with a cannula to permit intraoral infusion of a test solution.

presentation is made possible through the use of a cannula, adapted from procedures developed for rat pups (Hall & Rosenblatt, 1977; see also Phifer, chap. 13 this volume). During insertion of the wire involved in the installation of the cannula, care must be exercised to maintain the fetus under the surface of the water, to avoid crimping or pulling the umbilical cord, and to provide enough tubing so as not to restrict fetal movement. The cannula permits precise infusion of a test solution to the fetus without handling or otherwise interrupting ongoing fetal activity. We typically employ a protocol in which successive infusions are delivered in a 1–2 s pulse consisting of 20 μl of the test solution. Alternatively, continuous infusion of a test solution can be automatically controlled by a syringe pump and timer (Kashinsky et al., 1990). The only limitation on chemosensory solutions that may be delivered by this infusion system is fluid viscosity; we have infused solutions of isotonic saline; sucrose; quinine hydrochloride; novel olfactorants such as lemon and citral; and biologically relevant fluids such as milk, amniotic fluid, and dimethyl disulfide to fetuses as young as 17 days of gestation. If two cannulas are installed along the midline of the tongue (a procedure we have employed in conditioning experiments), it is possible to coordinate infusion of two different test solutions.

The three alternative methods for presenting chemical stimuli to the fetus offer unique advantages. The least invasive method of exposing fetuses to chemical stimuli is through manipulation of maternal diet. This approach also provides the least control over parameters of stimulus exposure to the fetus. Intra-amniotic injection results in chemical stimulation that persists within the closed environment of the amniotic sac for variable periods ranging from minutes to days. This method is appropriate for ensuring prenatal exposure to a stimulus in experiments calling for testing at a later point in gestation (Smotherman & Robinson, 1985) or after birth (Smotherman, 1982a; Stickrod, Kimble, & Smotherman, 1982) or for manipulating the chemical context in which other aspects of fetal behavior are assessed (Smotherman & Robinson, 1988a). Intraoral infusion of test solutions is better suited for direct measurement of immediate fetal responsiveness to chemical cues. The infusion technique has been applied to studies of fetal learning, habituation, sensory acuity, and stimulus-evoked motor development (Smotherman & Robinson, 1987a, 1988a). The evident drawback with intraoral infusion is the necessity to test fetuses Ex Utero, prohibiting longitudinal study.

MEASUREMENT OF FETAL BEHAVIOR

With the foregoing emphasis on technical procedures for preparing the pregnant rat and the fetus for observation, altering the sensory environment of the fetus, and manipulating fetal physiology, one must not neglect to devote equal consideration to problems of definition and measurement of behaviors produced by the fetus. We employ a system of observing and recording fetal behavior that exhibits the following key features. (1) Categories of fetal movement are defined in operational terms of the movements themselves. Specifically, behavioral categories are defined by the anatomical region of the body in which a particular movement occurs (e.g., foreleg, rearleg, head, mouth, trunk). Each movement is considered as an instantaneous point event.

(2) Every behavioral performance by the subject is scored, not just a focal subset of events relevant to a specific research question. Because categories are independently defined, it is possible for two or more categories to occur simultaneously. (3) Each act, comprising one or more categories of movement, is entered into an event recorder that preserves the time of data entry (± 1 s). The product is a continuous record of fetal motor activity during an observation session that preserves temporal and sequential information. (4) Fetuses are simultaneously videotaped during observation sessions. A permanent videographic record augments the basic event record and permits retrospective analysis of fetal behavior. This observation and recording protocol is consistent in its results: We have reported inter-rater reliabilities in excess of .90 when different observers are asked to detect and categorize fetal movement events. Successive observations of videographic records by the same observer result in higher reliability scores.

Clearly, an exhaustive behavioral record, even over a limited period of observation, must be reduced to describe fetal behavior or assess experimental effects. The strategy we have employed in data reduction is to construct several standard data summaries and then employ a battery of analytical techniques to address specific behavioral questions. The standard summaries provide information about the overall activity of the fetus, the frequency of events in specific behavioral categories, and the incidence of two or more independent categories of movement occurring synchronously as a single event (Smotherman, Richards, & Robinson, 1984). Summaries of overall activity, category frequency, and incidence of synchronous movements can be integrated to provide additional information about behavioral organization and repertoire diversity (Smotherman & Robinson, 1986a; Robinson & Smotherman, 1987, 1988). The frequency of activity, or of specific categories of movement, is further reduced by parsing events among successive intervals of time during an observation session; depending upon the interval selected (ranging from 5 to 60 s), relatively fine or coarse descriptions of temporal changes in behavior are obtained. Temporal summaries provide the basis for a variety of analytic procedures, including tests for bout structure (aperiodic temporal clustering of events), cyclic motor organization (Smotherman, Robinson, & Robertson, 1988), and sequential organization (Robinson & Smotherman, 1988). Temporal summaries also enable measurement of abrupt changes in fetal behavior coincident with acute changes in the fetal environment, such as produced by stimulus infusion or umbilical cord occlusion (Smotherman & Robinson, 1987b, 1988a). See Robinson and Smotherman (1988) for more detailed descriptions of these analytical procedures.

CONCLUDING REMARKS

Improved techniques for preparing the mother and fetus, manipulating the intra-uterine environment, and measuring behavior of the fetus are opening a window on a portion of the lifespan that has heretofore received comparatively little attention (Smotherman & Robinson, 1988c). Studies founded on this technology are continuing to document how some elements of the fetal repertoire promote adaptation of the fetus to changes in the intrauterine environment (Smotherman & Robinson, 1986a;

1987b) while other elements are coextensive with postnatal behavior (Smotherman & Robinson, 1987a; 1988d). As mentioned above, rat fetuses exhibit a remarkably consistent behavioral response to brief occlusion of the umbilical cord. The behavioral hyperactivity and cardiac deceleration that characterize the fetal response to hypoxia are exactly opposite to the response of newborn rat pups that are deprived of oxygen (Eden & Hanson, 1987; Smotherman & Robinson, 1988e). Because the fetal hypoxic response is evident only during the prenatal period upon manipulation of an anatomical feature unique to the fetus (the umbilical cord), we have interpreted this response as an ontogenetic adaptation.

Conversely, on day 20 of gestation rat fetuses respond to intraoral infusion of lemon with facial wiping, a stereotypic behavioral pattern that constitutes an important element of the aversive behavior of adult rats, but which, paradoxically, is not expressed during the early postnatal period. The apparent discontinuity in the development of facial wiping behavior has led us to investigate the reason for the absence of wiping during the neonatal period. Briefly, a series of experiments have now revealed that (1) fetuses near term perform facial wiping in response to lemon infusion, (2) newborn pups tested in an age-typical (terrestrial) environment do not perform wiping, (3) newborns immersed in a buoyant fluid medium do respond to infusions with facial wiping, and (4) pups immersed in a fluid but maintained in ventral contact with a submerged hard substrate do not exhibit wiping (Smotherman & Robinson, 1989). We have interpreted these findings as evidence that the early postnatal absence of facial wiping is not due to immaturity of neural substrates that subserve the wiping response or to a lack of continuity between fetal and adult aversion behavior but to the environmental constraint of the expression of facial wiping. These experiments, which not only describe the development of this form of behavior but also have implications for the conduct of developmental research in general, were predicated upon the advent of methods and research questions applied to the study of fetal behavior.

The fetus is an active organism that exists in a complex and changing environment. The fetus undergoes development in utero, but simultaneously interacts with its environment through its behavior and physiology. In this way the fetus may be viewed as an active participant in its own development and not merely a passive object in a static environment (Smotherman & Robinson, 1987c). We feel that this change in our view of the fetus is healthy. Renewed interest in fetal development is a logical extension of the epigenetic perspective of the prenatal period. Such a perspective, which constitutes the theoretical underpinnings of all developmental research, holds that ontogeny is a cumulative process. Thus, behavioral sophistication in the neonate implies that the organism begins the postnatal period with a rich behavioral and experiential history. It is our belief that more detailed knowledge of this prenatal history will foster a more complete understanding of behavioral development in general.

ACKNOWLEDGMENTS

This research is supported by a grant from the National Institute of Child Health and Human Development (NIH) HD 16102-07. William P. Smotherman is supported by

Research Career Development Award HD 00719-03 and a grant from NATO Collaborative Research Program (0551/88).

REFERENCES

Alberts, J. R., & Cramer, C. P. (1988). Ecology and experience: Sources of means and meaning of developmental change. In E. M. Blass (Ed.), *Handbook of behavioral neurobiology: Vol. 9. Developmental psychobiology and behavioral ecology* (pp. 1–39). New York: Plenum.

Angulo y Gonzalez, A. W. (1932). The prenatal development of behavior in the albino rat. *Journal of Comparative Neurology, 55*, 395–442.

Basmajian, J. V., & Ranney, D. A. (1961). Chemomyelotomy: Substitute for general anesthesia in experimental surgery. *Journal of Applied Physiology 16*, 386.

Bekoff, A., & Lau, B. (1980). Interlimb coordination in 20-day-old rat fetuses. *Journal of Experimental Zoology, 214*, 173–175.

Birnholz, J. C. (1984). Fetal neurology. In R. Saunders & M. Hill (Eds.), *Ultrasound annual* (pp. 139–160). New York: Raven Press.

Birnholz, J. C., & Benacerraf, B. R. (1983). The development of human fetal hearing. *Science, 222*, 516–518.

Blass, E. M., & Pedersen, P. E. (1980). Surgical manipulation of the uterine environment of rat fetuses. *Physiology and Behavior, 25*, 993–995.

Blass, E. M., & Teicher, M. H. (1980). Suckling. *Science, 210*, 15–22.

deVries, J. I. P., Visser, G. H. A., & Prechtl, H. F. R. (1982). The emergence of fetal behavior. I. Qualitative aspects. *Early Human Development, 7*, 301–322.

deVries, J. I. P., Visser, G. H. A., & Prechtl, H. F. R. (1985). The emergence of fetal behavior. II. Quantitative aspects. *Early Human Development, 12*, 99–120.

Eden, G. J., & Hanson, M. A. (1987). Maturation of the respiratory response to acute hypoxia in the newborn rat. *Journal of Physiology, 392*, 1–9.

Galigher, A. E., & Kozloff, E. N. (1971). *Essentials of practical microtechnique* (2nd ed.). Philadelphia: Lea and Febiger.

Hall, G. H., & Rosenblatt, J. S. (1977). Suckling behavior and intake control in the developing rat pup. *Journal of Comparative Physiology and Psychology, 91*, 1232–1247.

Hepper, P. G. (1986). Kin recognition: Functions and mechanisms: A review. *Biological Review, 61*, 63–93.

Hepper, P. G. (1987). The amniotic fluid: An important priming role in kin recognition. *Animal Behaviour, 35*, 1343–1346.

Hepper, P. G. (1988). Adaptive fetal learning: Prenatal exposure to garlic affects postnatal preferences. *Animal Behaviour, 36*, 935–936.

Humphrey, T. (1953). The relation of oxygen deprivation to fetal reflex arcs and the development of fetal behavior. *Journal of Psychology, 35*, 3–43.

Kashinsky, W. M. L., Rozboril, W., Robinson, S. R., & Smotherman, W. P. (1990). An inexpensive rotary infusion pump for delivering microliter volumes of fluids to animal subjects. *Physiology and Behavior, 47*, 1279–1281.

Kirby, M. L. (1979). A quantitative method for determining the effect of opiates on fetal rats in utero. In *Problems of drug dependence. NIDA Research Monographs, 27*, 191–197.

Kodama, N., & Sekiguchi, S. (1984). The development of spontaneous body movement in prenatal and perinatal mice. *Developmental Psychobiology, 17*, 139–150.

Leon, M. (1974). Maternal pheromone. *Physiology and Behavior, 13*, 441–453.

Mann, L. I. (1986). Pregnancy events and brain damage. *American Journal of Obstetrics and Gynecology, 155*, 6–9.

Maruniak, J. A., Silver, W. L., & Moulton, D. G. (1983). Olfactory receptors respond to blood-borne odorants. *Brain Research, 265*, 312–316.

Moessinger, A. C. (1983). Fetal akinesia deformation sequence: An animal model. *Pediatrics, 72*, 857–863.

Narayanan, C. H, Fox, M. W., & Hamburger, V. (1971). Prenatal development of spontaneous and evoked activity in the rat. *Behaviour, 40,* 100–134.

Narayanan, C. H., Narayanan, Y., & Browne, R. C. (1982). Effects of induced thyroid deficiency on the development of suckling behavior in rats. *Physiology and Behavior, 29,* 361–370.

Pedersen, P. E., & Blass, E. M. (1982). Prenatal and postnatal determinants of the 1st suckling episode in albino rats. *Developmental Psychobiology, 15,* 349–355.

Pedersen, P. E., Stewart, W. B., Greer, C. A., & Shepherd, G. M. (1983). Evidence for olfactory function in utero. *Science, 221,* 478–480.

Preyer, W. (1885). *Specielle Physiologie des Embryo. Untersuchungen uber die Lebenserscheinungen vor der Geburt.* Grieben: Leipzig.

Robinson, S. R., & Smotherman, W. P. (1987). Environmental determinants of behavior in the rat fetus. II. The emergence of synchronous movement. *Animal Behaviour, 35,* 1652–1662.

Robinson, S. R., & Smotherman, W. P. (1988). Chance and chunks in the ontogeny of fetal behavior. In W. P. Smotherman & S. R. Robinson (Eds.), *Behavior of the fetus* (pp. 95–115). Caldwell, NJ: Telford Press.

Smotherman, W. P. (1982a). Odor aversion learning by the rat fetus. *Physiology and Behavior, 29,* 769–771.

Smotherman, W. P. (1982b). In-utero chemosensory experience alters taste preferences and corticosterone responsiveness. *Behavioral and Neural Biology, 36,* 61–68.

Smotherman, W. P., Richards, L. S., & Robinson, S. R. (1984). Techniques for observing fetal behavior in utero: A comparison of chemomyelotomy and spinal transection. *Developmental Psychobiology, 17,* 661–674.

Smotherman, W. P., & Robinson, S. R. (1985). The rat fetus in its environment: Behavioral adjustments to novel, familiar, aversive and conditioned stimuli presented in utero. *Behavioral Neuroscience, 99,* 521–530.

Smotherman, W. P., & Robinson, S. R. (1986a). Environmental determinants of behaviour in the rat fetus. *Animal Behaviour, 34,* 1859–1873.

Smotherman, W. P., & Robinson, S. R. (1986b). A method for endoscopic visualization of rat fetuses in situ. *Physiology and Behavior, 37,* 663–665.

Smotherman, W. P., & Robinson, S. R. (1987a). Prenatal expression of species-typical action patterns in the rat fetus (*Rattus norvegicus*). *Journal of Comparative Psychology, 101,* 190–196.

Smotherman, W. P., & Robinson, S. R. (1987b). Stereotypic behavioral response of rat fetuses to acute hypoxia is altered by maternal alcohol consumption. *American Journal of Obstetrics and Gynecology, 157,* 982–986.

Smotherman, W. P., & Robinson, S. R. (1987c). Prenatal influences on development: Behavior is not a trivial aspect of fetal life. *Journal of Developmental and Behavioral Pediatrics, 8,* 171–176.

Smotherman, W. P., & Robinson, S. R. (1988a). Behavior of rat fetuses following chemical or tactile stimulation. *Behavioral Neuroscience, 102,* 24–34.

Smotherman, W. P., & Robinson, S. R. (1988b). The uterus as environment: The ecology of fetal behavior. In E. M. Blass (Ed.), *Handbook of behavioral neurobiology: Vol. 9. Developmental psychobiology and behavioral ecology* (pp. 149–196). New York: Plenum.

Smotherman, W. P., & Robinson, S. R. (Eds.). (1988c). *Behavior of the fetus.* Caldwell, NJ: Telford Press.

Smotherman, W. P., & Robinson, S. R. (1988d). Fetal expression of the leg extension response to anogenital stimualtion. *Physiology and Behavior, 43,* 243–244.

Smotherman, W. P., & Robinson, S. R. (1988e). Response of the rat fetus to acute umbilical cord occlusion: An ontogenetic adaptation? *Physiology and Behavior, 44,* 131–135.

Smotherman, W. P. & Robinson, S. R. (1989). Cryptopsychobiology: The appearance, disappearance and reappearance of a species-typical action pattern during early development. *Behavioral Neuroscience, 103,* 153–160.

Smotherman, W. P., Robinson, S. R., & Miller, B. J. (1986). A reversible preparation for observing the behavior of rat fetuses in utero: Spinal anesthesia with lidocaine. *Physiology and Behavior, 37,* 57–60.

Smotherman, W. P., Robinson, S. R., & Robertson, S. S. (1988). Cyclic motor activity in the rat fetus. *Journal of Comparative Psychology, 102,* 78–82.

Stickrod, G. (1981). In utero injection of rat fetuses. *Physiology and Behavior*, *27*, 557–558.

Stickrod, G., Kimble, D. P., & Smotherman, W. P. (1982). In utero taste/odor aversion conditioning in the rat. *Physiology and Behavior*, *28*, 5–7.

Swenson, E. A. (1926). *The Development of movement of the albino rat before birth*. Unpublished doctoral dissertation, University of Kansas, Lawrence.

Windle, W. F., & Becker, R. F. (1940). Relation of anoxemia to early activity in the fetal nervous system. *Archives of Neurology and Psychiatry*, *43*, 90–101.

II.
PHYSIOLOGICAL SYSTEMS

A.
REGULATION

The majority of studies on the physiological regulatory systems that support living organisms are done on adults. And when young organisms are studied, it has usually been at a single point in time, as if the subjects were a different subspecies. Through application of the methods described in these chapters, it has been discovered that physiological regulatory systems, like behaviors, have definable stages of development and characteristic patterns of function that change as the animal progresses from one stage to the next. Two decades ago, it was widely believed that early stages in physiological regulation were simply less precise, less efficient, and less complex forms of the adult system. Instead, application of novel methods designed to solve the problems presented by small, immature organisms with special housing and care requirements resulted in the discovery that infants' physiological systems are highly organized, efficient, and complex but are different from adults'. And these differences are related to the differences in the experiential demands and supports of the particular stage of the infant's development, the "ecological niche." In the case of many infant mammals in the postnatal period, its ecological niche is its relationship with its mother. These insights have had far-reaching implications for our understanding of the nature and function of early social relationships.

Satinoff, Chapter 10, begins by emphasizing the crucial importance of the thermoneutral zone for the study of infant behavior. Learning, memory, and even some simple behaviors do not occur when pups are slightly cool. The new methods Satinoff describes offer the potential for discovering why this should be so. Conversely, certain infant behaviors have been discovered to have thermoregulation as their goal. Physiological as well as behavioral thermoregulation have well-defined stages in development, and the application of novel methods such as miniaturized telemetry capsules have revealed a circadian organization in infant mammals with several unexpected features.

Phifer, in Chapter 11, summarizes a number of methods devised in W. G. Hall's laboratory for study of the gastrointestinal system in the control of early feeding. The application of these methods has led to major insights into how ingestive behavior develops. Before these methods were developed, it had been thought that nursing was the only way by which food intake could be regulated in infant mammals and that it was the necessary precursor stage for the development of independent ingestion. Very little was known about the nature of the infant's control over ingestion in nursing, and the major regulator was assumed to be the mother. Phifer describes the methods that revealed the presence of independent ingestive control in very young suckling infants, the differences in its organization and control from those of nursing, and the parallel rather than sequential nature of the development of the two systems.

In Chapter 12, Williams presents novel methods for the study of hormonal influences on sexual differentiation and describes how these methods led to the discovery of early forms of sexual responses during infancy. The behavioral methods discussed in this chapter are as important as the endocrine and neuroanatomical ones, leading the author to a new concept of organizational as opposed to activational processes in the development of sexual differentiation. The discovery of infantile sexual behavior in this chapter and infantile free feeding in Chapter 11 (which appear well before the ''need'' for these capabilities) raises interesting questions about the basic nature and evolution of developmental processes.

In Chapter 13, O'Grady and Hall introduce the application to early development of methods that have recently established a functional interrelationship in adults between the immune system and the nervous and endocrine systems. These methods have revealed unexpected influences on the developing immune system, acting through its autonomic neural and endocrine connections. The authors outline specific ways in which the mother-infant interaction influences the development of immune competence. They describe methods which have revealed critical periods for endocrine and early experiential effects on this system. The early development of these linked systems is just beginning to be studied, so the methods and approaches described in this chapter offer exceptional opportunities for new discoveries and conceptual insights.

These methods have revealed a number of ways in which physiological systems interact with behavior. Not only do these systems in some instances direct and control the infant's behavior; several examples are given of how the mother-infant behavioral interaction serves to modify and control levels of the pups' physiological function. Thus, these methods have revealed specific ways in which behavior and physiology are linked in the shaping of development through the course of the early mother-infant relationship.

10

Developmental Aspects of Behavioral and Reflexive Thermoregulation

EVELYN SATINOFF

Newborn rat pups are blind, hairless, and practically immobile at birth. Although their central nervous systems are immature, in the realm of learning and motivation they show several behaviors that are surprisingly sophisticated. Pups 1 to 3 days old can learn odor aversions (Rudy & Cheatle, 1977), perform an operant response for the reward of milk in their mouths (Johanson & Hall, 1979) or electrical stimulation of their brains (Moran, Lew, & Blass, 1981), feed independently of their mother (Hall, 1979), and behaviorally thermoregulate, either by negotiating a thermally graded alleyway (Kleitman & Satinoff, 1982) or by operant responding (Guenaire, Costa, & Delacour, 1982).

In fact, rat pups and neonates of other species can probably do much more than they are presently given credit for, but uncovering these abilities requires one indispensable experimental condition—that testing be done within the pups' thermoneutral zone. Newborns have poor or nonexistent physical and physiological mechanisms for heat conservation. Testing a newborn at the ordinary laboratory temperatures of 21–23°C is like taking an adult human, immersing him or her in ice water, and then testing the individual's cognitive capacities or motivational state.

In this chapter I will discuss, first, why an animal's thermoregulatory abilities and the environmental temperature at which testing is done are critically important for interpreting results. Second, I will review physiological and behavioral thermoregulation in infant rodents. Third, I will examine what is known about infant rodents' circadian rhythms of body temperature. Fourth, I will describe methods for measuring various aspects of thermoregulation. Finally, I will pose some questions for future research.

THERMOREGULATION AND DEVELOPMENT

In carrying out developmental studies it is perilous to ignore thermoregulation and thermoneutrality, both in the nest and in the isolated individual pup, because all behavior depends on movement, and body temperature affects arousal level and

169

locomotor ability. It is especially important for studies designed to uncover either the first appearance of a particular reflex or behavior or the age at which an infant achieves adult competence. If infants are not tested in their thermoneutral zones, wrong information may be obtained.

For example, several behaviors previously thought not to be in an infant's repertoire until 2 to 3 weeks of age, are clearly seen in rat pups during the first week or even first day of life, if the pups are tested within the environmental temperature range of 33–35°C. Behavioral thermoregulation is an obvious example: newborn rat pups can become hypothermic very quickly, and if they are placed in a thermal gradient at room temperature, they become immobile and will not move to a warmer spot (Fowler & Kellogg, 1975; Johanson, 1979). However, if they are placed at 30°C (still a cold temperature for rat pups, but not incapacitating), 1-day-old pups will eventually move up a thermal gradient (Kleitman & Satinoff, 1982).[1] One- and 2-day-old rat pups will even make operant responses to obtain heat (Guenaire, Costa, & Delacour, 1982).

When pups are tested at warm ambient temperatures one can see the earlier appearance of behaviors that seemingly have nothing to do with thermoregulation. For instance, there have been several reports that in both females and males primed with estradiol benzoate, female sexual behavior—lordosis and ear wiggling—occurs no earlier than 2–3 weeks of age. In fact, several hypotheses have been presented as to why these behaviors should appear so late in rats (see Williams, 1987, and chap. 12, this volume, for references). However, when Williams (1987) tested pups at 33–35°C, both responses were seen in pups of both sexes at 4–6 days of age, *without* steroid priming. High doses of estradiol facilitated these responses in the younger pups. "The finding of estrogen-activated sexual responses at 4 to 6 days of age in rats questions the concept of totally separable periods of organizational and activational actions of steroids as it applies to the development of female sexual behavior" (Williams, 1987). And all this simply by testing the pups in thermally comfortable surroundings.

Other behaviors that can be elicited in rat pups as young as 3 days old when they are tested at warm environmental temperatures include feeding in response to food deprivation (Hall, 1979) and to tail pinch (Szechtman & Hall, 1980), drinking in response to angiotensin injections (Misantone, Ellis, & Epstein, 1980), and odor discrimination (Alberts, 1981) and odor conditioning (Johanson & Hall, 1979; Pedersen, Williams, & Blass, 1982). In earlier studies in which pups were tested at about 22–23°C, olfactory orientation, liquid food intake, and, indeed, most sensorimotor and regulatory behaviors did not appear until much later (e.g., Altman & Sudarshan, 1975; Almli & Fisher, 1977).

Not only behavior but the development of neural pathways is affected by the temperature at which infants are maintained. Horwitz, Heller, and Hoffmann (1982), studying the functional development of the dopaminergic nigrostriatal pathway, originally found that transection of the pathway acutely elevated striatal dopamine levels in adult and 8- or 10-day-old rats, but not in 4- or 6-day-olds. They interpreted their results as indicating a lack of impulse traffic in this pathway in the younger animals. When they maintained the pups at ambient temperatures of 34–35°C, however, they found that 4-day-old pups showed qualitatively all of the biochemical responses seen in the older animals. What is special about 33–35°C? It is within the newborn pup's thermoneutral zone.

THE CONCEPT OF THERMONEUTRALITY

For every animal there is a range of ambient temperatures at which the basal rate of heat production equals the rate of heat lost to the environment, and a minimal amount of thermoregulatory effort is required to maintain a constant body temperature. The most accurate definitions of thermoneutrality are based on both heat loss and heat production responses. Thus, a "zone of least thermoregulatory effort" can be bounded on the low end by increased metabolic rate and on the high end by increased evaporative heat loss (Mount, 1974).

These criteria can define a precise thermoneutral zone, which in human infants is only 1–2°C wide (Hey, 1974). It is generally thought that this precision applies only to species with metabolically inexpensive, low-threshold heat loss responses such as panting or sweating. Adult rodents evaporate heat by grooming saliva onto their skin. Since this is not as efficient as panting or sweating (Dawson, 1973) and may appear late in the sequence of adaptive responses to heat stress (Roberts, Mooney, & Martin, 1974), rodents are assumed to have a wider thermoneutral zone. Heat loss considerations constrain the upper end of the thermoneutral zone for adult rats to about 30°C. The lower end, measured by increases in metabolic rate, is often found to be 26°C. However, one early paper places the thermoneutral zone of adult rats at 28–30°C (Benedict & MacLeod, 1928). This is also the range reported by Szymusiak and Satinoff (1981) using maximal rapid eye movement sleep time instead of metabolic rate as an index of thermoneutrality. This is as narrow a range as has been found in any other mammalian species.[2]

Infant mammals neither groom saliva on their (nonexistent) fur in the heat, although they do sprawl, nor do they shiver in the cold, although they can increase nonshivering thermogenesis. Under appropriate circumstances they have very good heat-producing and excellent behavioral thermoregulatory abilities, which they need, for they also have narrow thermoneutral zones, ranging from 34–35°C at birth and gradually approaching the adult range over 3–4 weeks. Infants are rarely confronted with the problem of being too warm, and most of the work on infant abilities is focused on how well they thermoregulate in the cold.

THERMOREGULATORY CAPABILITIES OF INFANT RODENTS

Since Adolph's comprehensive study of the ontogeny of physiological regulations in rats (1957) there have been several reviews centering on one or another aspect of thermoregulation (e.g., Hull, 1973; Leon, 1986; Schmidt, Kaul, & Heldmaier, 1986). In considering the question of how well an individual infant rodent thermoregulates, four variables must be taken into account: physiological heat production, vasomotor heat loss and heat conservation responses, ability to thermoregulate behaviorally, and morphology. First, a brief summary. Infant rats increase metabolic rate in response to cold stress, mostly or completely by generating heat in their brown adipose tissue. This response is as good or better than adults', but infants quickly become hypothermic at ambient temperatures below their thermoneutral zone, and metabolic responses fail. Very little is presently known about the adequacy of vasomotor heat loss and heat conservation responses. Pups can behaviorally thermoregulate if they do not become incapacitated by the cold. The key problems infant rodents face are a lack of thermal insulation, both in terms of fur and subcutaneous fat stores, and an unfavorable surface-to-mass[3] ratio. Since these latter problems resolve themselves simply with age, I will discuss physiological and behavioral thermoregulatory capacities in some detail.

Physiological Thermoregulation

There is a widespread impression that behavioral thermoregulatory responses appear before physiological responses in the development of altricial mammals. In fact, this is not the case. Although infant mammals respond behaviorally and not reflexively to low doses of pyrogens (fever-producing substances; Kleitman & Satinoff, 1980), the reflexive response of increasing metabolic rate in response to a cold stress appears as early as any behavioral response.

An individual newborn rat's rectal temperature closely approximates the surrounding temperature, and it is not until the pup is roughty 21 days old that it is homeothermic—that is, can maintain a constant body temperature in environments above about 10°C. The newborn pup's ectothermy (that is, its *inability* to maintain a constant body temperature) is not caused by an inability to increase oxygen consumption (Figure 10.1; see Taylor, 1960, for references to other species that also show metabolic responses to cold). As Taylor (1960) has shown,

> all rats, even on the first day of life, showed an increase in oxygen consumption on exposure to a cold environment. . . . The four-hr-old rats showed only a very small increase in oxygen consumption between [environmental temperatures of] 38 and 31°C. In the 23-hr-old rats there was a large increase in oxygen consumption when the temperature was reduced below 36°C, and there appeared to be a neutral zone between 36 and 38°C. In the 21-day-old rat the neutral zone extended from about 34 to 38°C, and below this there was a progressive increase in oxygen consumption down to 11°C, the lowest temperature studied.

The core temperatures of young rats can be severely affected by cold stress even though oxygen consumption increases. Conklin and Heggeness (1971) reported that 5-day-old rat pups had colonic temperatures of 37.5, 35.3, and 30.8°C after 90 min exposure to ambient temperatures of 35, 30, and 25°C, respectively. More recently, Spiers and Adair (1986) measured thermoregulatory responses in rats 5–19 days old and also found increases in oxygen consumption at all ages, with older animals showing larger rises.

All three studies make the point that in young rats the thermoneutral zone, where metabolic rate is minimal, is extremely small. In Taylor's study on Wistar rats, the lower limit of the zone was 34°C even in 21-day-old rats. In Conklin and Heggeness' and Spiers and Adair's studies on Sprague-Dawley rats, the lower critical temperature was lower, shifting from 34.5°C at 5–7 days of age to 32°C at 17–19 days to 29–31°C at 21 days. These ambient temperatures are much higher than the usual laboratory temperatures of 21–24°C. For an adult rat, they constitute a heat stress, but for rats as old as 5 days, anything much below about 34°C is a cold stress and will impair cognitive and motoric capacities. Therefore, to reiterate one of the most important messages in this chapter, any work on infant rodents directed to finding out the age at which particular abilities appear will necessarily estimate the age as later than it might actually be if the studies are done at ambient temperatures below the pups' lower limit of thermoneutrality.

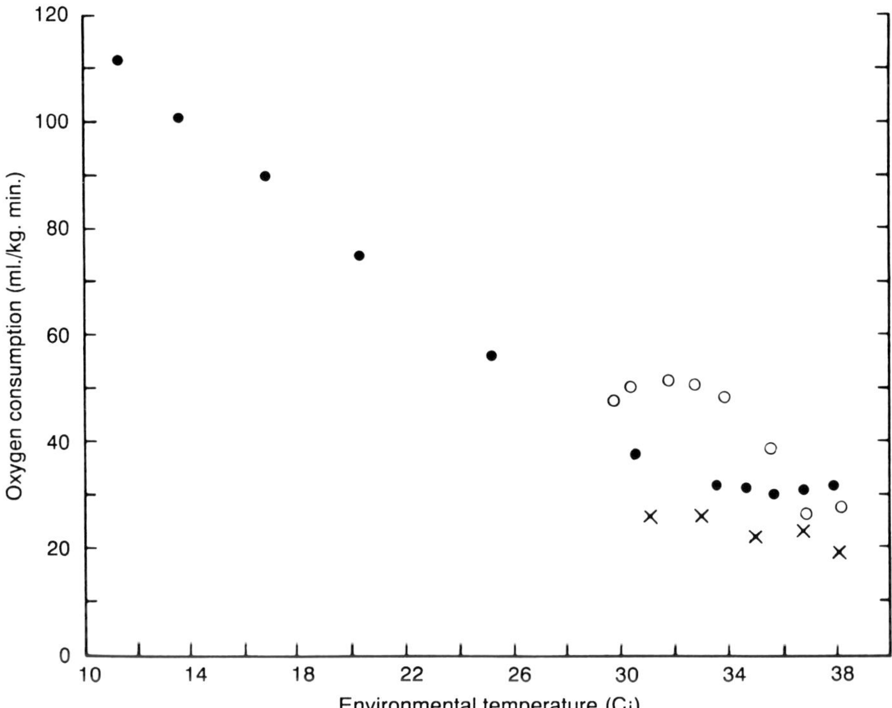

Figure 10.1. Measurement of oxygen consumption per unit body weight at different environmental temperatures in a group ($n = 6$–7) of rats 4 h old (x), in rats ($n = 6$–7) 23 h old ($\circ$) and one rat 21 days old ($\bullet$). (From "Oxygen Consumption in New-Born Rats" by P. M. Taylor, 1960, *Journal of Physiology (London)*, 154, p. 55.)

Another example of this key point is found in the infant response to injections of norepinephrine, which increase oxygen consumption by activating brown adipose tissue. Subcutaneous injections of norepinephrine in rats caused maximal increases in oxygen consumption, which remained constant between 3 and 30 days of age if the injections were made at the subneutral temperature of 30°C (Moore & Simmonds, 1966). However, if the injections were made at 35°C, maximum responses increased with age, even from Day 1 to Day 2 (Hsieh, Emery, & Carlson, 1971).

Much less is known about infant rats' ability to conserve and lose heat through vasoconstriction and vasodilation, respectively, than about their ability to produce heat internally. One reason is a technical problem—vasomotor tone is much more difficult to measure than is oxygen consumption. A direct evaluation of vasomotor tone requires measuring peripheral blood flow. Even if this were done, the handling, immobilization, and disturbance of the infants might seriously distort the measurement.

The standard way of assessing peripheral vasomotor tone in adult animals involves the indirect method of measuring some skin temperature—in adults, usually the tail. This creates another problem because it involves a confounding of skin temperature with core and air temperature: the higher the core or air temperature, the higher skin temperature is likely to be, and vice versa, depending on the strength of core-to-skin and air-to-skin temperature gradients.

Given these constraints, it is not surprising that very few studies have measured skin temperature in infants. Poczopko (1961) reported that newborn rats had no effective cutaneous vasoconstriction until postnatal Day 4, and Taylor (1960) thought this reponse did not appear until as late as Day 12. Conklin and Heggeness (1971), from determinations of thermal conductances between air and core temperature, concluded that vascular responses to cold become effective after 12 days of age.

These latter authors also concluded that "The improvement in the homeothermic ability of the rat during this period [5–19 days of age] appears to be primarily related to changes in thermal insulation." This conclusion is further buttressed by the observation that shaving does not reduce body temperature until the third postnatal week (Hahn, 1956; Poczopko, 1961).

In summary, rodents are endothermic from birth: that is, they are born with the capacity for altering internal heat production. Homeothermy, the ability to maintain a constant body temperature, arrives only after the development of fur.

Behavioral Thermoregulation

Although infant rodents can increase their metabolic rate from birth, this capability is practically useless because they cannot conserve heat. In any case, heat production, in the form of nonshivering thermogenesis initiated and maintained by brown adipose tissue, is a very costly use of metabolic energy that could better be used for growth and development. One might assume that infants would be better off drifting into hypothermia in the cold, but hypothermic rats develop much more slowly than do euthermic pups (Stone, Bonnet, & Hofer, 1976). Even premature human infants maintained in incubators that support slightly lower body temperatures have higher mortality rates and more metabolic disturbances than do euthermic infants (Beutow & Klein, 1964). Therefore, it is desirable for newborn mammals to maintain a relatively high body temperature.

All infant mammals studied seek heat as early as one day after birth and prefer warm fur or even warm metallic coils to cold fur (puppies and rabbits: Jeddi, 1970; monkeys: Harlow, 1971). In thermally graded alleys neonatal mice (Ogilvie & Stinson, 1966), pigs (Mount, 1963), rabbits (Baccino, 1935; Satinoff, McEwen, & Williams, 1976), gerbils (Eedy & Ogilvie, 1970), hamsters (Leonard, 1974), kittens (Olmstead, Villablanca, Torbiner, & Rhodes, 1979), and rats orient and move along the thermal gradient from cool to warm if they are not immobilized by the cold (Figure 10.2; Kleitman & Satinoff, 1982).

The above results are not surprising in light of work showing that several different species of mammals that are born in litters regulate their body temperatures by active group huddling. Rat pups in a huddle move from the top of the pile down into it when too much of their surface area is exposed and they get cool and move out to the edges when they get too warm (Figure 10.3; Alberts, 1978). In fact, huddling in rat pups is almost exclusively thermally directed during the first week of life (Alberts & Brunjes, 1978). Huddling leads to lower oxygen consumption in the individuals than if they are alone (mice: Stanier, 1975; rats: Alberts, 1978; puppies: Sugano & Nagasaka, 1979).[4] One reason for this is that huddled animals expose less surface area to the cool surround and thereby lose less heat.

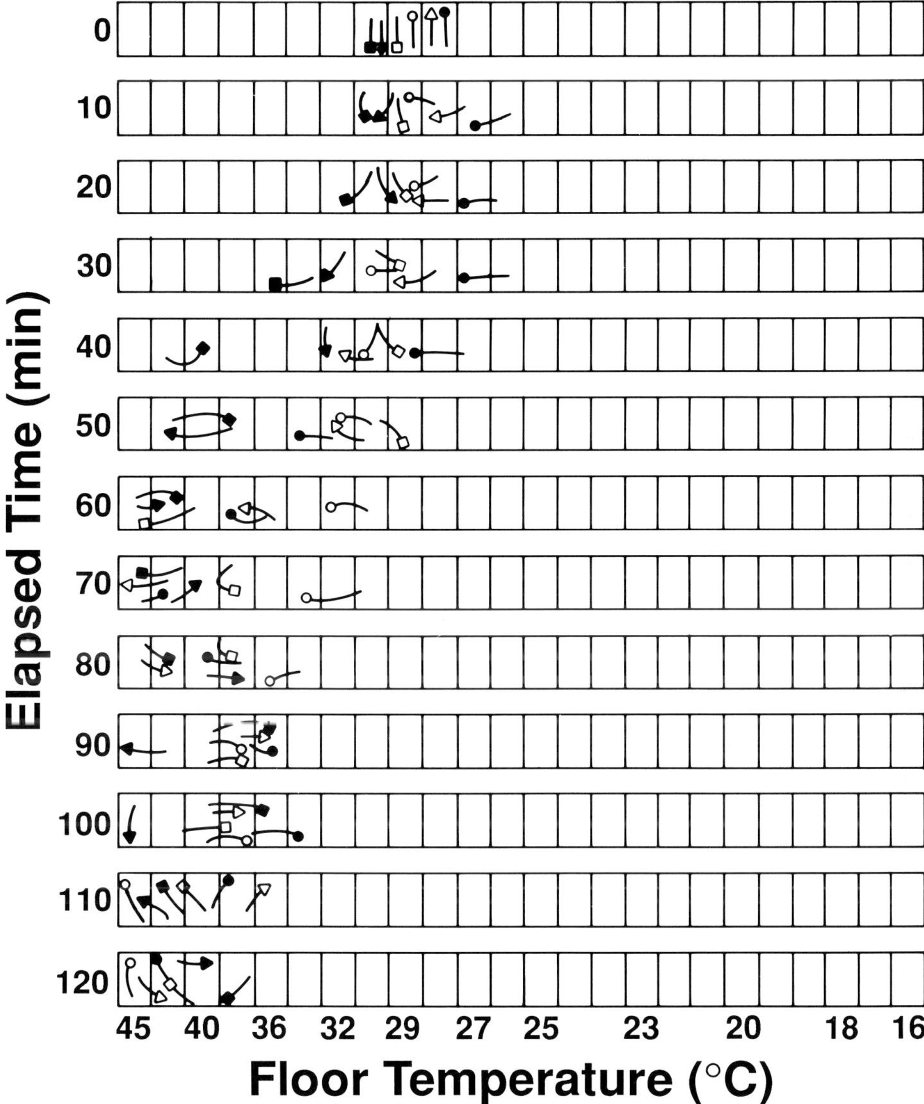

Figure 10.2. Positions of 1-day-old rat pups in a thermal gradient at 10-min intervals for 120 min. (From N. Kleitman & E. Satinoff, unpublished.)

Huddling makes good adaptive sense. Without a source of heat conservation, increased heat production is worse than useless. The main business of infants is to grow, and spending energy on achieving a stable body temperature that could not be maintained would be a waste. The only method newborn animals born in litters have of both producing and conserving heat is behavioral—by changing their position in the huddle or orienting and going toward other littermates when they are scattered around the nest site.

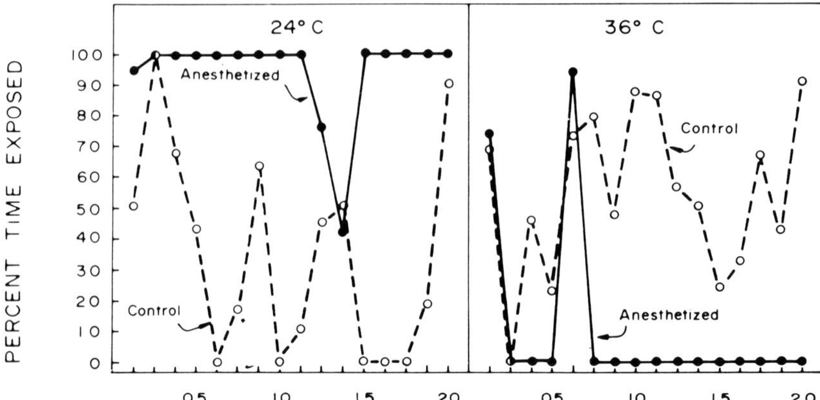

Figure 10.3. Percentage of time that individual anesthetized and control pups are on the surface of litters huddling at cool and at warm ambient temperatures. Position of pups was determined by time-lapse photography. (From "Ontogeny of Olfaction: Reciprocal Roles of Sensation and Behavior in the Development of Perception" by J. R. Alberts, 1978, in R. M. Aslin, J. R. Alberts, and M. R. Peterson (Eds.), *Development of Perception: Psychobiological Perspectives* [Vol. 1, pp. 321–357]. New York: Academic Press.

CIRCADIAN RHYTHMS OF BODY TEMPERATURE

This chapter has so far discussed reflexive and behavioral means by which an infant achieves thermal homeostasis. But in adults, body temperature is not constant over 24 hours but rather fluctuates in a precise manner. When these fluctuations occur in synchrony with the time cues in the environment, once every 24 h, they are called *diurnal* rhythms. If they still occur, or freerun, with a period of about 24 h, in the absence of any external time cues, they are called *circadian* rhythms. The development of circadian rhythmicity may lag by some weeks the development of the function that will later become rhythmic. Thus, for example, infant rats can respond to a stress with an increase in adrenal corticosterone, even though the daily adrenal rhythm itself becomes clear only at Day 16 (Levin & Levine, 1975). In thermoregulation also it appears that some homeostatic mechanisms, for example, metabolic changes in response to cold stress and behavioral responses, are in place well before diurnal or circadian rhythms of body temperature develop.

It is really impossible to determine a precise day on which the body temperature rhythm emerges, because one cannot tell which measurements should refer to circadian timing mechanisms and which to thermoregulatory mechanisms themselves. For instance, pups can show adultlike daily amplitudes of body temperature, that is, peaks as high and troughs as low as adults have, by days 19–20, but these bear no relation to the light/dark cycle. These changes may be due to the pups' maturing ability to thermoregulate reflexively (mainly because of increased heat conservation abilities). The two characteristics that are undeniably germane to the development of the circadian (or diurnal) temperature rhythm are the shape of the waveform and its timing relative to the light/dark cycle. Using these two criteria, and measuring body temperature by telemetry in pups as young as 10 days old, Kittrell and Satinoff (1986) found that the body temperature rhythm reached its adult timing between 31 and 36

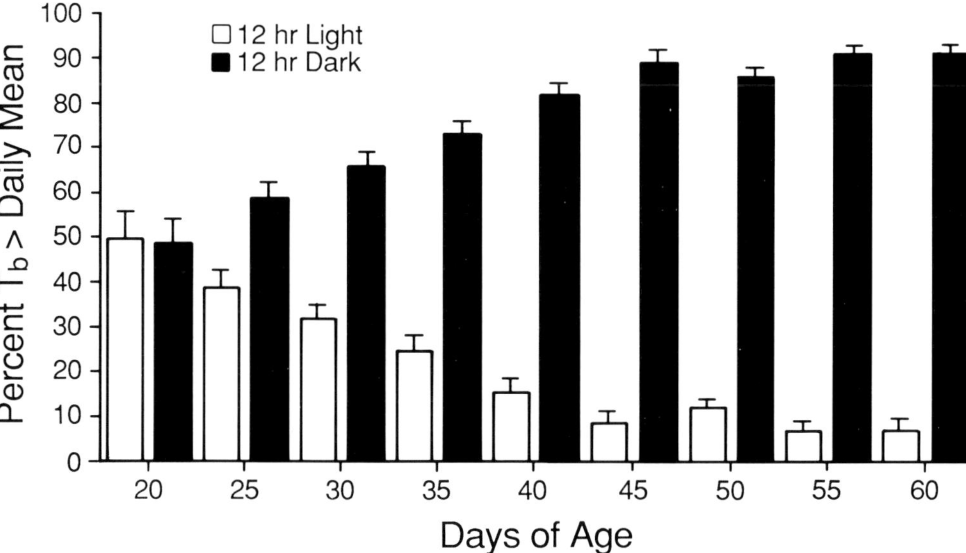

Figure 10.4. Mean percentages of body temperatures greater than the daily mean temperatures that occur in the light (open bar) and the dark (solid bar) period of a 12:12 LD cycle at 20–60 days of age in 5-day blocks. $n = 13$ or 14 up to Day 40 and 6–10 at later ages. (From "Development of the Circadian Rhythm of Body Temperature in Rats" by E. M. W. Kittrell and E. Satinoff, *Physiology and Behavior, 38,* p. 102.

days and attained a stable adult form, with over 90% of the body temperatures that were higher than the daily mean occurring at night, around Day 42 (Figure 10.4). The waveform continued to change slightly between 42 and 50 days of age for most pups.

Some recent preliminary work suggests that the body temperature rhythm may arise much earlier. Schmidt, Kaul, and Heldmaier (1986) reported that four isolated, artificially reared lean Zucker rat pups showed a diurnal temperature rhythm as early as 5 days of age, which had faded by the third postnatal week. If this phenomenon is replicated, and if it is generalizable to other strains, we would have the interesting situation of a diurnal rhythm that is present at birth and then disappears until sometime after weaning.

This prospect raises some fascinating questions about possible mechanisms. In adult rats, increased metabolic heat production due to general activity, feeding, grooming, and so forth, contributes greatly to the higher body temperature during the dark period (or, in animals living in constant dark, during the rat's "subjective night"). Adult rats may also have an endogenous rhythm of vasomotor tone, with an increased ability to conserve heat at night, just as do adult monkeys in the day (Fuller, Sulzman, & Moore-Ede, 1979). But what could the isolated rat pup be using to generate the rhythm? Behavioral responses and entrainment by the mother's rhythms are ruled out by the procedure. Pelage is inadequate. Heat-conservation (vasomotor) mechanisms are not functional, and obviously a rhythm cannot develop before the function that is rhythmic has developed. The only possibility left is a difference in day/night metabolic rates; and, in fact, Spiers (1988) found just this in rats 2–11 days old: their metabolic rate was higher at night. This difference had disappeared when the pups were 15 days old. The working out of this story may prove to be one of the most interesting aspects in the study of the ontogeny of thermoregulation.

What influences the development of infant rhythmicity? The most important biological clock in adults is the suprachiasmatic nucleus. Using labeled 2-deoxy-d-glucose, Reppert and Schwartz (1983) showed that there is a diurnal rhythm of glucose utilization in this nucleus even before birth, and its metabolic activity is sensitive to light immediately after birth (Fuchs & Moore, 1980). Thus, the neural mechanism for entrainment is in place very early on. In the nest, prior to weaning, the dam is the most obvious entraining agent. Levin and Stern (1975) reported that the dam exerts a strong influence upon suckling and feeding rhythms in her pups, and Deguchi (1975) suggested the same for the developing pineal NAT rhythm. Itoh, Hirota, and Katsura (1980) reported that the circadian periodicity of plasma corticosterone in neonatally blinded rats was similar to that of normal controls while they were in contact with their dams, but after weaning the rhythm of the blinded rats rapidly phase-shifted. The circadian adrenocortical rhythm in blinded pups is entrained by the nursing mother regardless of her genetic relationship to the pups (Yamazaki & Takahashi, 1983; Takahashi, Murakami, Hayafuji, & Sasaki, 1984). In summary, the major part of the brain responsible for rhythmicity is functional at birth or even before, although the rhythmicity may not manifest itself for some weeks after birth. When it does, it appears that the dam is responsible for the form it takes.

ONTOGENY OF FEVER

This chapter has concentrated on the necessity of taking thermoregulatory abilities into account when doing "nonthermoregulatory" experiments with developing organisms. This led to a discussion of the importance of the thermoneutral zone: Many behaviors that appear to mature with age are actually in place at birth or earlier than suspected if the pups are tested at nonstressful environmental temperatures. However, many questions remain unanswered about the ontogeny of thermoregulation itself. One of the most intriguing has to do with the development of fever.

Newborns of several species do not respond physiologically to injections of endotoxin[5] sufficient to cause fever in adults, and consequently they do not develop a fever. Yet newborns *do* shiver or increase heat production through brown fat thermogenesis in response to cold (guinea pigs: Blatteis, 1975; humans: Smith, Platou, & Good, 1956, Brück, 1961; lambs: Alexander, Thorburn, Nicol, & Bell, 1972; Kasting, Veale, & Cooper, 1979; rabbits, Satinoff, McEwen, & Williams, 1976; Figure 10.5). If infants can respond appropriately to cold from birth, and even before, why is there no reflexive response to endotoxin for several weeks?[6] We have here a situation in which infants respond both reflexively and behaviorally to cold stress, but only behaviorally to endotoxin injection. Either there are different thermoreceptors responding to the two stimuli, and they mature at different rates, or the behavioral threshold of response to a pyrogen is lower than the autonomic response. This is an issue that remains to be worked out.

Another puzzle in fever development involves species differences. As mentioned earlier, much larger doses of pyrogen will produce fever in guinea pigs and rabbits, but not in lambs. Lambs injected either 4 or 60 h after birth with *Salmonella* remained afebrile, regardless of how high the dose. However, if lambs that had previously been

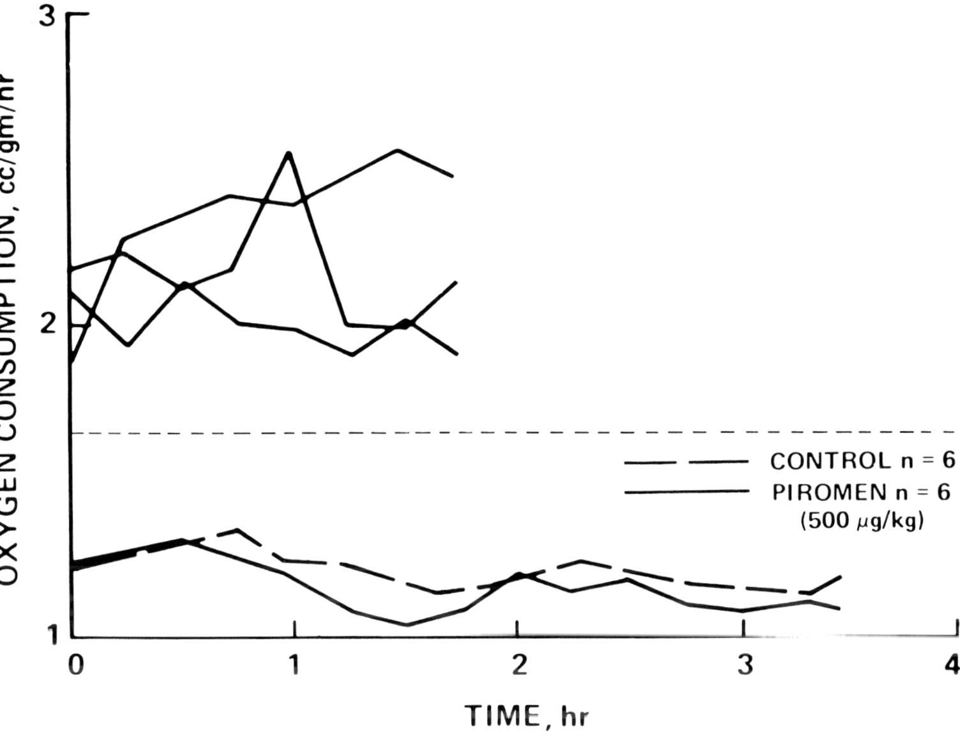

Figure 10.5. Oxygen consumption of 1-day-old rabbit pups. The three upper curves are individual animals at 22°C. The two lower curves are means of pups injected with a pyrogen or saline at 34°C. (From E. Satinoff, G. N. McEwen, and B. A. Williams, unpublished).

injected at 4 h were reinjected at 60 h, they did develop a fever (Veale, Cooper, & Pittman, 1978). If the second injection was given less than 40 h after the first, it usually did not produce fever. Thus, the febrile response seems to depend both on sensitization to the pyrogen and maturation of the lamb.

All of these results point to fascinating unsolved issues in fever and thermoregulation. For instance, infants may not develop fevers as easily as adults do because there may be substances (endogenous antipyretics) circulating in newborns that inhibit fever. Alternatively, the neural systems that release prostaglandins, the putative final common mediator of bacterial fever, may not be mature for some weeks. But since infants respond behaviorally to pyrogens, prostaglandins may not be involved in infant fever. Brain neurotransmitter systems involved in fever have hardly been examined in infants and are not well worked out in adults.

With respect to the last point, part of the confusion in assigning particular neurotransmitters to specific heat loss or heat production systems may have to do with the fact that animals tested as adults may have had different thermal experiences as infants. What is true for sexual behavior may be true for thermoregulation. Neural mechanisms of sexual behavior are sensitized by exposure to gonadal hormones during a critical period around birth and require these hormones later in life for their appropriate activation. Cooper, Ferguson, and Veale (1980) reported that adult rabbits raised from birth in a warm environment (33°C) showed significant drops in colonic temperature when exposed to cold. Control animals reared at 20°C had no trouble maintaining normal body temperatures. The warm-reared rabbits also had a reduced fever after endotoxin injection and a negligible rise in body temperature after intravenous noradrenaline infusion. These results were not seen in rabbits that had been placed in a warm room a few weeks before the tests were begun. Furthermore, hypothalamic injection of noradrenaline that produced hypothermia in rats that were warm acclimated as adults had the opposite effect in rats reared in a warm environment (Ferguson, Veale, & Cooper, 1981). Thus, exposure to appropriate thermal information early in life may sensitize neural mechanisms of thermoregulation.

METHODOLOGY

Body Temperature Measurement with Telemetry

The implantation method described below can be used when a rat pup is 10 days of age or older. Any earlier than that the transmitter is simply too big to fit into the peritoneal cavity (or even under the skin). The pup is anesthetized with an inhalable gas anesthetic and implanted with a biotelemetry thermistor weighing approximately 1.1 g (Model X-M, Mini-Mitter, Inc.). The transmitter is removed from its original plastic casing and is enclosed in a short section of ¼-in. heat-shrink tubing. It is then coated with a thin seal (Silastic Medical Adhesive; Dow Corning) to render it smooth and moisture resistant. Each transmitter is individually calibrated in a water bath, according to the manufacturer's recommendations, prior to implantation.

The transmitter is inserted into the peritoneal cavity through a 5–6 mm longitudinal incision on the ventral surface of the abdomen. It is important to displace this incision laterally to the midline, because we discovered in pilot work that dams tended to ignore sutures placed on pups in this region. Once inserted, the transmitter lies perpendicular to the midline in a very young pup, but will move about freely in an older animal. The peritoneal wall incision is closed with 4-0 chromic gut using a purse-string stitch. The skin incision is closed with 3–4 separate stitches using 6-0 chromic gut. One limitation of this technique is that because of the size of the coated transmitter it cannot be used on pups under about 20 g in weight, in whom it still comprises about 18% of total body weight.

Mini-Mitters emit beats that can be detected with any AM radio. Output is proportional to the surrounding temperature. Short-term measurement of body temperature can be made by using a stopwatch and handcounter to count beats or a frequency tuner to determine frequency. Long-term body temperature measurement can be accomplished with a computerized data collection system. The Mini-Mitter Company sells one called Dataquest, which can be used with Apple or IBM computers and can also monitor activity.[7] The signal is amplified and sent to a microcomputer where it is converted to a body temperature value using the signal rate and two predetermined calibration values. The battery life of the model X-M is about 6 weeks. A larger, longer-lasting transmitter (Model M) can then be inserted if more body temperature data are desired. Other model transmitters can also be used to measure brain or skin temperature of a pup. Body temperature can be determined for any preset interval and stored on disk for later analysis. The body temperature of only one pup per litter can be measured using the X-M, because these small transmitters produce frequencies that are broadband and can interfere with one another. The body temperature of a dam or the ambient temperature could be measured simultaneously with the body temperature of a pup by using a large frequency-tuned transmitter for the dam or surrounding area.

Rectal Temperature Measurement

Body temperatures can be measured in pups from birth, with either thermistors or thermocouples, after coating the tip with Vaseline. Thermistors are made of material whose resistance changes with its temperature. They can be bought in many sizes and shapes, already made to plug into temperature recorders (e.g., from Yellow Springs Instruments Co.), or they can be bought "raw" (e.g., from Victory Engineering Co.), the two wires emerging from the tip insulated by hand and plugged into a Wheatstone bridge. Thermocouples consist of two wires of dissimilar metals, such as platinum and iridium or iron and constantin, whose voltages vary differentially with temperature. They are cheaply and easily made by buying lengths of thermocouple wire that is already insulated. A short length of insulation is burned off at each end. For the measuring end, the two wires are wrapped tightly around each other, soldered together, and cut so that there is only a small, uninsulated tip. The other two ends of the wires are connected to a measuring device. This is a very easy technique, and if care is taken in smoothing the tip there are no problems. In previous work in rat pups we (Kleitman & Satinoff, 1982) used lengths of 36-gauge copper-constantin thermocouple wire to make probes that were inserted (depending on age) 0.5–1 cm into the anus. The thermocouples were connected to a Sentel digital thermometer. Spiers and Adair (1986) used smaller, 40-gauge thermocouples.[8]

There is no way of knowing, short of implanting a transmitter and simultaneously taking rectal temperatures, whether the handling involved in the rectal measurements are altering the pup's actual body temperature. This has, in fact, been done in adult rats. Figure 10.6 shows the results of taking rectal temperatures in rats that had previously been implanted with telemetry devices. Without the rectal measurements, the amplitude of the body temperature rhythm in unrestrained, unhandled rats was about 2°C. With insertion of a rectal probe, even at intervals as long as 4 h, the amplitude was reduced to about 1°C. With rectal measurements every hour, the rhythm is almost flat (Miles, 1962). Notice that it flattened at the *high* end of the rats' undisturbed body temperature rhythm, a sure sign of stress, since each exposure to insertion of a rectal probe raises rats' body temperature (Poole & Stephenson, 1977). Furthermore, the huge literature on early handling stress implies that rectal temperatures should probably not be taken on young pups if some other variable is to be measured later on and handling stress is not purposely included in the design.

Thermal Gradients

The following is an example of a thermally graded alleyway in use in my laboratory, but the main point is the principle. The alleyway consists of an aluminum floor and wall 75 cm long by 12 cm wide by 15 cm high with a hinged Plexiglas top. The floor extends 30.5 cm beyond the walls on each end and is wrapped with heating tape on one

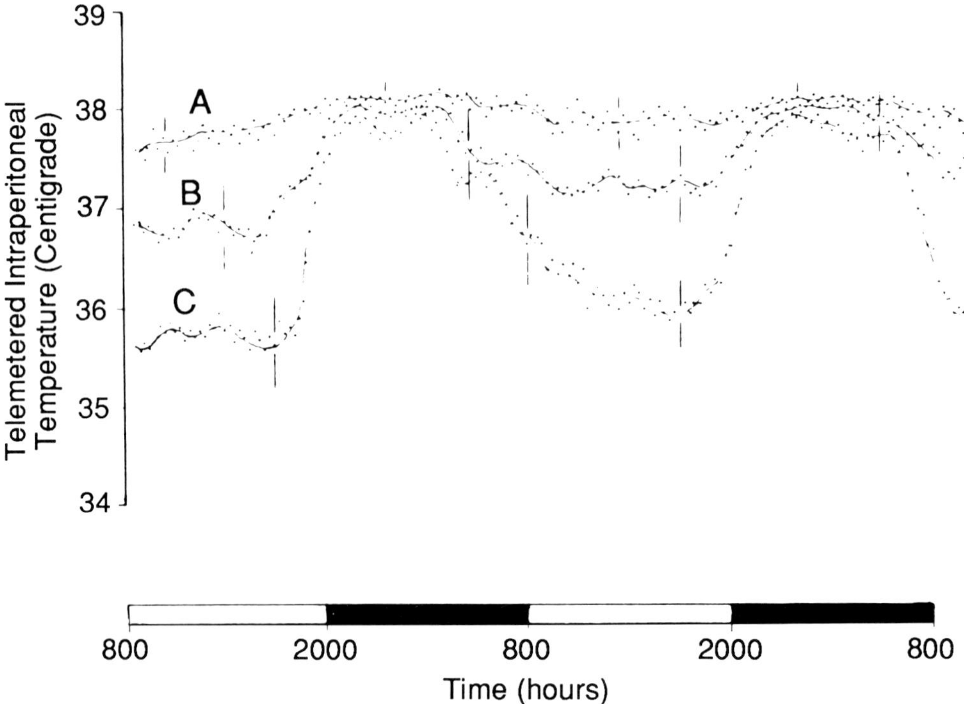

Figure 10.6. Effects of taking rectal temperature measurements on telemetered intraperitoneal temperatures of three groups of rats ($n = 4$/group). (A) Rectal measurement every hour. (B) Rectal measurement every 4 h . (C) Unhandled. Telemetered readings plotted every 15 min. Vertical lines are standard errors. (From "Telemetering Techniques for Periodicity Studies" by G. H. Miles, 1962. *Annals of the New York Academy of Sciences, 98*, p. 860.)

end and encased in ice packs on the other. The gradient temperatures vary from 45 to 17°C, although many ranges are possible by regulating the voltage to the heating tapes with step-down variable resistors (variacs). Floor temperatures are measured by thermocouples at the desired intervals, which should be at least a body length apart. The floor inside the walls is covered with one layer of drafting paper. Lines are drawn and numbered every 3 cm to record the positions of the pups in the alleyway (Figure 10.7). Of course, all this can be automated with photoelectric beams placed at intervals along the bottom of the sides of the gradient. After a pup has been in the gradient, it is necessary to remove any odor cues by wiping down the floor and walls with alcohol. Several questions involving the use of thermal gradients must be considered, because they will determine some of the experimental design. For instance, should the pups be placed in the gradient facing or away from the heat source? The most sensible initial position may be perpendicular to the gradient. At what floor temperature should the pups be placed? If they are positioned at 20°C, most rat pups will not move at all. Even at a floor temperature of 30°C, a pup may take 25–85 min to move 3 cm (mean 50 min), or it may not move at all. At a floor temperature of 45°C, all pups will move much more quickly—with a mean latency of 15 min and a narrower range of about 10–20 min. But this is very hot for a newborn, and some of them may heat up so fast that they will die.

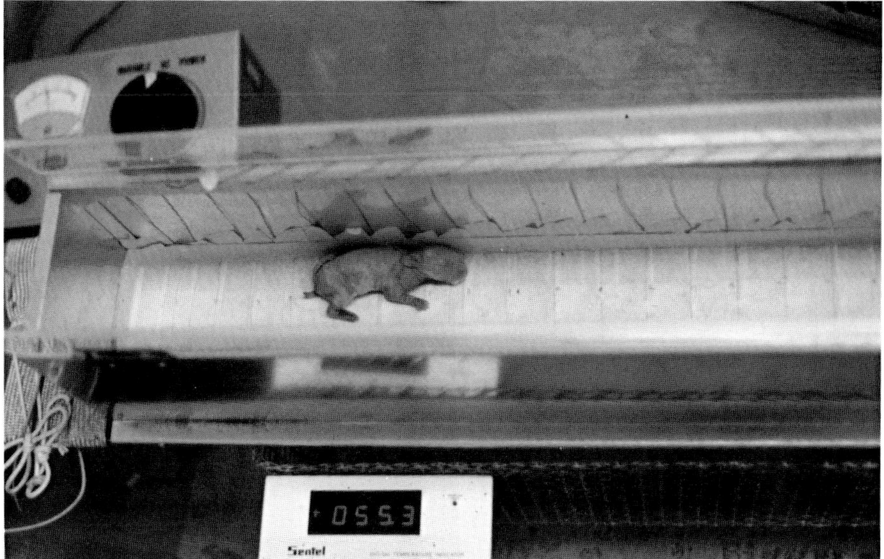

Figure 10.7. One-day-old rabbit pup in thermally graded alleyway.

Orientation, as opposed to movement, occurs much more rapidly. Rat pups placed at 30 or 20°C will position themselves toward, and pups placed at 45°C away from, the heat source as early as 5 min after being placed in the alleyway. Thus, if orientation is being studied, a trial may be as short as 5 min. If movement to a preferred floor temperature is the variable under investigation, since newborn rat pups move so slowly, a trial may have to last several hours.

If older pups are studied, other problems arise, such as what the dependent variable is to be. All 6-day-old pups placed at 20°C moved at least 10 cm in 2 hours, but this raised their body temperatures only to about 26°C. They did not move far enough to attain a final body temperature of 37°C until they were 12 days old. Latency to move away from 45°C was slowest in 12–17-day-old pups because other behaviors, like circling, interfered with progression along the alleyway (Kleitman & Satinoff, 1982). And, of course, there are the perennial problems of whether to run the pups in the day or the night and after or before nursing. Again, these decisions depend upon the nature of the experiment.

UNANSWERED QUESTIONS

How do we select which facts to study? The French mathematician Henri Poincaré (1914) wrote:

> the more general a law is, the greater is its value. This shows us how our selection should be made. The most interesting facts are those which can be used several times, those which have a chance of recurring. We have been fortunate enough to be born in a world where there are such facts.

Since thermoregulation is so important for understanding the ontogeny of almost all developing systems, the study of the maturation of its circadian rhythmicity should be important for understanding many other aspects of ontogeny. In this respect, many fruitful facts are waiting to be discovered, and in this final section I pose some questions that may be used to find them.

How do nonrhythmic functions become rhythmic? Do some rhythms appear earlier than others? Is there a necessary sequence of cyclicity? How do individual rhythms become synchronized with each other during ontogeny? If one could prevent, say, the diurnal fluctuations in body temperature, would that retard the development of other rhythmic functions? Just as exposure to hormones and thermal environments early in life affect sexual differentiation and ability to thermoregulate later on, would early exposure to aberrant light cycles affect later entrainment and circadian rhythmicity? How do the suprachiasmatic nuclei, which show circadian oscillations in rats at birth, synchronize rhythms such as plasma corticosterone, pineal N-acetyltransferase, and locomotor activity, which do not appear until some weeks after birth? Assuming that the mother's circadian system entrains (at least) some of the rhythms of the pups during gestation, what are the mediators that transfer the phase of the maternal circadian system to the pups' internal system? If body temperature rhythms are not entirely controlled by the suprachiasmatic nucleus (Satinoff & Prosser, 1988), what is the mechanism by which the temperature oscillator becomes synchronized with the rhythms that nucleus *does* supervise? These questions take only a few minutes to ask, but years to answer. Nevertheless, I believe they must be answered if we are to unravel the one great problem of how the various infantile behavioral and physiological responses become integrated into the mature adult pattern.

NOTES

1. They may take 30 min to start moving, whereas newborn hamsters may traverse 35 cm in under a minute to reach a warm area (Leonard, 1974), thus emphasizing the importance of species differences in assessing infant capabilities.

2. This means that when adult rats are housed at 21–22°C they are mildly cold-stressed during lights-on, which is when measures of thermoneutrality have been taken. However, 21°C may be thermoneutral for rats at night when they are active and producing more heat. No one has ever measured this, but it is a reasonable assumption, since thresholds for evaporative heat loss are lower, and for increases in metabolic rate higher, at night.

3. A body loses heat in proportion to its surface area. The higher the ratio of body surface to body mass, the more heat is lost. Small mammals in general, then, will lose more heat than large ones, and the newborns of small mammals are at an even greater disadvantage.

4. One paper reports that huddling elevated oxygen consumption in 3–4 day-old mice when compared with same-age isolated mice (Bryant & Hails, 1975). The authors attributed this to increased activity.

5. Many agents cause fever, including viruses, fungi, antigens, and bacteria. Endotoxin, also called lipopolysaccharide or bacterial pyrogen, is a constituent of the cell walls of gram-negative bacteria. They are often extracted from cultures of various forms of, for example, *Salmonella* or *E. coli* (see Nowotny, 1969, for discussion) and used to induce experimental fever. Bacterial pyrogens are not inactivated unless they are heated above 150°C for 2 hours. This means that normal autoclaving will not destroy them, and is why most papers in the thermoregulatory literature that involve injecting drugs into the subjects will state that the drug was dissolved in sterile, pyrogen-free saline.

6. If very high doses of endotoxin are administered, then some of these species, like guinea pigs and rabbits, will develop fevers reflexively, but, at least for rabbit pups, the fevers are monophasic, whereas in adult rabbits there are two febrile peaks (Székely, 1978).

7. The principle is elegant. If an antenna is wrapped around the animal's cage, the signal strength of the telemetry device will vary with its distance from the antenna. Since the telemetry unit is inside the animal whenever the animal moves relative to the antenna, signal strength changes. Thus, beat frequency measures temperature, and beat amplitude measures movement.

8. As with everything in life, each method has its advantages and disadvantages. Sufficiently miniature thermistors bought ready to use are expensive. Insulating them in one's laboratory is tedious; generally lead wires have to be attached, and they break easily. Furthermore, each one has to be calibrated individually in a water bath over the range of temperatures expected to be seen. On the other hand, any Wheatstone bridge will give an accurate measurement. Also, since the principal of temperature measurement is a resistance change, they can be connected through swivel joints if they are implanted for long-term monitoring.

Insulated thermocouple wire is cheap, is bought in long lengths, and individual thermocouples can be made in a minute. Since the thermocouples are all made from the same spool of wire, only one calibration is required (assuming that the lengths do not exceed 10 or 20 feet). On the other hand, since temperature is converted from a particular voltage difference, it is imperative that no other metal be in the path from the tip to the measuring device. This means that if one wants to have a swivel joint, it has to be made from the same metals as the thermocouple. Because the voltage difference is so small, it also means that there is no inexpensive way to measure it.

REFERENCES

Adolph, E. F. (1957). Ontogeny of physiological regulations in the rat. *Quarterly Review of Biology*, *32*, 89–137.

Alberts, J. R. (1978). Huddling by rat pups: Group behavioral mechanisms of temperature regulation and energy consumption. *Journal of Comparative and Physiological Psychology*, *92*, 231–235.

Alberts, J. R. (1981). Ontogeny of olfaction: Reciprocal roles of sensation and behavior in the development of perception. In R. M. Aslin, J. R. Alberts, & M. R. Peterson (Eds.), *Development of perception: Psychobiological perspectives* (Vol. 1, pp. 321–357). New York: Academic Press.

Alberts, J. R., & Brunjes, P. C. (1978). Ontogeny of thermal and olfactory determinants of huddling in the rat. *Journal of Comparative and Physiological and Psychology*, *92*, 897–906.

Alexander, G., Thorburn, G., Nicol, D., & Bell, A. W. (1972). Survival, growth and metabolic response to cold in prematurely delivered lambs. *Biology of the Neonate*, *20*, 1–8.

Almli, C. R., & Fisher, R. S. (1977). Infant rats: Sensorimotor, ontogeny and effects of substantia nigra destruction. *Brain Research Bulletin*, *2*, 425–459.

Altman, J., & Sudarshan, K. (1975). Postnatal development of locomotion in the laboratory rat. *Animal Behaviour*, *23*, 896–920.

Baccino, M. (1935). La température optimum de croissance des jeunes homéothermes. Diverses méthodes de détermination. *Comptes rendus de la Société de Biologie*, *119*, 1246–1248.

Benedict, F. S., & MacLeod, G. (1928). The heat production of the albino rat. II. Influence of environmental temperature, age and sex: Comparison with the basal metabolism of man. *Journal of Nutrition*, *1*, 367–398.

Beutow, K. C., & Klein, S. W. (1964). Effect of maintenance of "normal" skin temperature on survival of infants of low birth weight. (1964). *Pediatrics*, *34*, 163–170.

Blatteis, C. M. (1975). Postnatal development of pyrogenic sensitivity in guinea pigs. *Journal of Applied Physiology*, *39*, 251–257.

Brück, K. (1961). Temperature regulation in the newborn infant. *Biology of the Neonate*, *3*, 65–119.

Bryant, D. M., & Hails, C. J. (1975). Mechanisms of heat conservation in the litters of mice (*Mus musculus L.*). *Comparative Biochemistry and Physiology*, *50A*, 99–104.

Conklin, P., & Heggeness, F. W. (1971). Maturation of temperature homeostasis in the rat. *American Journal of Physiology*, *220*, 333–336.

Cooper, K. E., Ferguson, A. V., & Veale, W. L. (1980). Modification of thermoregulatory responses in rabbits reared at elevated environmental temperatures. *Journal of Physiology*, *303*, 165–172.

Dawson, T. J. (1973). Primitive mammals. In G. C. Whittow (Ed.), *Comparative physiology of thermoregulation* (Vol. III, pp. 1–46), New York: Academic Press.

Deguchi, T. (1975). Ontogenesis of a biological clock for serotonin: Acetylcoenzyme A N-acetyl-transferase in pineal gland of rat. *Proceedings of the National Academy of Sciences, 72,* 2814–2818.

Eedy, J. W., & Ogilvie, D. M. (1970). The effect of age on the thermal preference of white mice (*Mus musculus*) and gerbils (*Meriones unguiculatus*). *Canadian Journal of Zoology, 48,* 1303–1306.

Ferguson, A. V., Veale, W. L., & Cooper, K. E. (1981). Evidence of environmental influences on the development of thermoregulation in the rat. *Canadian Journal of Physiology and Pharmacology, 59,* 91–95.

Fowler, S. J., & Kellogg, C. (1975). Ontogeny of thermoregulatory mechanisms in the rat. *Journal of Comparative and Physiological Psychology, 89,* 738–746.

Fuchs, J. L., & Moore, R. Y. (1980). Development of circadian rhythmicity and light responsiveness in the rat suprachiasmatic nucleus: A study using the 2-deoxy [1-^{14}C] glucose method. *Proceedings of the National Academy of Sciences, 77,* 1204–1208.

Fuller, C. A., Sulzman, F. M., & Moore-Ede, M. C. (1979). Circadian control of thermoregulation in the squirrel moneky, *Saimiri sciureus*. *American Journal of Physiology, 236,* R153–R161.

Guenaire, C., Costa, J. C., & Delacour, J. (1982). Conditionnement opérant avec renforcement thermique chez le rat nouveau-né. *Physiology and Behavior, 29,* 419–424.

Hahn, P. (1956). The development of thermoregulation. III. The significance of fur in the development of thermoregulation in rats. *Physiologica Bohemoslavica, 5,* 428–431.

Hall, W. G. (1979). Feeding and behavioral activation in infant rats. *Science, 205,* 206–209.

Harlow, H. (1971). *Learning to love.* San Francisco: Albion.

Hey, E. N. (1974). Physiological control over body temperature. In J. L. Monteith & L. E. Mount (Eds.), *Heat loss from animals and man* (pp. 77–95). London: Butterworth.

Horwitz, J., Heller, A., & Hoffmann, P. C. (1982). The effect of development of thermoregulatory function on the biochemical assessment of the ontogeny of neonatal dopaminergic neuronal activity. *Brain Research, 235,* 245–252.

Hsieh, A. C. L., Emery, N., & Carlson, L. D. (1971). Calorigenic effect of norepinephrine in newborn rats. *American Journal of Physiology, 221,* 1568–1571.

Hull, D. (1973). Thermoregulation in young mammals. In G. C. Whittow (Ed.), *Comparative physiology of thermoregulation* (Vol. 3, pp. 167–200). New York: Academic Press.

Itoh, S., Hirota, R., & Katsura, K. (1980). The rate of phase shift of plasma corticosterone circadian rhythm during early developmental stages in neonatally blinded rats. *Japanese Journal of Physiology, 30,* 41–48.

Jeddi, E. (1970). Confort du contact et thermorégulation comportementale. *Physiology and Behavior, 5,* 1487–1493.

Johanson, I. B. (1979). Thermotaxis in neonatal rat pups. *Physiology and Behavior, 23,* 871–874.

Johanson, I. B., & Hall, W. G. (1979). Appetitive learning in 1-day-old rat pups. *Science, 205,* 419–421.

Kasting, N. W., Veale, W. L., & Cooper, K. E. (1979). Development of fever in the newborn lamb. *American Journal of Physiology, 236,* R184–R187.

Kittrell, E. M. W., & Satinoff, E. (1986). Development of the circadian rhythm of body temperature in rats. *Physiology and Behavior, 38,* 99–104.

Kleitman, N., & Satinoff, E. (1980). Fever in normal and maternally neglected newborn rabbits. In J. Lipton (Ed.), *Fever* (pp. 197–205). New York: Raven Press.

Kleitman, N., & Satinoff, E. (1982). Thermoregulatory behavior in rat pups from birth to weaning. *Physiology and Behavior, 29,* 537–541.

Leon, M. (1986). Development of thermoregulation. In E. M. Blass (Ed.), *Handbook of behavioral neurobiology* (Vol. 8, pp. 297–322). New York: Plenum.

Leonard, C. M. (1974). Thermotaxis in golden hamster pups. *Journal of Comparative and Physiological Psychology, 86,* 458–469.

Levin, R., & Levine, S. G. (1975). Development of circadian periodicity in base and stress levels of corticosterone. *American Journal of Physiology, 229,* 1397–1399.

Levin, R., & Stern, J. (1975). Maternal influences on ontogeny of suckling and feeding rhythms in the rat. *Journal of Comparative Physiology and Psychology, 89,* 711–721.

Miles, G. H. (1962). Telemetering techniques for periodicity studies. *Annals of the New York Academy of Sciences, 98,* 858–865.

Misantone, L. J., Ellis, S., & Epstein, A. N. (1980). Development of angiotensin-induced drinking in the rat. *Brain Research, 186,* 195–202.

Moore, R. E., & Simmonds, M. A. (1966). Decline with age in the thermogenic response of the young rat to l-noradrenaline. *Federation Proceedings, 25*, 1329–1331.

Moran, T. H., Lew, M. F., & Blass, E. M. (1981). Intracranial self-stimulation in 3-day-old rat pups. *Science, 214*, 1366–1368.

Mount, L. E. (1963). Environmental temperature preferred by the young pig. *Nature, 199*, 1212–1213.

Mount, L. E. (1974). The concept of thermal neutrality. In J. L. Monteith & L. E. Mount (Eds.) *Heat loss from animals and man* (pp. 425–439). London: Butterworth.

Nowotny, A. (1969). Molecular aspects of endotoxic reactions. *Bacteriology Review, 33*, 72–98.

Ogilvie, D. M., & Stinson, R. H. (1966). The effect of age on temperature selection by young mice (*Mus musculus*). *Canadian Journal of Zoology, 44*, 511–517.

Olmstead, C. E., Villablanca, J. R., Torbiner, M., & Rhodes, D. (1979). Development of thermoregulation in the kitten. *Physiology and Behavior, 23*, 4849–4895.

Pederson, P. E., Williams, C. L., & Blass, E. M. (1982). Activation and odor conditioning of suckling behavior in 3-day-old albino rats. *Journal of Experimental Psychology: Animal Behavior Processes, 8*, 329–341.

Poczopko, P. (1961). A contribution to the studies on changes of energy metabolism during postnatal development. I. Development of mechanisms of body temperature regulation in rats. *Journal of Cellular and Comparative Physiology, 57*, 175–184.

Poincaré, H. (1914). *Science and method.* New York: Dover (reprinted 1952).

Poole, S., & Stephenson, J. D. (1977). Core temperature: Some shortcomings of rectal temperature measurements. *Physiology and Behavior, 18*, 203–205.

Reppert, S. M., & Schwartz, W. J. (1983). Maternal coordination of the fetal biological clock in utero. *Science, 220*, 969–971.

Roberts, W. W., Mooney, R. D., & Martin, J. R. (1974). Thermoregulatory behaviors of laboratory rodents. *Journal of Comparative and Physiological Psychology, 86*, 693 699.

Rudy, J. W., & Cheatle, M. D. (1977). Odor aversion learning by neonatal rats. *Science, 198*, 845–846.

Satinoff, E., McEwen, G. N., & Williams, B. A. (1976). Behavioral fever in newborn rabbits. *Science, 193*, 1139–1140.

Satinoff, E., & Prosser, R. A. (1988). Suprachiasmatic nuclear lesions eliminate circadian rhythms of drinking and activity, but not of body temperature, in male rats. *Journal of Biological Rhythms, 3*, 1–22.

Schmidt, I., Kaul, R., & Heldmaier, G. (1986). Thermoregulation and diurnal rhythms in 1-week-old rat pups. *Canadian Journal of Physiology and Pharmacology, 65*, 1355–1364.

Smith, R. T., Platou, E. S., & Good, R. A. (1956). Septicemia of the newborn. *Pediatrics, 176*, 549–575.

Spiers, D. E. (1988). Nocturnal shifts in thermal and metabolic responses of the immature rat. *Journal of Applied Physiology, 64*, 2119–2124.

Spiers, D. E., & Adair, E. R. (1986). Ontogeny of homeothermy in the immature rat: Metabolic and thermal responses. *Journal of Applied Physiology, 60*, 1190–1197.

Stanier, M. W. (1975). Effect of body weight, ambient temperature and huddling on oxygen consumption and body temperature of young mice. *Comparative Biochemistry and Physiology, 51A*, 79–82.

Stone, E. A., Bonnet, K. A., & Hofer, M. A. (1976). Survival and development of maternally deprived rats: Role of body temperature. *Psychosomatic Medicine, 38*, 242–249.

Sugano, Y., & Nagasaka, T. (1979). Effect of huddling on heat losses in infant dogs. *Journal of the Physiological Society of Japan, 41*, 145–147.

Szechtman, H., & Hall, W. G. (1980). Ontogeny of oral behavior induced by tail pinch and electrical stimulation of the tail in rats. *Journal of Comparative and Physiological Psychology, 94*, 436–445.

Székely, M. (1978). Biphasic endotoxin fever in the newborn rabbit. *Acta Physiologica Academeiae Scientiarum Hungaricae, 51*, 389–392.

Szymusiak, R., & Satinoff, E. (1981). Maximal oxygen consumption defines a narrower thermoneutral zone than does minimal metabolic rate. *Physiology and Behavior, 26*, 687–690.

Takahashi, K., Murakami, N., Hayafuji, C., & Sasaki, Y. (1984). Further evidence that circadian rhythm of blinded rat pups is entrained by the nursing dam. *American Journal of Physiology, 246*, R359–R363.

Taylor, P. M. (1960). Oxygen consumption in new-born rats. *Journal of Physiology (London), 154*, 153–168.

Veale, W. L., Cooper, K. E., & Pittman, Q. J. (1978). Prostaglandin action on central thermoregulation and fever: Ontogenetic aspects. In F. Coceani & P. M. Olley (Eds.), *Advances in prostaglandin and thromboxane research* (Vol. 4, pp. 199–213). New York: Raven Press.

Williams, C. L. (1987). Estradiol benzoate facilitates lordosis and ear wiggling of 4- to 6-day-old rats. *Behavioral Neuroscience, 101*, 718–723.

Yamazaki, J., & Takahashi, K. (1983). Effects of change of mothers and lighting conditions on the development of the circadian adrenocortical rhythm in blinded rat pups. *Psychoneuroendocrinology, 8*, 237–244.

11

The Study of Early Feeding and Drinking Behaviors

C. B. PHIFER

STATE OF THE FIELD

Questions, Background, and Limitations of Available Methodology

Intensive research efforts, dating from the beginning of this century (e.g., Cannon & Washburn, 1912), have taught us a great deal about the control of ingestive behaviors in both adult and immature mammals. Many important questions related to the ontogeny of ingestive behavior have been addressed, and others are still being studied. For example, do newborn mammals have the ability to control their intake but simply do not express this ability until weaning, or do control mechanisms develop postnatally? If, as studies have indicated, the controls develop postnatally, what ontogenetic changes are responsible? Are the changes in controls the result of maturation of sensory systems for detection of nutrients or metabolites, maturation of new CNS structures or their connections, or simply changes in motor capabilities? And how vulnerable is the developing animal to perturbation? Can small changes in the preweanling's dietary constituents, quantity of nutrients, or pattern of delivery significantly affect the control of ingestion in the adult, and therefore be related to the development of pathological feeding disorders (e.g., obesity)?

In most of the attempts aimed at answering these and other questions about the ontogeny of ingestive behavior, mammals have been the preferred subjects. However, since infant mammals satisfy the majority of their nutritional and hydrational requirements by suckling and were considered incapable of any other form of ingestion, many of the earlier studies in the area of feeding behavior development were conducted on suckling animals (e.g., Drewett & Cordall, 1976; Friedman, 1975; Hall, Cramer, & Blass, 1977; Houpt & Epstein, 1973; Lytle, Moorecroft, & Campbell, 1971). Although our understanding of controls for early ingestive behavior grew as a result of these investigations, such studies were plagued by two important limitations that were associated with studying ingestion within the confines of the nest and maternally derived nutrition. First, suckling animals have less control over their own intake than they would in a free-feeding situation (Hall & Rosenblatt, 1977). Instead, intake is largely dependent on the presence of the mother and her supply of milk. Second, and more important, the range of experimental manipulations that can be utilized is severely restricted because of problems associated with maternal care (e.g., maternal grooming of infants disturbs surgical implants and stitches) and the nesting condition (e.g., problems with separating deprivational components). These and other difficulties associated with studying suckling as a model for ingestive behavior development led investigators to seek another form of ingestion in infant mammals. (See Brake, chap. 3, this volume, for recent progress in overcoming these difficulties.)

Then in the late 1970s infant rats where shown to be capable of ingestion completely independent of suckling and of the rat dam. Wirth and Epstein's (1976) experiments showed that rat pups can drink water when held to a flowing spout, and subsequently Hall (1979) demonstrated that pups can ingest liquid diet that is infused into their mouths. As a result of these studies, we now know that two forms of ingestion exist concurrently in rat pups. One of them, suckling, is the naturally expressed form of ingestion seen in the nest, whether in the laboratory or in the wild. The other, which has been called *independent ingestion*, is normally latent but can be elicited in rat pups simply by placing them in the appropriate environment (warm ambience) and offering them food in specific fashions.

The discovery of independent ingestion in rat pups has allowed investigators to overcome many of the limitations that plagued studies of ingestion in suckling animals. First, investigators were, for the first time, able to observe infant rats that were obtaining nourishment from a source other than the rat dam. Thus, pups were able to reveal their own system of controls, and these controls became amenable to study. Second, independent ingestion allowed investigators more freedom to manipulate the feeding experience. The types of manipulations that have been applied in feeding studies with infant rats include variations in the types of diet (e.g., taste or nutrient content), the delivery or availability of diet (e.g., continuous vs. intermittent, oral vs. gastric), and surgical procedures that restrict diet to specific regions of the digestive tract (e.g., sham feeding with gastric fistula, pyloric noose). These and other manipulations have been instrumental in studies designed to unravel the development of ingestive behavior controls. This chapter concentrates on descriptions of the techniques used in studying independent ingestion by rat pups. (For more information on studies of suckling behavior, nutrition during infancy, and the role of learning in the development of food intake control, see Brake, chap. 3, Diaz, chap 16, and Galef, chap. 7, this volume.)

DEVELOPMENT OF NEW METHODOLOGIES

Many investigators studying the control of feeding behaviors in adult animals have facilitated their analysis by imagining the animal as divided into compartments (e.g., Smith & Gibbs, 1979). Thus, they have started with the mouth and proceeded through the gastrointestinal tract to the stomach, the intestines, and postabsorptive sites. At the same time, they have progressed from the periphery, which includes sensory stimulation by food both externally (e.g., smell and sight) and internally (e.g., gastrointestinal chemoreceptors), to the level of the CNS. This division of intake-control sites has led to a better understanding of ingestion as a behavior with multiple controls and to a clearer understanding that ingestion is not a single behavior but rather a group of component behaviors with both separate and overlapping controls (Zeigler, 1976).

With a similar progression, the ensuing discussion of methodologies used in the study of independent ingestion by rat pups, after a brief discussion of deprivation techniques, proceeds from oral manipulations to gastric procedures to postgastric and postabsorptive loci of control. Space limitations prohibit a discussion of CNS involvement in ingestion.

General Procedures for Studying Independent Ingestion

Deprivation

Prior to being used in any tests of ingestive behavior, rat pups are deprived by removing them from the dam and placing them in a warm, humid incubator (Isolette, Air Shields, Inc.; surplus incubators can often be obtained at hospitals). Incubators are maintained at 70–90% RH, 33 + 0.5 °C for neonates to 9 days of age and 32°C for pups to 15 days; pups older than 15 days do best when kept in a tub cage on top of the incubator. As in adults, increased deprivation of pups results in increased intake (Hall, 1979), but 24 h is the maximum deprivation commonly used. Pups may be deprived with or without littermates; either condition leads to avid ingestion, but the former slightly reduces general activity during the ingestion tests.

Deprivation using a foster dam (i.e., females whose own pups have been removed; Rosenblatt & Siegal, 1981) in place of the natural mother provides important advantages in studies of ingestive behavior. These dams exhibit relatively normal maternal behavior (retrieving, licking, and grooming pups) and even allow nipple attachment, but do not lactate. Pups given oral infusions (see below) after being deprived with foster dams show avid ingestion, but do not exhibit the greatly enhanced activity characteristic of pups deprived without a dam (Bornstein, Terry, Browde, Assimon, & Hall, 1987). Although this enhanced activity is interesting for many reasons (e.g., it may be an externalized reflection of internal reward; Hall, 1987), it is not always desirable in studies of ingestion. Indeed, recent studies have indicated that less active pups are better at attending to a localized food source than more active pups (Phifer, Denzinger, & Hall, unpublished). Also, particular experimental designs may demand separation of the effects of maternal and nutritional deprivation.

In practice, experienced dams (i.e., mothers that have reared at least one litter) are commonly used as foster dams starting at 7–10 days after removal of their own litter and for 4–6 weeks thereafter. If the foster dams are used at least once every 2 weeks and no more than 3 days per week, they continue to exhibit maternal care, but lactation is not induced.

Oral Cannulas

In most independent ingestion tests a liquid diet is delivered by oral infusion; therefore, oral cannulas are installed in pups at least 60 min before testing. The cannula can be implanted in either the anterior or posterior oral cavity (Figure 11.1A and 11.1B), depending on the needs of the experiment (Hall, 1979). The anterior placement, which allows the infusion to enter the front of the mouth just behind the lower

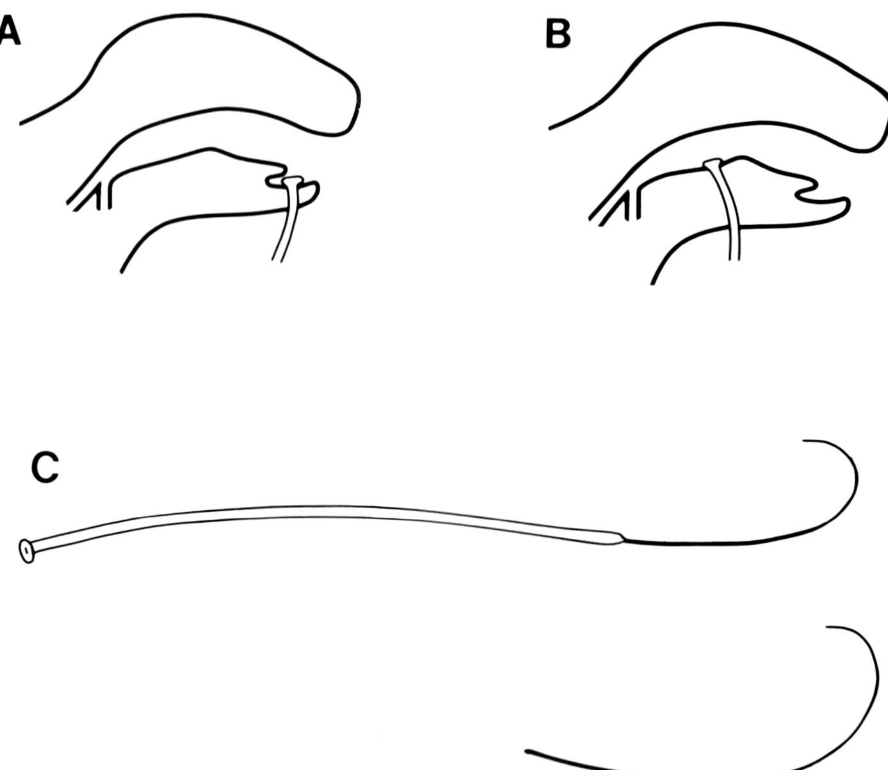

Figure 11.1. Sagittal sections of rat pup heads showing the anterior (A) and posterior (B) placements of the oral cannula used to provide infusions of liquid diet into the mouth. (C) Oral cannula and stainless steel wires used to install oral cannula in the anterior (upper wire) or posterior (lower wire) placements.

incisors, permits the pup to either accept the diet by moving it to the back of the mouth for swallowing or reject the diet by allowing it to spill out the front of the mouth. This placement is therefore useful in studies on factors that might affect initiation or termination of intake.

Alternatively, the posterior placement allows the infusion to enter the back of the mouth where it is reflexively swallowed; this placement is useful when the experimenter wishes to ensure uniform ingestion across experimental conditions. The posterior cannula placement also provides a useful control for assessing nonspecific or debilitative effects resulting from experimental treatments. Thus, in experiments on the effects of the satiety peptides cholecystokinin and bombesin, the latter placement was used as a control treatment to ensure that swallowing reflexes were not adversely affected by the peptides (Phifer & Hall, 1984).

The same cannula is used for either anterior or posterior placements, but the installation procedures differ. The cannula (Fig. 11.1C) is fashioned from a 10-cm length of PE-10 (Clay-Adams) tubing by heating one end of the tubing over a flame and pressing it against a flat surface to form a flange approximately 2 mm in diameter. A stainless steel wire (8–10 cm long, 0.28 mm in diameter; Small Parts, Inc.), shaped into the appropriate curve (Figure 11.1C) is used to install cannulas.

Pups are anesthetized before the cannula installation procedure. A 3–4 min exposure to methoxyflurane inhalation (Metofane, Pitman-Moore, Inc.) induces several minutes of anesthesia and does not appear to reduce ingestion or other activity when pups are tested 60 min later. Alternatively, a 15-s exposure to CO_2 (from vaporization of dry ice) induces approximately 60 s of anesthesia, which allows sufficient time to install a cannula. Prior to installing an anterior cannula, the nonflanged end of the PE tubing is fitted onto the noncurved end of the installation wire. Then, while the rat pup is grasped firmly but gently, the point of the wire is placed into the pup's mouth and pressed through the soft tissue just behind the symphysis of the mandibles. The wire should exit through the skin just posterior to the submental grouping of vibrissae. Lastly, with gentle traction on the wire, the cannula is pulled through the lower jaw until the flange rests against the floor of the mouth.

To install a posterior cannula, one starts with a bare wire, that is, without the tubing attached (Figure 11.1C). The point of the wire closest to the curve is pressed through the skin under the chin, and, as with the anterior placement, the wire should penetrate the skin just posterior to the submental grouping of vibrissae. The point of the wire is directed toward the top of the pup's head and toward a point about midway between the eyes and the pinnae; then it is advanced until its point just exits the top surface of the tongue. Next, the point is redirected forward and pushed until 2–3 cm of the wire extends from the open mouth. The nonflanged end of the cannula is attached to the projecting wire, and the wire is pulled until the cannula flange seats on the upper tongue surface. With proper procedure, the cannula opening should be just posterior to the intermolar eminence, a small bump on the dorsal surface of the tongue.

Ingestive Behavior Tests

In feeding tests with adult animals, intake is usually determined from weight or volume changes for a known amount of diet. But because rat pups receiving oral infusions generally spill some of the diet on the floor, the most reliable measure of intake is weight gain during ingestion tests. Also, because rat pups rarely exhibit spontaneous urination or defecation until close to weaning, weight change during an ingestion test represents intake alone. However, to ensure the reliability of this measure, the bladder and rectum of pups must be voided before testing by briskly stroking their anogenital region with a moist, soft, artist's brush. In addition, applying a drop of fast-drying acrylic cement (Superglue) to the anogenital region is useful for extended ingestion tests if pups are to be sacrificed after the experiments.

Rat pups exhibit independent ingestion only when they are in a warm environment (Johanson & Hall, 1980). Thus, ingestion tests are conducted in a special testing incubator, made by attaching a hinged Plexiglas cover to a 57-l glass aquarium (Figure 11.2). This incubator is warmed and humidified by a small electric fan (Pamotor; Model 8500C has a capacity of 27 cfm comparable to Dayton 4C596 and Rotron Suzei) that directs a flow of air over an adjustable 100-watt aquarium heater and a tray of water (Hall, 1979). This inexpensive testing incubator maintains a very stable temperature (32.5 ± 0.5°C for pups to 9 days of age and 1–2°C cooler for older pups) and constant humidity (75 ± 5% RH). Pups are separated from each other in the test

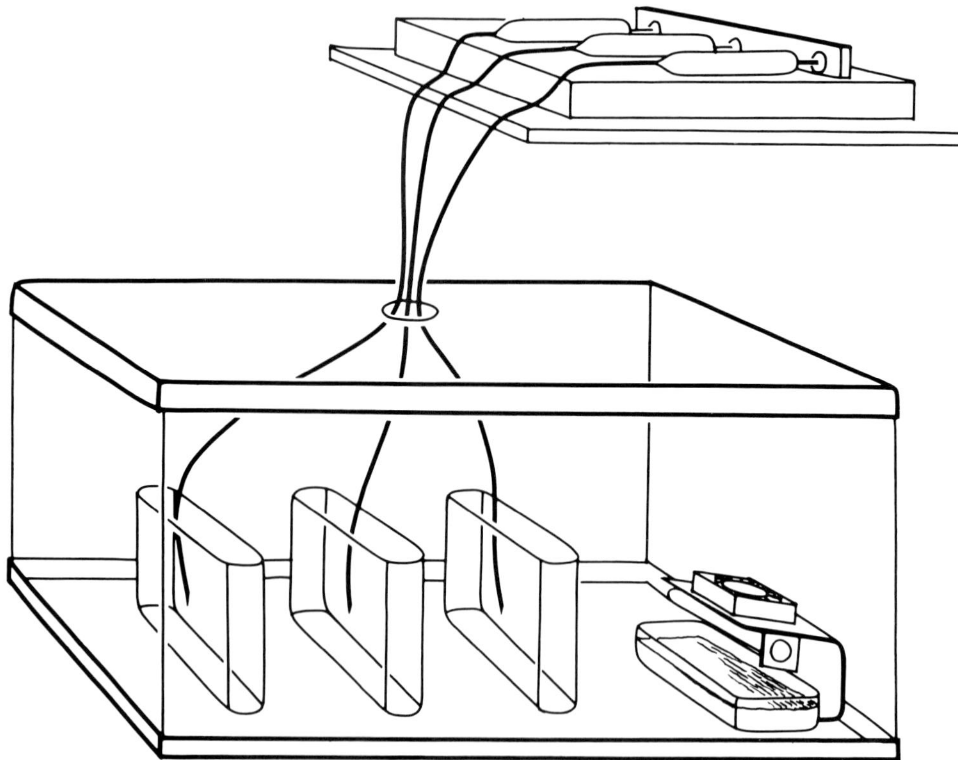

Figure 11.2. The testing incubator used to maintain a warm, humid environment for pups during independent ingestion tests. Transparent plastic containers for holding pups and the temperature/humidity maintenance device are shown inside the incubator; an infusion pump with syringes for delivery of diet is shown above the incubator.

incubator by placing them in clear plastic food containers (7 by 12 by 17.5 cm high) available from kitchen supply retailers.

To start an oral-infusion ingestion test, pups' cannulas are attached to flexible leads (Microrenathane Type MRE 040 tubing, Braintree Scientific, Inc.) that are connected to syringes with 23-gauge needles. The syringes are filled with liquid diet prewarmed to approximately body temperature and then driven by an infusion pump (e.g., Harvard Apparatus) that is controlled by a darkroom timer or other programmable timer (e.g., Chrontrol Timer; Harvard Apparatus also supplies pumps that allow computerized control of infusion parameters). The liquid diet most commonly used for ingestion tests with rat pups is a commercial mix of half bovine milk and half cream (Half and Half, 11% fat, 4.6% carbohydrate, 3.2% protein). Although this diet is not suitable for maintenance and growth (see Diaz, chap. 16, this volume), it is adequate for acute tests of ingestion. Other diets used in such tests include sucrose (0.15–$0.3\,M$), glucose (0.3–$0.6\,M$), or saccharin ($0.01\,M$) in distilled water.

Different infusion parameters are used depending on the type of ingestion test and on the component of ingestion being tested. The initial willingness to ingest and also the vigor of intake can best be assessed by providing pulses of diet at a rate of flow that is faster than the pup can swallow (15-s pulses containing 0.5% body weight of liquid diet delivered at 2-min intervals; these intervals allow the easily fatigued pup a chance to recover between pulses). Any excess diet spills out of the mouth onto the floor and is therefore not included in the pup's weight gain. This type of ingestion test usually lasts 10 min.

Termination of intake ("satiety") can be better assessed by giving pups continuous infusions that last 30 min and deliver diet at 0.25% body weight per minute. When 24-h-deprived pups are given this type of infusion, they typically ingest all the diet for 15–20 min and only then begin to spill diet from their mouths. Alternatively, in another assessment of intake termination, the pup terminates its own test; that is, the infusion is stopped when spilling of diet or the absence of mouthing reaches a criterion level. This assessment is particularly useful when ingestion is prolonged, for example, when an animal has an open gastric fistula (Phifer, Sikes, & Hall, 1986).

When diet is provided by infusion through an oral cannula, the pup must actively mouth and swallow (or just swallow with a posterior cannula placement) in order to ingest. Although this type of diet delivery is useful for assessing basic consummatory responses, the more appetitive aspects of intake control are not revealed (Hall & Bryan, 1980). In yet another variation of independent ingestion testing, pups are required to consume liquid or semisolid diet (e.g., moist rat chow) from the environment. This task requires the pup to recognize, contact, and maintain contact with a food source and, therefore, bears a closer resemblance to adult ingestion than does infusion feeding. In practice, a piece of terrycloth or several layers of paper toweling are placed on the bottom of the pup's test container and then soaked with diet. Because pups' poor thermoregulatory abilities are further compromised in this situation, the temperature of the incubator should be raised to 35–37°C to maintain normal body temperature. In a variation of this test, the ability of pups to localize diet is assessed by restricting diet to a fraction of the test container floor. Diet is limited to a small piece of terrycloth or folded toweling. Alternatively, the diet is placed in a shallow plastic tray (2 by 2 by 0.4 cm deep, cut from a plastic ice tray) that is secured to the floor of the test container.

Regardless of the way diet is provided to the pup, behavioral observations are essential for a detailed analysis of ingestion. Without actual observation of the pup's behavior, it is impossible to tell whether a pup ingested more diet because it mouthed more avidly or because it spilled less diet from its mouth. Therefore, the use of a clear glass aquarium as a test incubator, coupled with a mirror placed behind the containers that hold the pups, allows detailed observations of pup behavior. Various observation schedules are employed by different laboratories and by different investigators in the same laboratory, but schedules usually call for either continuous observation of every pup that is tested or a time-sampling technique in which each pup is observed at a different time (e.g., 10 s per pup every 2 min). Generally, the shorter the test, the more each pup is observed. The types of behaviors that are monitored and recorded also vary from one investigator to another, but the most frequently recorded behaviors are mouthing, probing with the snout or forepaws, stretching (whole body extension that sometimes accompanies delivery of diet), oral spilling of diet, and general activity level.

Surgical Manipulations in Infant Rodents

The manipulations described in this section are designed to deliver diet to or exclude diet from specific regions of the gastrointestinal tract. Most of these procedures require general anesthesia (see Hofer, chap. 2, this volume, for specific recommendations on anesthesia in young mammals).

Gastric Fistula for Sham Feeding

A gastric fistula is a chronically implanted conduit between the stomach and the skin that allows liquid or semiliquid diet to flow freely from the stomach. An animal feeding with an open gastric fistula experiences only limited gastrointestinal stimulation (Young, Gibbs, Antin, Holt, & Smith, 1974). Oroesophageal stimulation is normal, but although gastric chemoreceptors may be stimulated, gastric distension does not occur, and postgastric stimulation is largely prevented (but see Sclafani & Nissenbaum, 1985).

The basic design of the fistula used in rat pups is an adaptation of the fistula used in adults (Young, Gibbs, Antin, Holt, & Smith, 1974). The main differences are the use of Teflon tubing rather than the stainless steel tubing used in adults, and, of course, the tubing sizes are different. Teflon tubing (Small Parts, Inc.) is used for pup fistulas because this material is inert, easily molded with heat, and offers little resistance to flow. To date, the gastric fistula has been used in rat pups at only two different ages, 6 and 15 days (Phifer & Hall, 1988), but the size of the fistulas could easily be adapted to additional ages. The fistula is formed from a length of tubing that has been heat flared on both ends (tubing sizes are given in Figure 11.3). A sleeve of slightly larger tubing provides an anchor for a loop of 4.0 silk suture that is used to secure the fistula to the body wall. A replaceable plug is made by heating and then flattening the end of a short piece of Microrenathane Type MRE 080 (for 6-day-olds) or PE 320 tubing (for 15-day-olds).

Prior to receiving a fistula, a pup should be food deprived for 4–6 h, because a partially emptied stomach makes the surgical procedure much easier. Initially, a 1–1.5-cm horizontal laparotomy is made at the level of the stomach and extending from just right of the pup's midline to the right flank. Using 6.0 silk, a circular purse suture (3–5 mm in diameter) is made in the ventral antrum of the stomach; then, inside the circle, an incision is made that is slightly larger than the diameter of the fistula. One end of the fistula is inserted through this incision, the purse suture is tightened, and the laparotomy is closed around the sleeve of the fistula by using the suture that was looped through the sleeve as one of the closing stitches. The fistula is later opened by first stabilizing it with one pair of forceps and then withdrawing the plug with another pair of forceps. This fistula preparation allows the stomach to empty completely; pups that are ingesting while the fistula is open do not gain, and often even lose, weight as a result of spillage of residual stomach contents (Phifer, Sikes, & Hall, 1986).

Gavage Loading of Stomach

Although not a surgical procedure, gavage loading is a gastrointestinal manipulation that has been useful in feeding studies on both infant and adult mammals (for a review, see Houpt, 1982). When diet is delivered to the stomach by this technique the

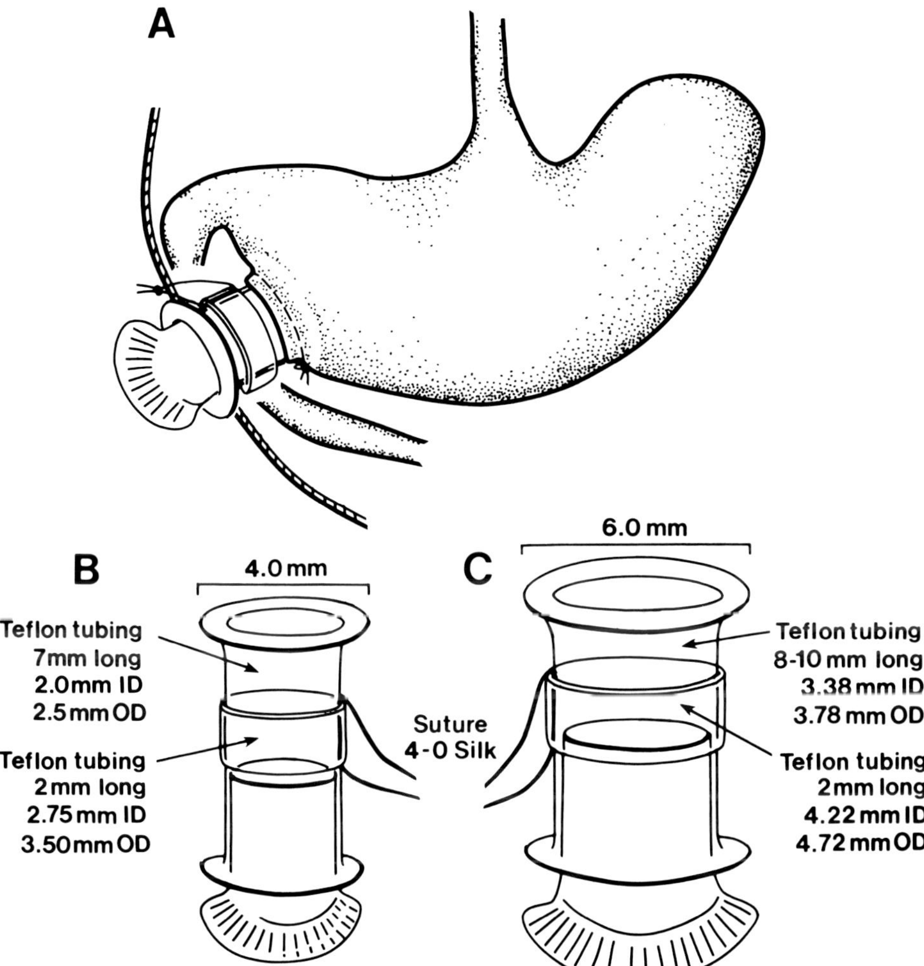

Figure 11.3. (A) The placement of a gastric fistula that permits sham feeding in rat pups. (B) Dimensions for the fistula used with 6-day-old pups and (C) 15-day-old pups.

oroesophageal stimulation that occurs during normal ingestion is prevented; also the rate of gastric filling is regulated by the experimenter rather than by the animal's feeding.

Anesthesia is not necessary prior to gavage loading of pups. An awake pup will usually swallow the gavage tube, thus making the procedure even simpler. The procedure for giving gavage loads is the same for pups at all ages, but the size of the tubing is different. For young pups, 3–9 days of age, a piece of Silastic tubing (Dow Corning Corporation; 0.508 mm ID, 0.94 mm OD) is attached to a 23-gauge needle; for older pups a piece of Silastic (0.635 mm ID, 1.19 mm OD) tubing is attached to a 21-gauge needle; the needle is then attached to a syringe of appropriate size to hold the solution that will be loaded. The larger size tubing is needed in older pups to provide the resistance necessary to push the tube through the mouth and down the esophagus.

In order to gauge how far to insert the gavage tube, a short sleeve of PE-100 (for the smaller-size tube) or Silastic (1.016 mm ID, 2.159 mm OD for the larger tube) is slid over the gavage tube to a point equal to the distance from the pup's stomach to its mouth. Lubricating the tube with vegetable oil facilitates passage down the esophagus.

To administer a gastric load, a pup is grasped with the head held fairly in line with the body axis. Next, the gavage tube is passed down through the throat and esophagus until the sleeve reaches the mouth; the required volume is then slowly emptied from the syringe. Care must be taken to avoid forcing the tube into the trachea instead of the esophagus, although this is rarely a problem. This procedure requires only 10–15 s for insertion of the gavage tube and another 45–60 s for actual loading; pups are then ready for testing immediately.

Intragastric Cannula for Gastric Loading
An intragastric cannula (Hall, 1975) is preferred over a gavage tube when the experimental design precludes gastric loading by gavage (e.g., when oral and intragastric inputs must be simultaneous), when repeated loadings are desired, or when chronic infusions are required. This procedure is also used for artificial rearing of infant rodents and is therefore described in detail by Diaz (chap. 16, this volume). The primary reasons for use of the gastric cannula in studies of ingestive behavior are the same as for the gavage tube: Both procedures allow diet to bypass the oroesophageal compartment and provide exact delivery of a specified volume of diet to the stomach.

Pyloric Noose
The pyloric noose provides complete occlusion of the pyloric sphincter and therefore prevents ingested diet from entering the intestines (Hall, 1973; Kraly & Smith, 1978; Phifer, Sikes, & Hall, 1986). Although the gastric fistula preparation (sham feeding) has previously been used to prevent intestinal stimulation (Young, Gibbs, Antin, Holt, & Smith, 1974), Sclafani and Nissenbaum (1985) have shown that some diet can enter the intestine of sham-feeding rats. But when the pyloric noose is combined with a gastric fistula, both postgastric stimulation and gastric distension cues are eliminated. Despite the usefulness of the pyloric noose, it has one drawback when used alone: Closing the pyloric noose simultaneously reduces intestinal stimulation and increases stimulation due to gastric distension. This opposition of actions can cause difficulty in identifying the particular mechanism or mechanisms involved in controlling intake.

The pyloric noose used in infant rats (Figure 11.4) consists of a 12–15-cm length of 4.0 silk suture with two overhand knots, one double thickness and one single, tied near the middle. When installed in a pup, the noose is formed into a loop that is held by a 3-mm length of Silastic tubing (0.635 mm ID, 1.19 mm OD). After a pup has been anesthetized, a 1.5-cm horizontal laparotomy is made just to the right of the stomach. The pylorus and duodenum are exposed; then the suture is passed around the pyloric junction, forming a loop, and the Silastic sleeve is threaded onto both ends of the suture. The sleeve is slid down the suture until the loop is about twice the diameter of the pylorus. It should rest against the double knot in the suture, and the single knot should be inside the sleeve to provide friction. The end of the suture proximal to the single knot is stitched to the body wall to provide a solid anchor for the noose. The other end of the suture is passed through the wound and trimmed to

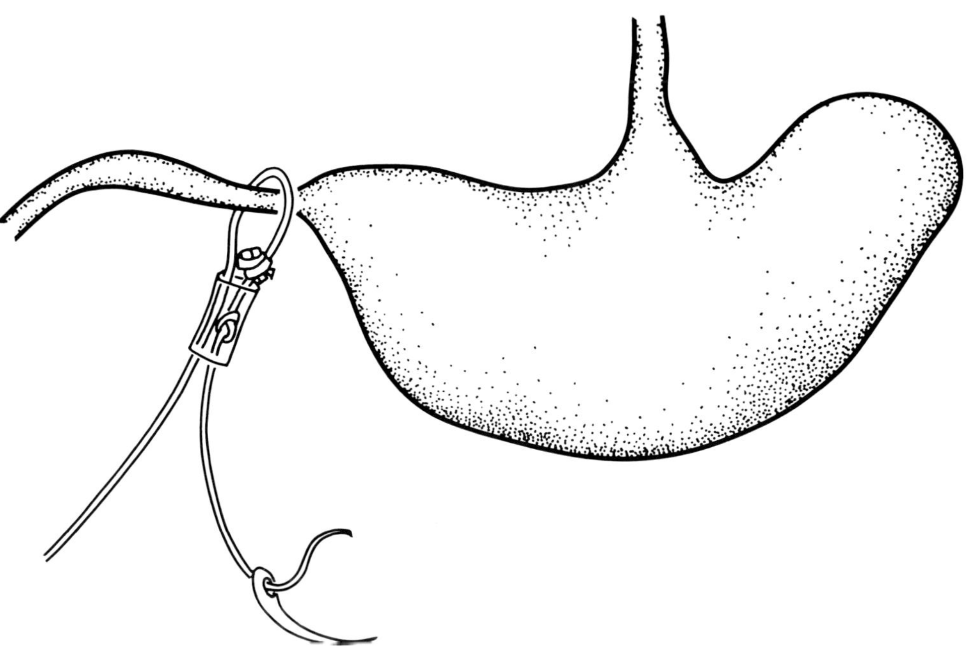

Figure 11.4. The pyloric noose (in relation to a rat pup's stomach) used to prevent gastric emptying and subsequent flow of diet into the intestines.

about 1.5 cm. This end of the suture is later pulled gently to tighten the noose and occlude the pylorus. After an experiment, proper closure of the noose can be verified by examining the intestine for the presence of diet (food coloring can be added to colorless diets to increase visibility.)

PERSPECTIVES

Advances in the Study of Ingestive Behavior Development

The techniques described in the previous sections have provided valuable tools for unraveling the developmental origins of feeding behavior and its control. Figure 11.5 summarizes some of the information we have obtained about the origins of this behavior in rodents.

The use of the oral cannula has allowed presentation of taste stimuli other than the milk provided to pups during normal suckling. As a result, we have learned that taste affects feeding responses quite early in the preweaning period. Pups as young as 1 day reject strong acid and quinine solutions (Johanson & Shapiro, 1986; compare Hall & Bryan, 1981; Moe, 1986); pups as young as 3 days reject strong salt solutions (Hall & Bryan, 1981; Moe, 1986); and pups as young as 6 days exhibit higher intakes of sweet solutions than water (Hall & Bryan, 1981). Together, these responses allow infant rats limited control over both the initiation and termination of ingestion.

Just as taste responsiveness increases with development, other types of controls are added during the maturation period, and this increase in complexity ultimately

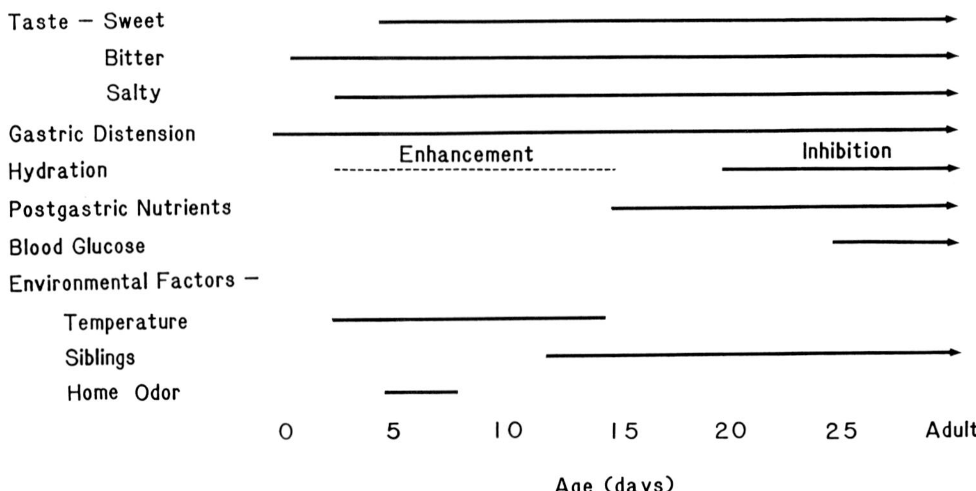

Figure 11.5. Summary diagram for the ontogeny of ingestive behavior controls in the young rat. See text for explanation and references.

leads to the feeding control system of adulthood. The techniques that deliver diet to limited regions of the gastrointestinal tract (e.g., gastric fistula, pyloric noose, gavage tube) have enabled us to determine the ontogenetic changes in responsiveness of specific regions or receptors. For example, we now know that, beginning at around 2 weeks of age, gastric fill is no longer the only intake-terminating cue; instead, the nutritive value of diet begins to affect intake (Phifer & Hall, 1988). This nutritive effect seems due to a postgastric but nonmetabolic mechanism because glucoprivation, as produced by 2-deoxyglucose or insulin, does not enhance intake by independent ingestion until at least a week later (Gisel & Henning, 1980; Williams & Blass, 1987). At about the same time that nutritive controls develop, the control mechanisms for hunger and thirst begin to differentiate. While dehydration leads to increased feeding in 6-day-old pups (Bruno, 1981; Hall & Bruno, 1984), food intake in 15-day-olds is not affected by dehydration (Phifer, Browde, & Hall, 1986), indicating the transition to the adult condition where dehydration is accompanied by reduced feeding (e.g., Hsiao & Trankina, 1969).

The independent ingestion techniques have only recently been applied to research on the origin of neurochemical controls for feeding. In numerous studies, the development of neurotransmitter control in both enhancement and reduction of ingestion has been demonstrated. And although our understanding of the mechanisms underlying these controls is far from complete, the techniques described here have been applied by many investigators in studies on both the classical neurotransmitters (adrenergic and serotonergic) and peptides (e.g., Areoyewun & Barr, 1982; Capuano, Barr, & Leibowitz, 1985; Ellis, Axt, & Epstein, 1984).

Although most of the efforts in developmental studies of ingestive behavior controls have been directed toward finding the onset of new controls, some intake-modulating factors are present during infancy but then disappear or decrease in effectiveness as pups mature. Perhaps the most obvious of these is temperature (another factor that was difficult to study before pups could be induced to feed in the absence of the rat dam and her body heat). Very young pups will not ingest unless they are placed in a warm ambience, regardless of their own body temperature, but temperature has much less influence over feeding in pups older than 15 days (Johanson & Hall, 1980). Even though other environmental cues, including substrate texture, home odor, and presence of siblings, begin to affect intake as pups mature, none match the absolute permissive role played by temperature (Johanson & Hall, 1981; see also Satinoff, chap. 10, this volume).

Future Research Directions

Although we have learned a great deal about ontogenetic changes in ingestive behavior controls over the past 15 to 20 years, to a major extent we are still working at the phenomenological level. We are now ready to begin investigating the physiological mechanisms for these controls, how the physiology changes with maturation, and how abnormalities in ingestive behavior controls might develop.

In order to accomplish the goals described above, future studies on the development of feeding controls, as well as other areas of developmental psychobiology, must make better use of ontogeny as a tool for separating, or "dissecting," control mechanisms. To use this "natural experiment" approach to analyze a behavioral or physiological mechanism, one must first find that animals at one age possess the trait to be studied and animals at another age either lack the trait or possess it in a modified form. Then, by comparing the two ages, one can search for differences in neural structure or function that may account for the difference in behavior or physiology. This natural experiment approach has long been regarded as an important advantage of developmental studies. However, we are just now reaching the point at which our understanding of developmental sequences in ingestive behavior and its controls will allow us to use this approach.

Such an approach could now be used to examine the similarities and differences between feeding and drinking behaviors, since these behaviors appear to differentiate during the preweaning period. Presently we do not know whether this differentiation reflects the separation of a single control system or the maturation of a new system at the level of the CNS or, alternatively, whether it simply reflects a differentiation of two peripheral sensory mechanisms, one that becomes specialized for nutrient detection and the other for hydration level.

The natural experiment approach can likewise be used to analyze the development of nutritive effects on intake, because this control mechanism also appears to mature during the preweaning period. Although we know that this mechanism operates postgastrically, we still do not know whether it is a preabsorptive or postabsorptive mechanism, or if intestinal nutrients reduce intake by inhibiting gastric emptying (as is the case in adults; McHuge, Moran, & Wirth, 1982). This brings up one of the most fascinating issues of food-intake control, both in developing and adult animals, and that is the interaction between different control mechanisms. Early in this chapter I stated that many investigators in the field of feeding behavior controls have found it convenient to divide the animal into figurative compartments for purposes of analysis. Although this approach has been useful, it has sometimes concealed important interactions between different mechanisms. We now know that the control of ingestion involves interactions between oroesophageal cues and both gastric and postgastric cues, and also that gastric cues interact with postgastric cues (Antin, Gibbs, & Smith, 1977; McCann & Stricker, 1986; McHugh, Moran, & Wirth, 1982; Ramirez, 1985). Thus, only by examining how each mechanism affects the others can we truly understand the entire complex of intake control.

Just as we are only beginning to understand the origins of peripheral controls for ingestion, we are also just beginning to study the mechanisms responsible for attracting young mammals to food and water. For example, we know that taste and olfactory cues are important for the initiation of ingestion by young rat pups (Hall & Bryan, 1981), but little is known concerning the development of other appetitive aspects of ingestion. Recent studies have revealed that pups as young as 3 days can maintain their position at a localized food source, but we still do not know what types of orientation cues they are using, nor do we understand their orienting mechanisms (Phifer, Denzinger, & Hall, unpublished). Such studies are prerequisite to investigating the development of food-searching and food-locating behaviors. We are also just beginning to investigate the temporal patterning of ingestion in rat pups that are given constant access to food. Preliminary results (also unpublished) suggest that even 6-day-old pups show definite rhythmicity in their feeding, but the relationship between this rhythmicity and the meal patterns of adults is totally unknown.

Another underutilized benefit of the developmental approach is the availability of simpler, and therefore more easily understood, mechanisms in immature animals. The young rat pup, which has relatively few controls of ingestive behavior, provides an excellent model for examining the mechanism that underlies the inhibition of intake by gastric fill. This aspect of feeding behavior control has seen a recent resurgence of interest (e.g., McHugh & Moran, 1985), and the relatively simpler system in the young pup has several advantages over the adult system. For example, many variables that confound the analysis of control by gastric fill in adulthood either have not yet developed (e.g., nutritive effects of diet) or have not yet had an opportunity to be formed (e.g., learned preferences and aversions to specific foods).

Just as infant rats' ingestive behavior is simpler because it has fewer peripheral controls, the neural control of ingestion is likely simpler and easier to understand in the pup. We have already seen that the neurochemical correlates of ingestive behavior controls show developmental changes, but we are only just beginning to investigate changes in the neuroanatomical basis of early ingestion (Almli, 1978; Hall, 1989; Kornblith & Hall, 1979).

Yet another advantage of the developmental approach, and indeed one for which development is the only approach, is the opportunity to investigate how manipulations during infancy affect animals later in life (e.g., Harlow, Dodsworth, & Harlow, 1965; Wiesel & Hubel, 1963). Such studies provide useful models for understanding the relationships between events in early life and behavioral abnormalities in adulthood. Although this type of study has often been fraught with problems caused by uncertainties as to whether the later effects were caused by the original manipulation or some intervening variable, the technique is still quite powerful and should not be neglected. As clinical research pushes our knowledge concerning the ages of onset for serious human feeding disorders earlier into childhood, we need to increase our understanding of how early ingestive experiences affect feeding in adults. The ability to rear rat pups away from their dam has made it possible to regulate the type, and to a smaller extent, the quantity, of diet that a pup receives from birth until weaning (e.g., West, Diaz, & Woods, 1982). Investigations of the resulting effects of these and other manipulations may provide valuable insights into the developmental bases of obesity, anorexia nervosa, and bulimia.

In conclusion, feeding and drinking behaviors, and the mechanisms responsible for their control, have been a topic of interest and curiosity for centuries. Research on the development of these behaviors has a shorter history but one that has been very fruitful. Continued use of the developmental approach for the analysis of feeding and drinking control systems should greatly increase our understanding of these behaviors as well as of regulatory mechanisms in general.

ACKNOWLEDGMENTS

I would like to thank W. G. Hall, D. J. Kucharski, D. Phifer, and L. M. Terry for critical reading of this manuscript. Support was provided by a grant from the National Institution for Child Health and Human Development (HD 17457).

REFERENCES

Almli, C. R. (1978). The ontogeny of feeding and drinking: Effects of early brain damage. *Neuroscience and Biobehavioral Reviews, 2*, 281–300.

Antin, J., Gibbs, J., & Smith, G. P. (1977). Intestinal satiety requires pregastric food stimulation. *Physiology and Behavior, 18*, 421–425.

Aroyewun, O., & Barr, G. A. (1982). The effects of opiate antagonists on milk intake of preweanling rats. *Neuropharmacology, 21*, 757–762.

Bornstein, B. H., Terry, L. M., Browde, J. A., Jr., Assimon, S. A., & Hall, W. G. (1987). Maternal and nutritional contributions to infant rats' activational responses to ingestion. *Developmental Psychobiology, 20*, 147–163.

Bruno, J. P. (1981). Development of drinking behavior in preweanling rats. *Journal of Comparative Physiology and Psychology, 95*, 1016–1027.

Cannon, W. B., & Washburn, A. L. (1912). An explanation of hunger. *American Journal of Physiology, 29*, 441–455.

Capuano, C. A., Barr, G. A., & Leibowitz, S. F. (1985). Independent feeding elicited by noradrenergic stimulation of the periventricular nucleus in the developing rat. Paper presented at the Meeting of the *International Society for Developmental Psychobiology*, p. 77.

Drewett, R. F., & Cordall, K. M. (1976). Control of feeding in suckling rats: Effects of glucose and osmotic stimuli. *Physiology and Behavior, 16*, 711–717.

Ellis, S., Axt, K., & Epstein, A. N. (1984). The arousal of ingestive behaviors by the injection of chemical substances into the brain of the suckling rat. *Journal of Neuroscience, 4*, 945–955.

Friedman, M. I. (1975). Some determinants of milk ingestion in suckling rats. *Journal of Comparative Physiology and Psychology, 89*, 636–647.

Gisel, E. G., & Henning, S. J. (1980). Appearance of glucoprivic control of feeding behavior in the developing rat. *Physiology and Behavior, 24*, 313–318.

Hall, W. G. (1973). A remote stomach clamp to evaluate oral and gastric controls of drinking in the rat. *Physiology and Behavior, 11*, 897–901.

Hall, W. G. (1975). Weaning and growth of artificially reared rats. *Science, 190*, 1313–1315.

Hall, W. G. (1979). The ontogeny of feeding in rats: I. Ingestive and behavioral responses to oral infusions. *Journal of Comparative Physiology and Psychology, 93*, 977–1000.

Hall, W. G. (1987). Early motivation, reward, learning, and their neural bases: Developmental revelations and simplifications. In N. Krasnegor, E. Blass, M. Hofer, & W. Smotherman (Eds.), *Perinatal development: A psychobiological perspective* (pp. 169–193). New York: Academic Press.

Hall, W. G. (1989). Neural systems for early independent ingestion: Regional metabolic changes during ingestive responding and dehydration. *Behavioral Neuroscience, 103,* 386–411.

Hall, W. G., & Bruno, J. P. (1984). Inhibitory controls of ingestion in 6-day-old rat pups. *Physiology and Behavior, 32,* 831–841.

Hall, W. G., & Bryan, T. E. (1980). The ontogeny of feeding in rats: II. Independent ingestive behavior. *Journal of Comparative Physiology and Psychology, 94,* 746–756.

Hall, W. G., & Bryan, T. E. (1981). The ontogeny of feeding in rats: IV. Taste development as measured by intake and behavioral responses to oral infusions of sucrose and quinine. *Journal of Comparative Physiology and Psychology, 95,* 240–251.

Hall, W. G., Cramer, C. P., & Blass, E. M. (1977). Ontogeny of suckling in the rat: Transitions toward adult feeding. *Journal of Comparative Physiology and Psychology, 91,* 1141–1155.

Hall, W. G., & Rosenblatt, J. S. (1977). Suckling behavior and intake control in the developing rat pup. *Journal of Comparative Physiology and Psychology, 91,* 1232–1247.

Harlow, H. F., Dodsworth, R. O., & Harlow, M. K. (1965). Total social isolation in monkeys. *Proceedings of the National Academy of Science, U. S. A., 54,* 90–97.

Houpt, K. A. (1982). Gastrointestinal factors in hunger and satiety. *Neuroscience and Biobehavioral Reviews, 6,* 145–164.

Houpt, K. A., & Epstein, A. N. (1973). Ontogeny of controls of food intake in the rat: GI fill and glucoprivation. *American Journal of Physiology, 225,* 58–66.

Hsiao, S., & Trankina, F. (1969). Thirst-hunger interaction: I. Effects of body-fluid restoration on food and water intake in water-deprived rats. *Journal of Comparative Physiology and Psychology, 69,* 448–453.

Johanson, I. B., & Hall, W. G. (1980). The ontogeny of feeding in rats. III. Thermal determinants of early ingestive responding. *Journal of Comparative Physiology and Psychology, 94,* 977–992.

Johanson, I. B., & Hall, W. G. (1981). The ontogeny of feeding in rats: V. Influence of texture, home odor, and sibling presence on ingestive behavior. *Journal of Comparative Physiology and Psychology, 95,* 837–847.

Johanson, I. B., & Shapiro, E. G. (1986). Intake and behavioral responsiveness to taste stimuli in infant rats from 1 to 15 days of age. *Developmental Psychobiology, 19,* 593–606.

Kornblith, C. L., & Hall, W. G. (1979). Brain transections selectively alter ingestion and behavioral activation in neonatal rats. *Journal of Comparative Physiology and Psychology, 93,* 1109–1117.

Kraly, F. S., & Smith, G. P. (1978). Combined pregastric and gastric stimulation by food is sufficient for normal meal size. *Physiology and Behavior, 21,* 405–408.

Lytle, L. D., Moorecroft, W. H., & Campbell, B. A. (1971). Ontogeny of amphetamine anorexia and insulin hyperphagia in the rat. *Journal of Comparative Physiology and Psychology, 77,* 388–393.

McCann, M. J., & Stricker, E. M. (1986). Gastric emptying of glucose loads in rats: Effects of insulin-induced hypoglycemia. *American Journal of Physiology, 25, (Regulatory Integrative Comparative Physiology, 20),* R609–R613.

McHugh, P. R., & Moran, T. H. (1985). The stomach: A conception of its dynamic role in satiety. *Progress in Psychobiology and Physiological Psychology, 11,* 197–232.

McHugh, P. R., Moran, T. H., & Wirth, J. B. (1982). Postpyloric regulation of gastric emptying in rhesus monkeys. *American Journal of Physiology, 243, (Regulatory Integrative Comparative Physiology, 12),* R408–R415.

Moe, K. (1986). The ontogeny of salt preference in rats. *Developmental Psychobiology, 19,* 185–196.

Phifer, C. B., Browde, J. A., Jr., & Hall, W. G. (1986). Ontogeny of glucose inhibition of independent ingestion in preweanling rats. *Brain Research Bulletin, 17,* 673–679.

Phifer, C. B., & Hall, W. G. (1984). Cholecystokinin and bombesin: Differential effects on ingestion in young and preweanling rat pups. Paper presented at the Meeting of the International Society for Developmental Psychobiology, p. 77.

Phifer, C. B., & Hall, W. G. (1988). Ingestive behavior in preweanling rats: The emergence of postgastric controls. *American Journal of Physiology, 255 (Regulatory Integrative Comperative Physiology, 24),* R191–R199.

Phifer, C. B., Sikes, C. R., & Hall, W. G. (1986). Control of ingestion in 6-day-old rat pups: Termination of intake by gastric fill alone? *American Journal of Physiology, 250, (Regulatory Integrative Comparative Physiology, 19),* R807–R814.

Ramirez, I. (1985). Oral stimulation alters digestion of intragastric oil meals in rats. *American Journal of Physiology, 248* (*Regulatory Integrative Comparative Physiology, 17*), R459–R463.

Rosenblatt, J. S., & Siegel, H. I. (1981). Factors governing the onset and maintenance of maternal behavior among nonprimate mammals. The role of hormonal and nonhormonal factors. In D. J. Gubernick & P. H. Klopfer (Eds.), *Parental care in mammals* (pp. 13–76). New York: Plenum.

Sclafani, A., & Nissenbaum, J. W. (1985). Is gastric sham feeding really sham feeding? *American Journal of Physiology, 248,* (*Regulatory Integrative Comparative Physiology, 17*), R387–R390.

Smith, G. P., & Gibbs, J. (1979). Postprandial satiety. *Progress in Psychobiology and Physiological Psychology, 8,* 180–242.

West, D., Diaz, J., & Woods, S. (1982). Infant gastrostomy and chronic infusion as a technique to overfeed and accelerate weight gain of neonatal rats. *Journal of Nutrition, 112,* 1339–1343.

Wiesel, T. N., & Hubel, D. H. (1963). Single-cell responses in striate cortex of kittens deprived of vision in one eye. *Journal of Neurophysiology, 26,* 1003–1017.

Williams, C. L., & Blass, E. M. (1987). The development of postglucoprivic insulin-induced suckling and independent feeding in infant rats. *American Journal of Physiology, 253,* (*Regulatory Integrative Comparative Physiology, 22*), R121–R127.

Wirth, J. B., & Epstein, A. N. (1976). The ontogeny of thirst in the infant rat. *American Journal of Physiology, 230,* 188–198.

Young, R. C., Gibbs, J., Antin, J., Holt, J., & Smith, G. P. (1974). Absence of satiety during sham feeding in the rat. *Journal of Comparative Physiology and Psychology, 87,* 795–800.

Zeigler, H. P. (1976). Feeding behavior of the pigeon. In J. S. Rosenblatt, C. Beer, & R. A. Hinde (Eds.), *Advances in the study of behavior* (Vol. 7, pp. 285–389). New York: Academic Press.

12

Development of a Sexually Dimorphic Behavior: Hormonal and Neural Controls

CHRISTINA L. WILLIAMS

It is well accepted, at least for mammalian species, that the sex chromosomes initiate the process of sexual differentiation. However, it was discovered early in this century that chemicals released by the fetal testes at a sensitive period in early development direct the differentiation of the body. In the presence of testicular secretions, male physiology develops; in their absence, a female body type appears (see Jost, 1985, for review). In 1959, a classic paper by Phoenix, Goy, Gerall, and Young provided the first evidence that gonadal hormones also direct the differentiation of sexual behavior, probably by acting on the neural tissue that mediated the sexually dimorphic responses. In this study, the sexual behavior of male and female guinea pigs born to mothers receiving testosterone propionate (TP) during most of pregnancy was investigated. As adults, these prenatally androgenized guinea pigs were gonadectomized and treated with estradiol and progesterone to stimulate female receptive behavior, or treated with TP to stimulate male sexual behaviors like intromission and the ejaculatory response. The results of this experiment led to the discovery that females which had been exposed to testosterone prenatally were less likely to show receptive behavior in response to female hormones and much more likely to show male sexual responses. That is, females' sexual behaviors were permanently defeminized and masculinized by exposure to testosterone during a critical period of early development.

The questions posed and the methodology used by this important paper directed research in the field for many years following its publication. Phoenix and colleagues (1959) were the first to differentiate between the effects of hormones early in development and in adult animals. Specifically, they called attention to the difference between their modes of action. Early effects were permanent, occurred only during a limited sensitive period, and might not be manifest until adulthood. In contrast, hormones circulating in adulthood have effects only while they are present. For example, female rodents display soliciting and receptive responses to a male when ovarian estrogens and progesterone are high, while gonadectomized females will not display these behaviors. Phoenix et al. (1959) termed the early effects *organizational* and the effects in adulthood *activational*. This framework has focused research on the differences between these modes of actions of steroid hormones and has led to the assumption that these different actions depend on the state of the neural tissue at the time of exposure. Thus, until recently, the prevailing view was that organizational and activational actions of steroids are separable in time.

206

The paper also served as a model for how research in the field should be done: Manipulate the hormonal system early in development and, after puberty, examine the result of the manipulation on the adult system. This is a reasonable protocol for examining a phenomenon that is dependent on some early sensitive period and for which the adult form is not displayed until adulthood. For example, this technique has been used extensively to examine the effects of selective deprivation on response characteristics of visual cortical neurons (see Pettigrew, 1978). In this type of analysis, the focus is on the long-term effects of a perturbation in a developing system.

The work described in this chapter is an attempt to study the process of sexual differentiation of reproductive behavior from a different perspective. This research was initiated because I wanted to examine the process while it is occurring, rather than examining the outcome of an early manipulation. This approach has been used successfully for the developmental analysis of behavioral systems that appear at birth and continue throughout the lifespan of an organism. The ontogeny of ingestive behavior and of various types of learning have been examined in this manner (for review, see Hall & Williams, 1983, and Spear & Campbell, 1979). Since sexual behavior does not normally appear in an animal's behavioral repertoire until after it is activated by gonadal hormones at puberty, its prepubertal development in nonhuman species has been generally ignored. In fact, there has been little research done to determine whether sexual behavior is not displayed prior to puberty simply because the gonadal steroids necessary to activate sexual behavior are not yet available, or because the neural and behavioral systems are not yet assembled.

Over the last 7 years my laboratory has identified a sexually dimorphic behavior in the neonatal rat that can be used to trace its prepubertal ontogeny. This strategy, of examining a system as it develops to see how it gets assembled, may be particularly revealing with regard to the development of a sexually dimorphic response. Within the same species, without any exogenous intervention, one can see the physiological and behavioral consequences of nature's own pharmacological manipulation. That is, at a critical period in early ontogeny male testes begin to secrete large quantities of the hormone testosterone, and development progresses in one direction. In contrast, female ovaries are relatively quiescent until puberty, and without exposure to gonadal steroids development progresses in another direction. In altricial species like the rat, much of the sensitive period for hormone-induced sexual differentiation of behavior occurs postnatally. This means that at birth male and female brains are not fully differentiated. The possibility of tapping into a behavioral system as it undergoes the process of sexual differentiation is the focus of this chapter. Do males and females at some point in development show no sexually dimorphic behaviors? Do males begin with a female brain that is defeminized and masculinized through the action of hormones? Do females become more sensitive or specific in their stimulus requirements for the display of sexual behavior? My hope is that this analysis may be able to provide answers to these questions.

TECHNIQUES FOR ELICITING AND QUANTIFYING
A SEXUALLY DIMORPHIC RESPONSE

I began my analysis of the development of a sexually dimorphic behavior because of several reports that lordosis, or behaviors that resemble lordosis, could be elicited from rodents well before puberty without the activational influences of gonadal steroids. *Lordosis* is a word of Greek origin, meaning *a bending*, and it is commonly used in medicine to describe a curvature of the spine producing a hollow in the back (Oxford English Dictionary). In the field of animal behavior, lordosis refers to the mating posture assumed by the estrous female mammal when she is mounted by the male. This motor response is extremely stereotyped in the female rodent, and its display is both sexually dimorphic (i.e., males do not normally display the response) and hormone dependent (i.e., priming by estrogens and progesterone are required). Thus, the term lordosis has been reserved for this hormonally controlled female mating behavior rather than referring, more generally, to an arched back posture.

Boling, Blandau, Wilson, and Young (1939) and Beach (1966) described a behavior in newborn guinea pigs in response to maternal licking of the anogenital region that looked like lordosis. Since the behavior appeared to be hormone independent and not sexually dimorphic, they concluded that back arching accompanied the micturition reflex in neonatal rats and that the behavior was not precocious lordosis. During the first week of life, rat pups also display a behavior that resembles lordosis when the dam licks their anogenital region to stimulate urination and defecation (Moore & Chadwick-Dias, 1986). Other work indicated that not only could infants display a lordosislike response without activating steroids, they also did not need anogenital stimulation. For example, 3–10-day-old rats will arch their backs and display immobility when the medial forebrain bundle is electrically stimulated (Moran, Schwartz, & Blass, 1983), and food- and suckling-deprived 3-day-old rats will arch their backs when milk is delivered to the backs of their mouths through a tongue cannula (Hall, 1979b). The appearance of these precocious lordosislike responses in very young infants suggested that it might be possible to trace their development. Several important questions directed the initial research effort: (1) Can lordosis be elicited in neonatal rats with stimuli similar to that which elicits lordosis in adult females? That is, could a technique be developed that delivered mounting stimulation to a week-old rat? (2) If week-old rats display lordosis to mounting stimulation, suggesting that the neural circuitry for lordosis was functional in rats of this age, then when did these circuits become sensitive to the activating influences of steroids? (3) Did males and females both possess the capacity for showing lordosis as infants? If so, how was this capacity lost, suppressed, or inhibited in the male and/or elaborated in the female?

As I began this work, I found several papers by Sodersten (1975, 1978), which were an attempt to examine lordosis development in prepubertal rats, to be quite perplexing, in light of the work (described above) that clearly demonstrates newborn rodents have the capacity to show lordosislike motor responses. Sodersten's studies reported that even with large doses of activating steroids (i.e., estradiol benzoate and progesterone), manual stimulation of the flanks and rump of rats did not cause them to display lordosis until they reached 15–18 days of age. Sodersten concluded that the neural and/or hormonal mechanisms underlying lordosis did not mature until after the second postnatal week. These data had been accepted because they supported the prevailing view: that activational actions of estrogens could not be demonstrated until after the period for estrogen's organizational actions had ended (see Williams, 1987, for a review).

The clue that was needed to reconcile Sodersten's work with other demonstrations of lordosislike behavior in neonatal rats was provided by Hall (1979a). He showed that if newborn pups are tested at nest temperature (34°C) rather than room temperature, they will lick and lap milk placed in a puddle on the floor. Since the commonly held belief was that infant rats did not show adultlike feeding responses until they reached 15 days of age, Hall's finding suggested that temperature might have a general permissive effect on the behavior of the neonate (see Satinoff, chap. 10, this volume). In all the previous reports of precocious lordosislike behavior from neonates, the infants were always tested at warm temperatures, either in an incubator (Moran, Schwartz, & Blass, 1983; Hall, 1979b) or with the dam at nest temperature (Boling et al., 1939; Moore, & Chadwick-Dias, 1986). When I repeated the Sodersten study, this time testing pups in an incubator, pups showed lordosis (see Figure 12.1). This was a clear demonstration that infant rats could respond to manual stimulation of the flanks and rump with a behavior resembling that normally displayed by a sexually receptive adult female rat to the mounting of a male.

The methods that allow us to elicit reliably this precocious behavior and to quantify its output have been developed in my laboratory. In order to support my contention that this response is an infantile form of lordosis, I have also demonstrated that the hormonal control and the neural substrates of the response are continuous with the adult form.

Procedure and Description of Behavior

In adult rats, it has been shown that flank stimulation is the necessary and sufficient cutaneous stimulation required to elicit lordosis from a receptive adult female (Pfaff, Montgomery, & Lewis, 1977). Therefore, I attempted to develop a method of giving pups tactile stimulation, in a fashion that resembled the flank stimulation that is provided to an adult rat during mating. After trying several "male substitutes," I discovered that a foam artist's brush (Polybrush, 3 cm wide, available at hardware stores) with an arch cut from its bottom edge would fit snugly over the rump and flanks of an infant rat. The modified brush can be seen just above the pup in Figure 12.2. To stimulate the pup, the experimenter places the arch over the pup's flanks and gently saddles the pup's lower back. To stimulate the infant, the wooden handle of the brush is rotated back and forth between the thumb and first two fingers. The foam portion of the brush massages the flanks alternately. The rate of rolling should be as fast as possible, about 2–4 rotations/s.

In response to this type of tactile stimulation of the back, rump, and flanks, 6-day-old rats clearly display the immobility, vertebral dorsiflexion, head and rump elevation, and tail deflection characteristic of the adult female mating response, lordosis. The response of an adult and 6-day-old infant rat to stimulation of the flank, back, and rump can be seen in Figure 12.2. The adult rat is being stimulated by a human index and middle fingers, forming an arch, while the infant is stimulated by the arched foam brush.

With continuous stimulation, pups will hold this position from 3 to 30 s. Often this response is accompanied by a rapid head rotation. Adult female rats display a proceptive behavior called ear wiggling, and the head rotation of the 6-day-old rat appears to be its precursor. Two forms of evidence support this view: (1) A cinegraphic analysis of ear wiggling in the adult rat has shown that in order to wiggle their ears rats vibrate their heads along the longitudinal axis so that each ear is

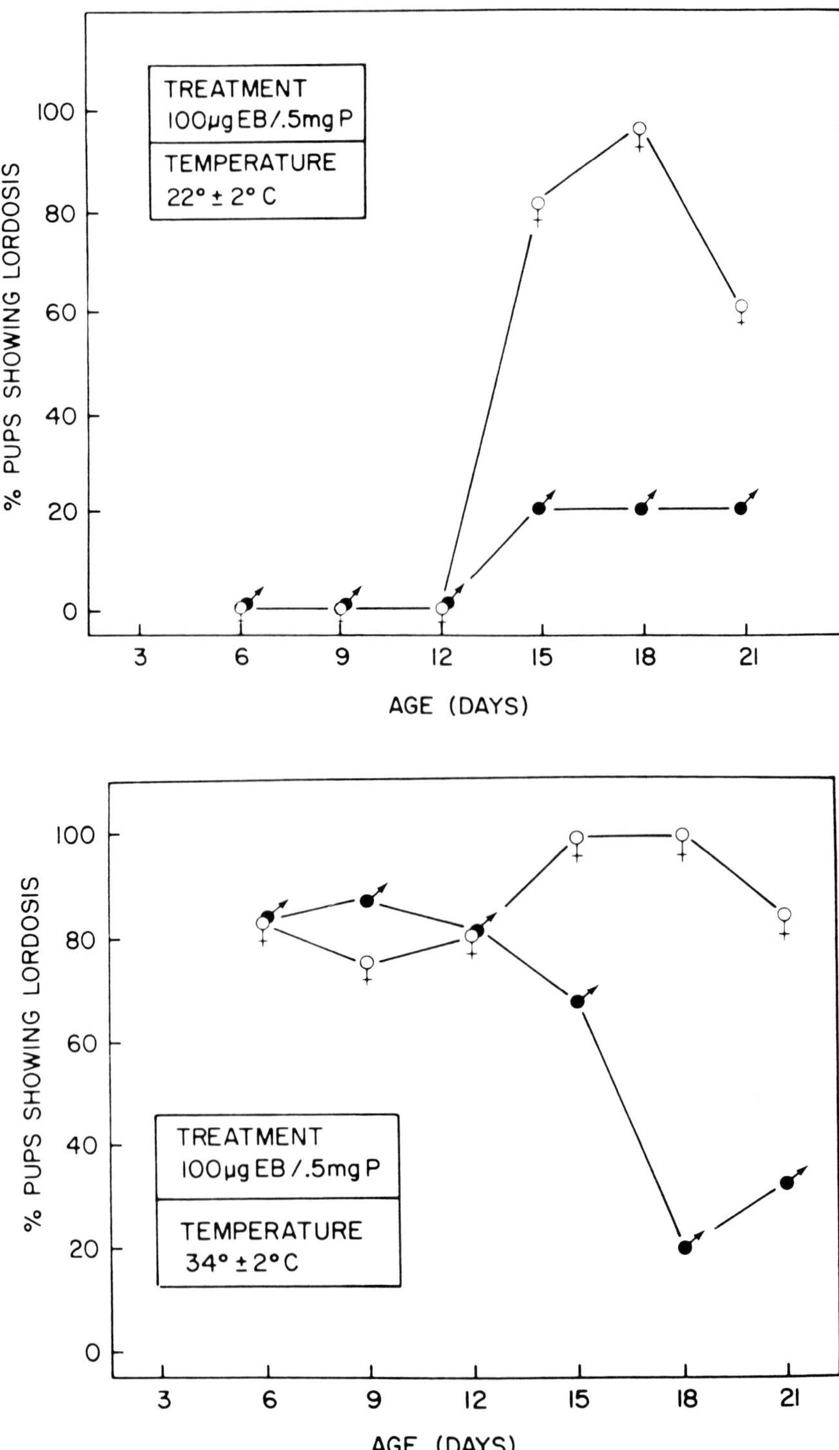

Figure 12.1. Percentage of steroid-primed 6- to 21-day-old male and female rats that displayed lordosis in response to tactile stimulation when they were tested at room temperature (top) and at incubator temperature (bottom).

Figure 12.2. A steroid-primed 6-day-old female rat (top) and a steroid-primed adult female rat (bottom) displaying lordosis to tactile stimulation of the flanks and rump. The infant is being stimulated with a foam brush and the adult with the standard manual stimulation technique.

alternately higher, then lower, about 25 times per second (Diakow, 1975). (2) As pups mature, and the ears separate from the skull, the behavior looks less like a head rotation and more like ear wiggling.

Special Conditions for the Response

Preliminary studies have shown that special conditions are critical for the display of these responses by pups younger than 15 days. As can be seen in Figure 12.1, rats younger than 15 days show lordosis only when they are tested in a warm (34 ± 2°C), moist (> 70%) incubator. The construction of these incubators is described in detail elsewhere (Hall, 1979b, and Phifer, chap. 11, this volume). Pups kept in these incubators maintain a consistent body temperature of about 34°C.

When young pups are tested at room temperature (26 ± 2°C) they rarely respond to stimulation by showing lordosis. In this study all pups were treated with 100 μg of estradiol benzoate (EB) to facilitate maximal response rates. Forty hours after EB administration, all pups were given 0.5 mg of progesterone and were placed in an incubator (see "Steroid Facilitation," below). Four hours after isolation, half of the pups were removed from the incubator and immediately stroked; the other half were stroked in the incubator. Even though their body temperatures were not lowered significantly compared with the incubator-tested pups, pups younger than 15 days did not show lordosis or ear wiggling. Therefore, the display of lordosis in the infant appears to be due to ambient temperature rather than to core temperature. This phenomenon has also been found to be true of precocious feeding responses of the rat (see Johanson & Hall, 1980).

Another necessary condition for the display of precocious lordosis and ear wiggling appears to be a period of isolation from maternal care and food. At this time, we do not know exactly how long this period must be; however, pilot work in our laboratory has shown that pups just taken from the litter and placed in an incubator do not reliably display these responses. In contrast, 3–4 h of deprivation in an incubator allows for reliable elicitation of lordosis. A short period of food, suckling, and/or somatosensory deprivation may arouse pups and change their response to tactile stimulation. A number of investigations have demonstrated multiple maternal separation effects on developing rat pups' behavior (Hofer, 1975, 1981; Moorcroft, Lytle, & Campbell, 1973). Without further research, it is impossible to know which aspect of maternal care is the critical factor. However, Bornstein, Terry, Browde, Assimon, and Hall (1987) have shown that an active maternal presence, rather than gastric fill, or suckling per se, determines whether or not pups will become behaviorally activated by oral infusions of milk. These data suggest that maternal licking, handling, and/or contact may modulate pups' later response to various forms of tactile stimulation.

Recording and Quantifying Behavior

Over the last 5 years, I have developed a system for accurate, reliable, and unbiased observation and quantification of the precocious behaviors of the infant rat. The strength of this system lies in the fact that all behavioral tests are videotaped, and all testing and scoring of the behaviors are performed by an assistant who is naive to the pup's experimental condition.

Using a macro and zoom lens on a color video camera (Pentax Saticon PC K1500A), pups are taped during behavioral tests. The camera is mounted so that it can be focused on a single marked area in the incubator. Pups are housed individually in small Plexiglas containers sized according to pups' ages (7.5 by 4.5 by 7 cm for 6-day-olds, 9 by 4.5 by 7 cm for 9-day-olds, etc.) The floors of the containers are lined with felt. Containers are placed on a focal spot in the incubator, so that pups rest perpendicular to the camera and fill the entire monitor screen. Pups are stroked continuously, regardless of their response to the stimulation, for 30 s. Tests are repeated every 30 min for a total of 4 tests. The person providing the stroking stimulation is always blind to the pup's treatment; therefore, variations in stimulus strength or vigor are randomized across conditions.

It is important that the behaviors of the pups be videotaped using a good-quality recording system, one that allows the experimenter to score the tapes in slow motion and use stop action. Most video recorders for home use do not provide a clear, sharp image during slow motion playback, and the edges of frames can be seen at stop action. I have found that the Panasonic Video Cassette Recorder (model number NV8950; possible vendors include MPCS Video and Tallahassee Camera Center) is an excellent system. It allows variable-speed playback in both forward and reverse modes, and the picture is always clear. Any video recorder used for film editing would provide these important features.

The 4 typical responses of the 6-day-old to stimulation by the foam brush are shown in Figure 12.3. When lordosis is displayed, both quantitative (frequency) and qualitative (duration, intensity) aspects of lordosis can be recorded. Frequency is calculated by the formula (number of lordosis responses/4 test periods) $\times$ 100%. This measure is comparable to the lordosis quotient score standardly used to measure

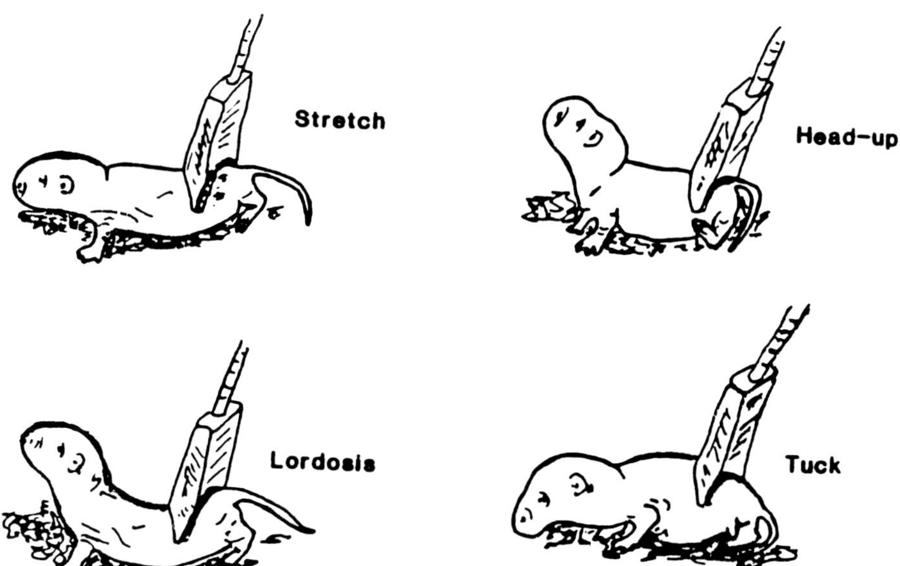

Figure 12.3. The four typical responses of the 6-day-old rat to stimulation by a foam brush (see text for descriptions) are pictured here. These line drawings were made by tracing frames from the video record.

adult receptivity. Ear wiggling frequency can be quantified in the same fashion. Duration of lordosis can be determined by the output of a time-date generator superimposed on the video record. Many good-quality video cameras include a time-date generator as a feature. The Pentax camera that I use monitors time in hundredths of a second. Intensity (maximal spinal depression) is easily viewed in stop action on the video monitor and measured by recording the distance from a straight line connecting the top of the head to the top of the rump to the lowest part of the back. This procedure has been used successfully to measure lordosis intensity in adult rats (Komisaruk & Diakow, 1973).

The duration of other behaviors that the pups display when they are not showing lordosis can also be recorded (see Figure 12.3). "Tuck" is the typical (about 80% of the time) response shown following lordosis. Pups' heads and rumps are in contact with the floor of the test container. "Head up" appears to be an active avoidance of the stimulation. Pups lift their heads, push forward with their hind limbs, and move away from the brush. Pups that display this behavior usually (75% of the time) maintain this response for the entire 30-s test period. "Stretch" is an immobility posture without the vertebral dorsiflexion or head and rump elevation that characterizes lordosis. The back is flattened rather than concave, as in a lordosis response. Ear wiggling almost always (90% of the time) accompanies "lordosis" and "stretch." By scoring these other behaviors, the range and sequencing of behavior that are shown by pups can be determined.

For all of our studies, pups remain with their dams until the day of testing. In each experiment, litters are culled to 10 pups, and not more than 2 pups of each sex are used for each treatment condition. At least 5 litters contribute pups to each treatment condition. This procedure assures that group vairances are not artificially depressed by litter homogeneity. In most experiments a sex × condition × age analysis of variance has been used, with littermates treated as nested factor.

TECHNIQUES USED FOR DEMONSTRATING CONTINUITY OF INFANTILE AND ADULT LORDOSIS

One of the first things that I discovered about infantile lordosis was that unlike the adult response, it could be elicited (1) in males and (2) without estrogen or progesterone priming (Williams, 1987). Thus, although the motor pattern and the tactile stimuli initiating the response were similar to those of the adult, it was possible that there was little functional continuity between the two behaviors. In order to use infant lordosis as the starting point for a developmental analysis of a hormonally controlled sexually dimorphic response, I needed some confirmation that what we were studying was the infantile form of a behavior that would gradually be elaborated into the adult pattern. To this end, two studies were initiated to examine the neuroanatomical and hormonal controls of lordosis and ear wiggling in the 6-day-old rat.

Neuroanatomical Pathways

To get a broad view of the neural circuitry underlying lordosis and ear wiggling of 6-day-old male and female rats, we performed a series of neural transections to see which brain regions were critical to support the behaviors in the neonate (Williams & Lorang, 1987). The procedure used for this study was one first utilized by Kornblith and Hall (1979) to examine infantile ingestive responses. Since this neurosurgery technique is quite different from anything that is used with adults, I will describe it in detail.

Essentially, a series of brain transections is made along the neuraxis from the caudal extent of the olfactory bulbs to the medulla in order to determine the amount and localization of neural tissue required for lordosis. Pups are removed from the dam and anesthetized in an ice bath (Phifer & Terry, 1986; Hofer, chap. 2, this volume). This procedure is particularly good for neurosurgery because the cooling minimizes hemmorhage caused by the brain cuts. Within each litter, cuts should be aimed at several different sites, and a control pup of each sex should always be included. Data can then be examined by comparing the behavior of the transected pups to control pups that were raised in the same litter and operated on at the same time.

For surgery a pup is placed in a prone position with its head held firmly between two fingers, and the surgeon makes a small incision in the scalp at the site of the transection using a pair of microscissors. With a 22-gauge needle, a hole is bored in the skull just lateral to the midsagittal suture at the exact level of the transection. This is done to avoid puncturing the midsagittal sinus. Then, a 30-gauge needle with a blunted tip is inserted at a 60° angle into the hole to the base of the brain and is swept across the brain in a single lateral movement (like a windshield wiper). To prevent injury to the base of the brain, we predetermined the length of the needle needed by using the skull and brain of a sacrificed littermate. Then, a small piece of rubber was used as a stopper so that the needle could be inserted only to specified depth. After, the needle is withdrawn and the wound is dried and closed with a very small amount of cyanoacrylate cement (Crazy Glue works just fine). A control pup receives the same treatment, except that after the hole is made in the skull the blunted needle is inserted to the base of the brain and slowly withdrawn, without any lateral movements. Pups are gradually warmed to room temperature (approximately 30 min) and then are placed in Plexiglas tubs in the incubator for 4 h prior to testing. Pups recover quickly from this major surgery, and even those with very caudal lesions still show a righting response, as has been shown (Bignall, 1974; Kornblith & Hall, 1979; Williams, 1987). In addition, all pups are capable of actively avoiding the brush. That is, "head up" behavior is exhibited by pups in all transection conditions. Since this surgery is an acute preparation, it does not suffer from the drawbacks that would be encountered with chronically transected animals.

Verification of the transections after behavioral testing is done by fixing the brains in a 20% formalin solution with added sucrose for 2 to 3 weeks. The brains are then frozen on the stage of a microtome and sliced sagittally in 50μ sections. Staining can be done with a standard cresyl violet preparation. The brain cuts are quite easily detectable, and they can be drawn onto sagittal figures of a 10-day-old rat brain from the atlas of Sherwood and Timiras (1970).

The results of this study demonstrated several very exciting findings: (1) Only very small sex differences appeared in the effects of our transections, suggesting that the neural control of infantile lordosis is similar in males and females. (2) Transections through the hindbrain eliminated lordosis, indicating that in the infant, as in the adult

(Pfaff, 1980), supraspinal facilitation is critical for the display of lordosis by the 6-day-old rat. That is, pups' response to tactile stimulation of the flanks and rump does not appear to be a spinal reflex. Since lesions from the caudal edge of the anterior hypothalamus to the mammillary bodies rarely displayed lordosis, whereas cuts about the anterior hypothalamus had no effect on lordosis, the ventral medial hypothalamus (VMN) and its descending connections appear to be the source of a behavior-facilitating signal in the infant as in the adult. Moreover, the lateral descending tracts are probably the critical pathway. (3) Unlike the adult, however, the VMN does not require estradiol to be active and facilitating. Electrophysiological studies are required for verification, but it appears that the VMN of the infant is active in the absence of estradiol. This suggests that as development progresses and the process of differentiation is completed, VMN cells may be inhibited by other neurons, and their stimulus requirements may become more specific. At this time estrogen and progesterone priming may be required to overcome this inhibition. Most important, this study demonstrates that parts of the central nervous system which mediate some components of adult female sexual behavior are functional soon after birth in the rat, even though these systems are not obviously useful to the organism until later in development.

Transection studies are useful as a first step in the analysis of the neural pathways that control a complex behavior. The problems of interpretation of such gross brain cuts are myriad. Potentially, there is interference with blood supply to structures caudal to cuts as well as interference with inhibitory and/or excitatory systems secondary to local tissue injury effects (similar to spinal shock). Follow-up studies using electrolytic lesions, or more limited knife cuts, would be a next step to delineate carefully the pathways involved in lordosis and ear wiggling.

Steroid Facilitation

Although the infantile lordosis response does not appear to require estradiol or progesterone priming, I wondered whether activational or facilitative effects of hormones on the behavior could be demonstrated. This is a critical experiment for several reasons. It would add to evidence suggesting a continuity in the infant and adult forms of the behavior, particularly if hormonal actions were localized to specific hypothalamic sites. Second, it would cause a reevaluation of the current concept of organizational and activational effects of estrogens. This is the belief that activational effects of steroid hormones cannot occur until after the process of steroid organization or differentiation is completed (for reviews, see McEwen, 1981, and MacLusky & Naftolin, 1981). In order to address these issues, I have done a number of studies to examine the effects of endogenously and exogenously administered steroid hormones on lordosis in infant rats (Williams, 1986, 1987; Williams & Blaustein, 1988). There are a number of different methods for administering hormones to infant rats. Three of the most useful techniques are described here.

The simplest method of treating infants with hormones is to dissolve the crystalline hormone in oil (sesame oil is available in pure form from Fisher Scientific) and inject subcutaneously. For pups less than 12 days of age, the volume injected should not be more than 0.05 ml. To avoid leakage of hormone in infants, a 27-gauge needle is inserted under the skin at the nape of the neck to a point on the rump, so that the oil-steroid bubble will not be near the skin puncture. Once the needle is withdrawn, the injection site is sealed with collodion (Fisher Scientific). In studies of adult rats, steroids are rarely administered by rats' body weight. However, since infants gain 50–60 g during the first 3 weeks of life, I believe that for all developmental work steroids should be administered to infants on a body weight basis. This precaution prevents a wrongful interpretation based on a gradually diminishing dose.

Using this technique, I have found that high doses of estradiol benzoate (EB) (10 and 100 μg/10 g body weight) administered 40 h prior to behavioral testing facilitate the frequency, duration, and intensity of lordosis and the frequency of displaying ear wiggling responses in infant females, while in infant males only duration of lordosis is facilitated. These data suggest that 4–6-day-old rats of both sexes possess portions of the estrogen-sensitive neural system for female sexual behavior and that under certain experimental conditions this neural system can be activated. There are, however, several notable differences between the infant and adult response: (1) Infants appear to require more EB per body weight to facilitate lordosis, possibly because they are less sensitive to EB than the adult or because some exogenous EB may be sequestered, and prevented from acting on CNS tissue, by an estrogen-binding protein, alpha-fetoprotein, which is in high concentrations in fetal blood. (2) In contrast to the adult, infant males display lordosis, and EB priming facilitates at least one component of the behavior.

The only major drawback to subcutaneous injection of hormones in an oil solution is that it is difficult to control the exact timing of hormone administration. Hormones, given in this fashion, are released slowly into circulation and may elevate hormone levels for many days after a single injection. Therefore, for some types of studies it is necessary to use a different means of hormone delivery. One method that is currently in use in my laboratory is to implant a small capsule of steroid under the skin and then remove it several hours or days later. Capsules are made out of Silastic tubing (Silastic medical-grade tubing, Dow Corning) packed with crystalline hormone. The packing can be easily done by tamping the open end of tubing into a small mound of crystalline hormone. The rate of hormone release into the circulation is controlled by surface area; thus, a shorter length or smaller diameter tubing provides comparably less hormone. I have found that tubing 0.058 mm i.d. and 0.077 mm o.d. is a good size for infants less than 12 days of age. Capsules should be sealed at both ends with Silastic adhesive (Dow Corning), soaked in absolute alcohol for 1 h to clean them, and checked for leakage. This is a modification of the procedures described by Meaney and Stewart (1981) and McEwen, Lieberberg, Chaptal, and Krey (1977). These capsules can then be implanted under the skin on the back using hypothermia as anesthesia and cyanoacrylate cement to close the wound.

A third method we have used to administer hormones to infants is to implant the crystalline hormone directly into the brain (Williams, 1986). The technique we have used is a modification of the intracerebral implant procedure described by Christensen and Gorski (1978) and Nordeen and Yahr (1982). For this procedure, pups are anesthetized using hypothermia and placed securely into a wax body and head mold attached to the platform of a Kopf sterotaxic apparatus. Coordinates for implant locations can be taken from the atlas of Sherwood and Timiras (1970). Implants are made by tapping 30-gauge stainless steel tubing (Small Parts, Inc.) into a small amount of crystalline hormone diluted with cholesterol and a small amount of carbon black. By using the dilution technique, it is possible to obtain implants that contain very small amounts of estradiol (i.e., in the 5–25 ng range), which do not facilitate lordosis if administered systemically. The carbon black serves as a marker during histology, since cannula tracts are very difficult to localize in the infant brain (see Barr, chap. 19, this volume). Although no work has examined this issue in the infant rat, Davis, Krieger, Barfield, McEwen, and Pfaff (1982) have provided convincing data using an autoradiographic analysis, indicating that 5–7 ng of estradiol implanted into the ventromedial hypothalamus of an adult female rat spreads less than 0.5 mm after 5–7 days. However, spread is dependent upon numerous factors: implant size, hormone dose, duration of implant, and implant location. Therefore, it is nearly impossible to estimate the exact diffusion hormone for any given study. The most valuable control to determine diffusion from implant sites is to examine neural sites at which estradiol does not act to facilitate lordosis. If implants into adjacent brain regions have no effect on behavior, then the implant's effects are restricted.

To implant a hormone pellet, the cannula tip is inserted just above the desired location. The implant is extruded by pushing a stylet through the cannula. The cannula and stylet should be constructed prior to implantation so that the stylet tip is just visible when it is inserted fully. Using this technique, I have found that estradiol in a dose as low as 10 ng is sufficient to facilitate lordosis in 4–6-day-old female rats when it is implanted into the region of the anterior hypothalamus-ventromedial hypothalamus, but not to the preoptic region (Williams, 1986). These data suggest that in the infant, as in the adult (Davis, McEwen, & Pfaff, 1979), estradiol stimulation of the ventromedial hypothalamus is sufficient to elicit lordosis. Further work is needed to examine other potential sites of action of estradiol in infant females, to examine whether similar sites activate lordosis in infant males, and to determine whether, and at what sites, progesterone facilitates lordosis and ear wiggling in infants.

In my studies of hormone facilitation of lordosis in the infant, one important finding causes some rethinking of our current notion of the organizational and activational actions of hormones and also suggests a caution to those interested in studying hormonal facilitation of behavior in week-old rats. The normally high serum testosterone levels of week-old male rats activates lordosis through its aromatization into estradiol at specific hypothalamic sites. Three lines of evidence support our view that endogenous testosterone activates female responses in the not-yet-defeminized infant rat brain: (1) Williams (1987) found that castration of newborn male rats decreases the frequency of lordosis compared with their sham-castrated littermates and castrated littermates treated with testosterone. (2) Williams and Blaustein (1987) demonstrated that endogenous testicular secretions of male infants increase the concentration of progestin-binding sites in the preoptic-medial basal hypothalamic area, whereas castration significantly decreases estradiol-induced progestin receptor formation. (3) Progesterone treatment to intact 6-day-old rats increases lordosis frequency of male infants but not female infants, suggesting that endogenous testicular secretions converted to estradiol in the hypothalamus of infant males acts synergistically with exogenously administered progesterone to facilitate lordosis. Together, these data suggest that in week-old male rats, endogenous testosterone can activate their not-yet-defeminized brain; at the same time, it is causing long-term effects that will lead to a permanent masculinization and defeminization of brain regions involved in reproductive behavior. Perhaps more important for the purposes of this chapter is that week-old male infants are normally providing endogenous testosterone (and aromatized estradiol) to their neural circuits. Researchers interested in comparing hormone effects on male and female infants should take this into account when doses are being compared.

PRECOCIOUS BEHAVIORS AND THE STUDY OF DEVELOPMENT

The discovery of hormonally modulated responses in the infant rat that appear to be continuous with the lordosis and ear wiggling behaviors normally displayed by receptive adult female rats should give us a starting point to trace the development of a sexually dimorphic behavior. Initial work has demonstrated that although the behaviors of the infant and adult rat share a common neural substrate and are both hormone sensitive, there are some intriguing differences between the control of the behaviors in infants and adults: The behaviors in the infant are not dependent on hormonal facilitation, and males are just as likely as females to show the responses. Our current work focuses on using lordosis and ear wiggling as windows on the development of neural, neurochemical, and neurohormonal functioning of the brain. Current work in our laboratory indicates that males lose the lordosis and ear wiggling responses, and females become completely dependent upon hormonal facilitation between 12 and 18 days of age (Williams & Blaustein, 1987). These data suggest that organizational effects of hormones occurring in the first week of life may set in motion a chain of events that are not functional until after the second week of life. Several important questions about the development of this sexually dimorphic behavior still need to be answered: (1) In males, what neural system is inhibited, and by what process is this inhibition occurring? (2) In females, what causes the sensory and hormonal requirements to become more specific?

Our analysis of infant lordosis complements other work demonstrating that under certain experimental conditions, components of the neural and behavioral systems underlying certain motivated behaviors are functional soon after birth in the rat, even though these systems are not obviously useful to the organism until later in development (see Stehower, chap. 8, this volume, for another example). Hall (1979a, 1979b) was the first to demonstrate that a newborn rat has neural and behavioral capacities to feed independently of its dam and that this feeding ability has some rudiments of physiological control. Development in this system adds additional capabilities and organizational complexity (Hall, 1985). In contrast, lordosis is lost in the male as development progresses and comes under strong hormonal control in the female. The analysis of a sexually dimorphic system may provide some interesting insights into a different sort of behavioral process. A seemingly analogous system may be the functional development of binocular vision in visual cortical cells (see Hubel & Wiesel, 1977). Infant kittens have cortical cells that are born with some intrinsic properties or predispositions toward monocularity or binocularity. If the kitten has early experience with reduced patterned visual input to a single eye, there is a dramatic decrease in the number of binocularly driven cells, and the eye is functionally blind. However, removal of the other eye rapidly restores function to the blind eye, suggesting that there is an active inhibition caused by experience. Perhaps testicular hormones cause a similar sort of inhibition to occur on the neural systems that subserve female sexual behavior.

It may be possible to trace the development of other sexually dimorphic behaviors. A report by Brunelli, Shindledecker, and Hofer (1987) shows that 18–30-day-old male and female rats show components of maternal behaviors when confronted with week-old pups. At 30 days of age, maternal blood plasma facilitates maternal behaviors in female but not male infants. These data clearly indicate that prior to puberty infant males have the capacity to show several behaviors that are hormonally controlled and hard to elicit in males when they are adult. Again, it would be interesting to determine what causes the inhibition of maternal responses in males.

One final consideration is that we do not really know why infant animals have the neural and behavioral capacities to display behaviors involved with reproduction and parental care prior to puberty. Two possible explanations have been offered, but as yet no research has addressed these important issues. These experimentally induced behaviors may used by the infant rat, but in a context quite different from the one in which it is expressed in the adult. For instance, a lordosis-type response accompanies the micturition reflex in the infant rat, and this response is elicited by maternal licking of the anogenital region and by maternal grasping of the pup's midriff region (Moore et al., 1986). Perhaps, as has been suggested by Komisaruk (1978), lordosis is simply a motor pattern that is part of a very general neural system. In the infant this pattern is elicited by a variety of stimuli. As development progresses, the neural system is actively inhibited in both sexes, but in the female hormonal activation can overcome the inhibition. This view of the development of a sexually dimorphic behavior suggests that one should examine whether lesions or transections that disrupt lordosis in the infant might also disrupt the lordosislike behavior that accompanies the micturition reflex.

An alternate view of this research is that the precocious abilities and their adultlike neural controls may serve no immediate function for the pup. These are systems that have evolved and develop in advance of function because it is adaptive for them to do so. Anokhin (1974) has argued that this may be the normal fashion in which neural systems develop. If this is the case, these adultlike behaviors that infants are capable of showing could be used as assays of the development of function of neural mechanisms important to the display of adult-motivated behavior. Experiments and clever techniques that ask the brain what it can do, rather than what it normally does, might reveal that an infant brain has many more capabilities than we commonly believe.

ACKNOWLEDGMENTS

This research was supported by NS 20671 from the National Institute of Neurological and Communicative Disorders and Stoke of the National Institutes of Health. The methods described here have been tested and refined by my students: Andrea Baron, Cynthia Keller, Odine Kleiner, Dominique Lorang, Lena Shnayder, and Susan Yoo. I thank them for their input. Comments by G. Benedict, R. Cohen, M. Hofer, and H. Shair were greatly appreciated.

REFERENCES

Anokhin, P. V. (1974). *Biology and neurology of the conditioned reflex and its role in adaptive behavior*. Oxford, England: Pergamon Press.

Bignall, K. E. (1974). Ontogeny of levels of neural organization: The righting reflex as a model. *Experimental Neurology, 42*, 566–573.

Beach, F. A. (1966). Ontogeny of "coitus-related" reflexes in the female guinea pig. *Proceedings of the National Academy of Science, 56*, 526–533.

Boling, J. L., Blandau, R. J., Wilson, J. G., & Young, W. C. (1939). Post-parturitional heat responses of newborn and adult guinea pigs. *Proceedings of the Society of Experimental Biological Medicine, 42*, 128–132.

Bornstein, B. H., Terry, L. M., Browde, J. A., Jr., Assimon, S. A., & Hall, W. G. (1987). Maternal and nutritional contributions to infant rats' activational responses to ingestion. *Developmental Psychobiology, 20*, 147–164.

Brunelli, S. A., Shindledecker, R. D., & Hofer, M. A. (1987). Behavioral responses of juvenile rats to neonates after infusion of maternal blood plasma. *Journal of Comparative Psychology, 101*, 47–59.

Christensen, L. W., & Gorski, R. A. (1978). Independent masculinization of neuroendocrine systems by intracerebral implantation of testosterone or estradiol in neonatal female rat. *Brain Research, 146*, 325–332.

Davis, P. G., Krieger, M. S., Barfield, R. J., McEwen, B. S., & Pfaff, D. W. (1982). The site of action of intrahypothalamic estrogen implants in feminine sexual behavior: An autoradiographic analysis. *Endocrinology, 111*, 1581–1586.

Davis, P. G., McEwen, B. S., & Pfaff, D. W. (1979). Localized behavioral effects of tritiated estradiol implants in the ventromedial hypothalamus of female rats. *Endocrinology, 104*, 898–903.

Diakow, C. (1975). Motion picture analysis of rat mating behavior. *Journal of Comparative and Physiological Psychology, 88*, 704–712.

Hall, W. G. (1979a). Feeding and behavioral activation in infant rats. *Science, 205*, 206–209.

Hall, W. G. (1979b). The ontogeny of feeding in rats: I. Ingestive and behavioral responses to oral infusion. *Journal of Comparative and Physiological Psychology, 93*, 977–1000.

Hall, W. G. (1985). What we know and don't know about the development of independent ingestion in rats. *Appetite, 6*, 333–356.

Hall, W. G., & Williams, C. L. (1983). Suckling isn't feeding or is it? A search for developmental continuities. *Advances in the Study of Behavior, 13*, 219–254.

Hofer, M. A. (1975). Studies on how early maternal separation produces behavioral change in young rats. *Psychosomatic Medicine, 37*, 245–264.

Hofer, M. A. (1981). Parental contributions to the development of their offspring. In D. J. Gubernick & P. H. Klopfer (Eds.), *Parental care in mammals*. New York: Plenum.

Hubel, D. H., & Wiesel, T. N. (1977). Grass Foundation Lecture. In *Society for Neruoscience, Sixth annual meeting, 1976, B.I.S. Conference Report No. 45* (pp. 9–24). Los Angeles: Brain Information Service.

Johanson, I. B., & Hall, W. G. (1980). The ontogeny of feeding in rats: III. Thermal determinants of early ingestive responding. *Journal of Comparative and Physiological Psychology, 94*, 977–992.

Jost, A. (1985). Sexual organogenesis. In N. Adler et al. (Eds.), *Handbook of behavioral neurobiology* (Vol. 7, pp. 3–19). New York: Plenum.

Komisaruk, B. K. (1978). The nature of the neural substrate of female sexual behavior in mammals and its hormone sensitivity: Review and speculation. In J. B. Hutchinson (Ed.), *Biological determinants of sexual behavior* (pp. 349–394). New York: Wiley.

Komisaruk B. K., & Diakow, C. (1973). Lordosis reflex intensity in rats in relation to the estrous cycle, ovariectomy, estrogen administration and mating behavior. *Endocrinology, 93*, 548–557.

Kornblith, C. L., & Hall, W. G. (1979). Brain transections selectively alter ingestion and behavioral activation in neonatal rats. *Journal of Comparative and Physiological Psychology, 93*, 1109–1117.

MacLusky, N., & Naftolin, F. (1981). Sexual differentiation of the central nervous system. *Science, 211*, 1294–1303.

McEwen, B. S. (1981). Cellular biochemistry of hormone action in brain and pituitary. In N. T. Adler (Ed.), *Neuroendocrinology of reproduction* (pp. 485–518). New York: Plenum.

McEwen, B. S., Lieberberg, I., Chaptal, C., & Krey, L. C. (1977). Aromatization: Important for sexual differentiation of neonatal rat brain. *Hormones and Behavior, 9,* 249–263.

Meaney, M. J., & Stewart, J. (1981). Neonatal androgens influence the social play of prepubescent rats. *Hormones and Behavior, 15,* 197–213.

Moorcroft, W. H., Lytle, L. D., & Campbell, B. A. (1973). Ontogeny of starvation-induced behavioral arousal in the rat. *Journal of Comparative and Physiological Psychology, 75,* 50–67.

Moore, C. L., & Chadwick-Dias, A. (1986). Behavioral responses of infant rats to maternal licking: Variations with age and sex. *Developmental Psychobiology, 19,* 427–438.

Moran, T. H., Schwartz, G. J., & Blass, E. M. (1983). Organized behavioral responses to lateral hypothalamic electrical stimulation in infant rats. *Journal of Neuroscience, 3,* 10–19.

Nordeen, E. J., & Yahr, P. (1982). Hemispheric asymmetries in the behavioral and hormonal effects of sexually differentiating mammalian brain. *Science, 218,* 391–394.

Pettigrew, J. D. (1978). The paradox of the critical period for striate cortex. In C. W. Cotman (Ed.), *Neuronal plasticity*. New York: Raven Press.

Pfaff, D. W. (1980). *Estrogens and brain function*. New York: Springer.

Pfaff, D. W., Montgomery, M., & Lewis, C. (1977). Somatosensory determinants of lordosis in female rates: Behavioral definition of the estrogen effect. *Journal of Comparative and Physiological Psychology, 91,* 134–145.

Phifer, C. B., & Terry, L. M. (1986). Use of hypothermia for general anesthesia in preweanling rodents. *Physiology and Behavior, 38,* 887–890.

Phoenix, C. H., Goy, R. W., Gerall, A. A., & Young, W. C. (1959). Organizing action of prenatally administered testosterone propionate on the tissues mediating mating behavior in the guinea pig. *Endocrinology, 65,* 369–382.

Sherwood, N. M., & Timiras, P. S. (1970). *A stereotaxic atlas of the developing rat brain*. Berkeley: University of California Press.

Sodersten, P. (1975). Receptive behavior in developing female rats. *Hormones and Behavior, 6,* 307–317.

Sodersten, P. (1978). Lordosis behavior in immature male rats. *Journal of Endocrinology, 76,* 233–240.

Spear, N. E., & Campbell, B. A. (Eds.) (1979). *The ontogeny of learning and memory*. Hillsdale, NJ: Erlbaum.

Williams, C. L. (1986). A reevaluation of the concept of separable periods of organizational and activational actions of estrogens in development of brain and behavior. *Annals of the New York Academy of Sciences, 74,* 282–292.

Williams, C. L. (1987). Estradiol benzoate facilitates lordosis and ear wiggling of 4- to 6-day-old rats. *Behavioral Neuroscience, 101,* 718–723.

Williams, C. L., & Blaustein, J. D. (1987). Development of hormonal control of lordosis and ear wiggling in male and female rats: Relation to hypotnalamic progestin receptor concentration. *Conference on Reproductive Behavior Abstracts, 36.*

Williams, C. L., & Blaustein, J. D. (1988). Steroids induce hypothalamic progestin receptors and facilitate female sexual behavior in neonatal rats. *Brain Research, 449,* 403–407.

Williams, C. L., & Lorang, D. (1987). Brain transections differentially alter lordosis and ear wiggling of 6-day-old rats. *Behavioral Neuroscience, 101,* 819–826.

13

Interactions Between the Developing Immune and Neuroendocrine Systems

MAUREEN P. O'GRADY AND NICHOLAS R. S. HALL

Less than a decade ago, the immune system was considered to carry out its functions in relative isolation from the known neural and hormonal regulatory activities that linked other physiologic systems to the adaptive needs of the organism in its environment. The immune system had its own developmental plan and responded, it was thought, only to certain molecules that entered the body. Most of its functioning could be demonstrated to occur in vitro and had, in fact, been elucidated in highly simplified laboratory systems extracted from the living organism.

The application of new methods forced a change in this view of the immune system and led to the discovery of the neuroendocrine and immune interactions described in this chapter. Some of the first studies, done in the 1960s and 1970s, employed the classic methods of early developmental psychobiology: contrived alterations of early experience (e.g., early weaning, crowding, handling) and testing in adulthood for modification of the course of laboratory model diseases (e.g., bacterial and parasitic infections, transplanted tumors) in which the immune system played a central role. These studies made it seem highly probable that development of the immune system could be modified by experiences other than exposure to antigens and that neural and/or endocrine influences were the most likely pathways. Other work demonstrating changes in immune function after hypothalamic lesions and stress-inducing protocols, as well as Ader's demonstration that an immune response could be classically conditioned, launched a new field, psychoneuroimmunology. These and other pioneering lines of work are summarized in Ader (1981).

In the last decade, new methods for neuroanatomical identification of innervation in lymphoid tissue, for identification of protein messenger molecules and their membrane receptors on elements of the immune and nervous systems, and for analysis of steroid hormones and their receptors have together opened up a new territory of interactions in which it is clear that the central nervous system (CNS) is influenced by immune function as well as the reverse. As usual, studies of early stages in the development of these interactions have lagged behind studies of the mature system.

This chapter addresses issues relating to the development of the immune system. First, we discuss briefly the ontogeny of the immune system and the role of maternal factors in the ability of the fetus/newborn to ward off infection. Next, we examine the interaction between the central nervous system and the immune system. Evidence describing the malleability of the developing immune and endocrine systems is then discussed. Finally, keenly aware that the immune system, the endocrine system, and the nervous system are not developing in isolation but in the context of the whole body, which is constantly bombarded by antigenic, sensory, and physiological stimuli, the last section discusses the methodological limitations of the study of developmental neuroimmunology and suggests areas for future research.

IMMUNE SYSTEM

A brief overview of the immune system is first given to orient readers who are not immunologists to the terms, concepts, and major constituents of the system and its links to neural and endocrine systems. A more detailed review for the nonspecialist may be found in Guillemin, Cohn, and Melnechuk (1985).

During the embryonic period, cells migrate from the yolk sac to the developing bone marrow and spleen, where they become precursors of immune system elements. These stem cells continue to be produced in the bone marrow throughout life. Some migrate to the thymus, where they differentiate, in response to local tissue influences, into T lymphocytes, one of the major components of the immune system. Other stem cells differentiate into another major branch comprised of B lymphocytes. Once mature, these cells migrate to lymph nodes and other lymphoid tissue throughout the body where they are ready to carry out their functions in interaction with local cells and tissue factors.

On these T and B lymphocytes are specific surface receptors that respond to molecules of a certain configuration—antigens. B cells respond to antigens by producing antibodies, a humoral immune response consisting of one or another type of immunoglobulin (Ig). These antibodies bind specifically to the antigen and promote its removal and destruction by phagocytic cells. T cells, in contrast, are the agents of the cell-mediated immune response. There are four major types of T-cell responses: helper T cells are essential for the production of humoral antibody by the B cells, and they produce lymphokines, which regulate B- and T-cell proliferation as well. Cytotoxic T cells recognize pathogenic cells and destroy them. Suppressor T cells interact with other cell types in a modulatory or regulatory manner, preventing excess activity. Finally, delayed hypersensitivity responses, as in tissue graft rejection and responses to chronic infection, are mediated by another subpopulation of T cells interacting with cytokines.

The capacity of the immune system to recognize a large number of antigens is made possible by the range of distinct receptor types present on lymphocytes, estimated in excess of one million. Only a few cells within this huge range respond to a given antigen. A proliferative response follows in which cells with this particular receptor type multiply. This provides a greater number of cells to respond in the future if the antigen is encountered a second time, as well as producing quantities of antibody and/or T-cell types in response to the initial encounter. Gradually individuals develop an immune "library" that reflects the history of their encounters with different specific antigens. To the extent that the recent past is a good predictor of the future, this prepares the organism for the next environmental challenges.

The thymus gland, in addition to its role in T-cell differentiation, releases a number of peptide products, including the thymosins. These not only play a role in

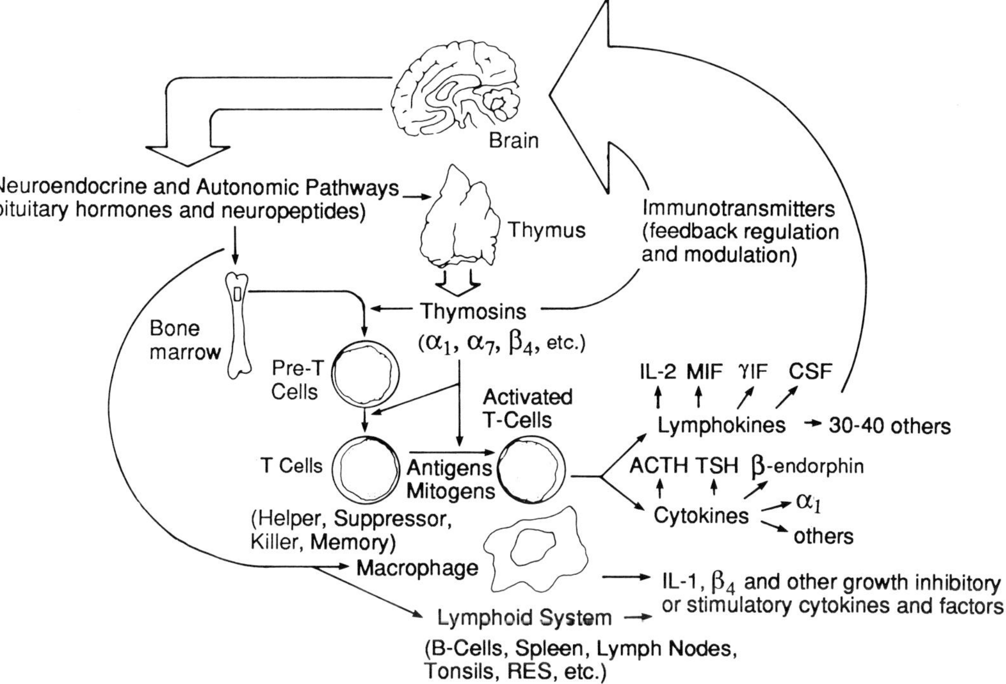

Figure 13.1. Schematic diagram of interactions among immune, endocrine, and neural systems. (Reproduced with permission from "Evidence That Thymosins and Other Biologic Response Modifiers Can Function as Neuroactive Immunotransmitters" by N. R. Hall, J. P. McGillis, B. L. Spangelo, and A. L. Goldstein, 1985, *Journal of Immunology, 135*(2), p. 807s.)

activating T cells but also transmit messages to the CNS and periphery. Together with the soluble products of the cellular immune response (lymphokines and cytokines) and the products of T-cell interaction with other cells such as macrophages (interferons, prostaglandins, and leukotrienes), these form a novel class of neuromodulators capable of influencing CNS and endocrine systems as well as the local cellular events of the immune response.

Steroid hormone interactions are somewhat better known, particularly the suppression of immune function by glucocorticoids. It is less well known that gonadal steroids can influence the development of immune function and that activated lymphocytes produce small amounts of adrenocorticotrophic hormone, beta-endorphin, thyroid-stimulating hormone, and human gonadotrophinlike peptides. Clearly, there are opportunities for immune regulation of endocrine and CNS function as well for regulation of the immune system by the other two. Figure 13.1 illustrates and summarizes these interactions.

Ontogeny of the Immune System

A review of the development of the immune system reveals that this process begins in the uterus, so that at birth the infant is already able to perform some immune functions. The findings to date show that in the human fetus, as in other species, the individual immune tissues and functions make their appearances at different times. The appearance of immunocompetence is dissociated in time across branches of the immune system (e.g., T-cell function vs. immunoglobulin synthesis) as well as within branches of the immune system. For example, B lymphocytes appear sequentially in a specific order by a process independent of antigenic stimulation.

The immune response can be divided into "innate" and "adaptive" components. Thus, when an infant is newly exposed to a pathogenic microorganism such as *Bordatella pertussis* (the agent for whooping cough), the innate immune system reacts and processes the antigen, but its response is too slow to prevent illness and risk of death. In contrast, the infant who has been immunized has had the benefit of activation and "education" of the adaptive immune system. Thus, immunologic memory allows for protection from reinfection by an antigen that has been processed previously. The individual has a repertoire of different immune responses available, though in the infant these are not all fully functional.

While some immune functions develop to effective levels during intrauterine life, others are incapable of full development because of inadequate antigenic stimulation. From an immunologic standpoint, birth represents an end to maternal protection and immediate exposure to many antigens. The exposure of the newborn to a host of antigens (primarily microbial) provides powerful antigenic stimulation of the immune system and furthers it subsequent development. The change is largely a quantitative one; although the uterus is not an antigen-free environment, the range and quantity of antigens to which the newborn infant is exposed is much greater than in gestation. Likewise, once the main immune reactions listed above first appear, it is quite some time before they reach adult levels. For instance, the ability to synthesize immunoglobulins appears in utero in the following order: IgM, IgG, and IgA, but adult serum concentrations of these immunoglobulins occur much later: IgM reaches adult levels by 12 months of age, IgG by 5–6 years, and IgA at about puberty. (These kinetics cannot be explained by lack of precursor B lymphocytes in utero since cells bearing membrane IgG and IgA are present in adult frequencies by midgestation [Gathings, Lawton, & Cooper, 1977].)

The sequential development of the various components (i.e., humoral vs. cell mediated; adaptive vs. innate immunity) of the immune system is complex; for details the reader should see Toivanen et al. (1981), Alford, Cartwright, and Sell (1976), or Stites, Caldwell, Carr, and Fudenberg (1975).

Interactions between T and B cells in the neonate are dominated by suppressor activity. There is evidence that the suppressor T cells present in the newborn may selectively inhibit responses by adult cells, and Olding, Benirsch, and Oldstone (1974) have suggested that the suppression of maternal immunologic functions by fetal cells might serve an important physiological role in the maintenance of pregnancy. This suppression has subsequently been shown to be mediated by T cells and is present in peripheral blood as early as 12 weeks of gestation and persists up until one year after birth (Unander & Olding, 1981).

Maternal-Placental-Fetal Interactions

For many years, pediatricians have observed that certain infections such as measles, chickenpox, and hepatitis A are rarely seen in infants less than 4–6 months of age. Conversely, infections caused by pathogenic strains of *E. coli*, staphylococci, and certain respiratory viruses are both more common and more severe in the postnatal period than in older children. These differences in the susceptibility of newborn and older infants to these types of infections are primarily determined by the passive transfer of immunoglobulins from the mother to the fetus during intrauterine life and by the short half-lives of certain antibody types.

The fetus possesses an adequate number of B lymphocytes with surface membrane immunoglobulin receptors, and these cells are capable of differentiation into antibody-producing cells (given adequate antigenic stimulation) in the first trimester of intrauterine life. However, the placenta protects the fetus from much antigenic exposure, so that the level of immunoglobulin synthesis by the fetus is rather low. To compensate partially for this immunoglobulin deficiency in humans, IgG antibodies in maternal serum are transmitted across the placenta into the fetal bloodstream. The primary biological purpose of this mechanism seems to be that of immediate passive protection for the fetus and newborn against infection by pathogenic microorganisms against which normal immunity is effective. For example, Solomon (1970) reports that there are several antibodies transmitted from the mother to the fetus in intra-uterine life that will protect the neonate against various infections. These include antibodies against measles, hepatitis A, German measles, poliomyelitis, diphtheria, and varicella.

However, antibodies against streptococci, staphylococci, and *E. coli* are not transmitted across the placenta. Not all the immunoglobulin classes can cross the placenta from mother to fetus during pregnancy. The impermeability of the human placental barrier to IgM, IgA (Stiehm & Fudenberg, 1966), and IgD immunoglobulins (Leslie & Swate, 1972) is well documented. Likewise, Ishizaka and Ishizaka (1967) found that IgE did not cross the placenta in amounts that could protect the neonate. Only IgG immunoglobulins are transmitted from mother to fetus (in humans) such that cord blood levels of IgG equals that of maternal serum.

There are species differences regarding the transmission of maternal antibodies to the fetus. In humans and monkeys, this transmission takes place for IgG antibodies in intrauterine life, as described above. However, in other species (e.g., pigs), the placenta is impermeable to maternal antibodies throughout gestation, and the young receive maternal antibodies solely in the colostrum and milk (see below).

Finally, the events surrounding delivery can modify newborn immune status. Cord blood was studied in newborns delivered after maternal labor and compared to neonates delivered by caesarean section without preceding labor. The data indicated that vaginal delivery decreased the total number of T lymphocytes and the subpopulation of T helper cells in cord blood (Pittard, Schleich, Geddes, & Sorensen, 1989). Maternal anesthesia is not likely to underlie these differences, since the same drug was administered by the same route to all the mothers. This observation is strengthened by previous reports by the same group on functional differences in immune responses involving both B and T lympocytes between vaginally and caesarean-delivered children (Pittard, Miller, & Sorensen, 1984, 1985). Thus, these studies suggest that the mode of delivery can impact on the immune system of the newborn.

Immunologically Active Factors in Colostrum and Milk

In addition to the IgG antibodies transmitted across the placenta and to its own defenses built up during gestation and after birth, the newborn infant receives further defense factors in maternal milk and colostrum, which aid the infant in its resistance to gastrointestinal infection. The significance of maternal colostrum and milk is manifested most dramatically in some animals (especially pigs) in which there is no transplacental antibody transmission during gestation: The first protective antibodies are not acquired until after birth, in colostrum and milk, when they are reabsorbed from the intestine into the bloodstream. In such species, survival of the neonate literally depends on the ingestion of colostrum in the first hours after birth.

The human neonate's survival is not directly dependent on breastfeeding, but infants fed breast milk have a reduced incidence of severe gastroenteritis and respiratory infections compared with those fed nonhuman milk or milk substitutes.

Analysis of human milk and colostrum shows that they contain cellular and humoral, specific and nonspecific immune products. Human milk and colostrum contain all the immunoglobulin classes (Bennich & Johansson, 1971). The most important component in the newborn is IgA (Ogra & Ogra, 1978), for which maternal milk is the main source, the amount of IgA produced locally by the lymphoid tissue of the infant's intestinal mucosa being rather small at this time (Gunther, 1975). IgA antibodies found in milk possess specificity for infectious agents endemic to or pathogenic for the intestinal or respiratory tracts (e.g., *E. coli*, *Salmonella*, *Streptococcus pneumoniae*, herpes simplex 1, Coxsackie viruses A and B, influenza A virus, poliomyelitis virus types 1, 2, and 3). The highest IgA concentration is found in the colostrum in the first days of lactation; later, in the milk, it falls to one tenth of the original value and remains at this level until lactation ends (Carlsson, Gotheford, Ahlstedt, Hanson, & Winberg, 1976). IgM levels show a similar pattern to that of IgA, though the levels are four to five times less than those of IgA. The concentration of IgG remains rather low in colostrum and milk throughout lactation.

In addition to the immunoglobulins discussed above, a host of other soluble factors and cellular components as well as contaminants are contained in human colostrum and milk. Briefly, these include immunologically specific agents such as T-lymphocyte products and histocompatibility antigen. There are a multitude of nonspecific soluble products in human milk and colostrum, including all the complement components (C1–C9), chemotactic factors, interferon (IFN), alphafetoprotein, bifidus factor, lactoferrin, lysozome, migration-inhibition factor, antistaphylococcal factor(s), and antiadherence substances. Different cells of the leukocyte line, which are plentiful in the mammary gland, pass into the colostrum and milk in relatively large amounts. T and B lymphocytes, neutrophils, macrophages, and epithelial cells are all found in human milk and colostrum. Lastly, several contaminants can sometimes be found in human milk and colostrum: bacterial antigens, rubella, HIV-1, herpes simplex, pharmacologic products (drugs, antibiotics, and diagnostic radioisotopes), environmental pollutants, and toxins.

Although living under the protection of the maternal immune system throughout pregnancy, the fetus is subjected to several types of immunogenic stimuli, including viral pathogens. The majority of infections in pregnant women occur in the gastrointestinal and upper respiratory tracts and are either treated with antimicrobial agents or resolve themselves spontaneously without treatment. These infections are usually localized and do not affect the developing fetus. However, the infectious agent may invade the bloodstream, and the fetus can become infected. Since its own immune system is not yet fully developed, the fetus has an increased susceptibility to infection.

A list of some microorganisms that have deleterious effects on the developing fetus via transplacental fetal infection are the following: viruses such as rubella, cytomegalovirus (CMV), Coxsackie B, hepatitis B; bacteria such as *Mycobacterium tuberculosis* and *Listeria monocytogenes*; and protozoa such as *Toxoplasma gondii* (Klein, Remington, & Marcy, 1983). In addition, an alarmingly high number of children born in some regions of the United States (i.e., 1 of 60 births in New York, according to a recent state survey) have been exposed to the human immunodeficiency virus (HIV). The impact of the viral infection as well as the subsequent immune response upon developmental processes has far-reaching ramifications, some of which will be discussed later.

Despite the fact that the immune system is still developing, the fetus is not without some protection. As discussed in the previous section, fetal cells are capable of certain antibody responses. In addition, infants are capable of producing some of the factors involved in cell-mediated immunity. Macrophages have not yet reached full functional capacity in the neonate, but are able to synthesize and secrete interleukin-1 (IL-1) in amounts equal to that of macrophages from adults (Lu & Unanue, 1982). Newborns are also capable of synthesizing IFN in response to a virus infection (Vilcek, 1964).

The immunologic capacity of the neonate has evolved to protect the organism from pathogens, but there is also evidence that certain products of the immune system may have deleterious effects in addition to beneficial ones. Gresser and his colleagues have performed a series of experiments in neonatal mice to test the role of IFN in the pathogenesis of lymphocytic choriomeningitis virus (LCM). To summarize their findings, Swiss mice inoculated at birth with LCM virus develop glomerulonephritis. Administration of antiinterferon globulin at the time of inoculation protects the animal from the glomerulonephritis. If IFN alone is administered to mice during the first week of life, glomerulonephritis develops just as in LCM virus infection (Gresser, Torrey, Maury, & Chouroulinkov, 1975; Gresser, Morlmar, Verroust, Riviere, & Guillon, 1978). Nephritis results from IFN treatment only in the first week of life; no adverse effects are observed when mice are treated from days 8 to 15. Based upon these results, Gresser extrapolates his findings to humans (with caution) to suggest that some cases of nephritis of unknown etiology could be due to the delayed effect of virus-induced IFN production when the patient was young. In more general terms, there seems to be a critical period in development when the induction of endogenous IFN (e.g., from exposure to a virus) can contribute to a disease process that manifests itself later in life.

CNS-IMMUNE SYSTEM INTERACTIONS

Only in the last 10 years has it become widely recognized that the immune system, like other physiological systems contributing to the maintenance of homeostatic balance, is subject to regulation or modulation by the CNS. The CNS is capable of modulating immune function through both neuroendocrine output and autonomic nervous system projections. In support of the latter, Felten, Felten, Carlson, Olschowka, and Livnat (1985) have demonstrated both noradrenergic and peptidergic innervation of lymphoid tissue. These findings provide evidence for a structural link between the nervous and immune systems. In addition, many hormones under the control of the brain have been found capable of modulating functions of the immune system. Perhaps the best characterized immunomodulatory hormones are the adrenocortical and gonadal steroids. Gonadal hormones are considered to be, at least partially, responsible for the observed sex difference in the ability to resist disease. Females are generally less prone to infection and have been found to produce higher antibody titers compared with males (Goble & Konopka, 1973; Krzych, Strausser, Bressler, & Goldstein, 1979). Adrenal corticosteroids have differential effects upon the various cell types involved in host defense. At low concentrations they have been reported to have stimulatory influences upon certain immunologic responses (Ambrose, 1964). Ritter (1977) found that low doses of corticosterone (0.2 and 1μg/ml) potentiate embryonic thymus lymphopoiesis in vitro. However, under stressful conditions or following exogenous administration, there is a wealth of information suggesting that glucocorticoids can inhibit immune functions (Monjan, 1981). Stressed individuals representing numerous species have been found to be more susceptible to infections and have a higher incidence of neoplastic diseases (see Hall & Goldstein, 1984). This has been attributed, in part, to the release of adrenal corticosteroids.

The lymphoid system can also influence neuroendocrine systems. Products of the immune system, such as thymosins and lymphokines, can alter endocrine function as well as neural activity. Thymosin fraction 5 (TF5), a partially purified bovine thymic extract, has several neuroendocrine actions. In addition to its reported immunostimulatory effects (such as increasing the rate of graft rejection and stimulating maturation of T lymphocytes), TF5 increases secretion of adrenocorticotrophin hormone (ACTH) (McGillis, Hall, Vahouny, & Goldstein, 1985), beta-endorphin (Farah, Hall, Bishop, Goldstein, & O'Donohue, (1987), growth hormone and prolactin (Spangelo et al., 1987), and luteinizing hormone-releasing hormone (LHRH) (Rebar, Miyake, Low, & Goldstein, 1981) and reduces serum thyroid-stimulating hormone (TSH) levels in young but not in old rats (Goya, Sosa, & Meites, 1987). Thymosin is also capable of blocking the binding of corticosteroids to the lymphocytic receptor, thereby modulating the immunoregulatory effects of the hormone at the molecular level (Osheroff, 1981). Several types of evidence support these concepts. These include in vivo studies in rodents, monkeys, and humans as well as in vitro studies.

The immune response appears to influence circulating steroid levels (Besedovsky, Sorkin, Keller, & Muller, 1975), and several soluble factors produced by the thymus gland or leukocytes have been reported to produce increases in serum corticosteroid levels. These include TF5 and thymosin alpha-1, a purified component peptide of TF5 (Hall et al., 1982), and supernatant fluids from mitogen-stimulated lymphocytes (Besedovsky, Del Ray, & Sorkin, 1981), which contain lymphokines. Neonatal mice have elevated corticosteroid levels during acute reovirus type 3 infection. These elevated corticosteroid levels are restored to normal by treatment with TF5 (Wiley & Ushijima, 1981). Considering that TF5 elevates resting corticosterone levels in normal animals while restoring corticosterone to baseline in reovirus-infected neonates, it may be that TF5 plays a regulatory role in the release of glucocorticoids, which, in turn, can play an immunomodulatory role.

In addition to its antiviral, antiproliferative, and immunoregulatory effects, it seems that IFN also has the ability to modulate endocrine activity. IFN mimics ACTH through a cyclic AMP-mediated increase in murine adrenal cells (Chany, Mathieu, & Gregoire, 1980). Estradiol, progesterone (Kauppila, Cantell, Janne, Kokko, & Vihko,

1982), and thyroxine (Orava, Cantell, Kauppila, & Vihko, 1983) are suppressed in normal women after IFN administration. IFN also inhibits testosterone secretion from Leydig cells in vitro (Orava, Cantell, & Vihko, 1985).

Recent evidence from three different groups suggests that IL-1, a polypeptide produced primarily by activated macrophages and monocytes that mediate the host's response to infection and inflammation, is capable of activating the hypothalamic-pituitary-adrenal (HPA) axis. There remains some controversy whether IL-1's effect is on the pituitary (Bernton, Beach, Holaday, Smallridge, & Fein, 1987) or the hypothalamus (Sapolsky, Riviere, Yamamota, Plotsky, & Vale, 1987; Berkenbosch, Van Oers, Del Ray, Tilders, & Besedovsky, 1987) and whether IL-1 alpha and/or beta is active (Uehara, Gottschall, Dahl, & Arimura, 1987). In addition, recombinant IL-1 administered to mice (i.p.) stimulates cerebral noradrenergic neurons, as indicated by a marked elevation in levels of the norepinephrine catabolite MHPG in several brain regions, but especially the hypothalamus (Dunn, 1988). The kinetics of this effect parallel the increase in plasma corticosterone seen after IL-1 administration.

Immune system products, and the cytokines in particular, are able to alter other CNS functions. IL-1 induces slow-wave sleep (Krueger, Dinarello, Wolff, Chedid, & Walter, 1984), as do IFN and tumor necrosis factor (TNF) (Shoham, Davenne, Cady, Dinarello, & Krueger, 1987). Brain norepinephrine levels are reduced by lymphokine-containing supernatant fluids (Besedovsky et al., 1983). Activated complement proteins exert dopaminergic effects (Schupf, Williams, Hugh, & Cox, 1983). IFN has norepinephrine-like actions on myocardial cells (Blalock & Stanton, 1980). IL-1 stimulates the proliferation of astroglia isolated from brains of newborn rats and adult brain tissue that has been damaged, but not healthy tissue (Giulian & Lachman, 1985). In fact, astrocytes are capable of synthesizing both IL-1 and IL-3-like factors (Fontana, Hengartner, de Tribolet, & Weber, 1984). The fact that the brain synthesizes both interleukins and interferons leads some investigators to challenge the classical view that the brain is an immunologically privileged tissue (Frohman, Vayuuegula, Van der Noort, & Gupta, 1988). The brain seems to be capable of synthesizing and secreting IFN. Indeed, the ability of viruses to induce IFN production in the brain is not limited to adult animals. Levels of IFN in the brains of suckling mice were in direct relationship to titers of Sindbis virus (Vilcek, 1964). The levels of IFN and virus were both higher in suckling than in adult mice.

These observations are important because they point out that (1) fetuses and newborns are capable of synthesizing and secreting these cytokines and (2) immune system cytokines have biological effects outside the immune system.

CRITICAL PERIODS OF IMMUNE AND ENDOCRINE DEVELOPMENT

A multidirectional interaction among the immune, endocrine, and nervous systems is quite apparent. However, almost all of the research investigating the effects of immune products on CNS and endocrine function has been done in adults or using in vitro systems. This has not been the case with endocrine modulation of immunity and CNS activity, since there have been numerous studies examining the effects of hormones administered in early development. Alterations in the hormonal status of the newborn have profound and permanent effects on subsequent biological functioning of the host. The observation that these hormonal alterations in the neonate can produce changes that remain throughout adulthood supports the notion of "critical periods" in development when the organism is maximally susceptible to internal or external influences.

Exposure of newborns or fetuses to exogenously administered hormones can have lasting effects on immunity. Gunn, Reece, Metrakos, and Colle (1981) found a diminished percentage of T lymphocytes in 5-year-old children who had received cortisol on the first day of life, for treatment of respiratory distress syndrome, compared with placebo-treated controls. They also noted an increased number of infections in these children compared with controls during the five years between birth and the follow-up study. Similar effects can be demonstrated experimentally in rabbits. The immunosuppressive effects of hormones are not limited to cortisol. Asklen and Kalland (1982) observed reduced lymphoproliferation in response to T-lymphocyte (Con A) and B-lymphocyte (LPS) mitogens and a reduction in natural killer cell activity in adult mice that were exposed to diethylstilbestrol (DES) for the first five days of life.

These studies demonstrate that gonadal and adrenocortical hormones, which are known to have temporary immunomodulatory effects in adults, can cause permanent immune disturbances if given early in life. These observations have tremendous implications for prenatal and postnatal care, since lasting changes in immune function in these populations seems to be associated with deleterious consequences in health history (respiratory infections, cervical cancer, etc.).

The amplitude and pattern of hormone secretions as well as hormone-dependent behaviors can be altered by events occurring in neonatal life. Ward (1972) observed that the male offspring of rats stressed during pregnancy showed low levels of copulatory behavior when tested as adults. Ward hypothesized that stress-induced maternal hormonal changes altered the testosterone-dependent process of sexual differentiation of the male fetuses. Increased estrous cycle length is manifested in adult females that were stressed in utero (Herrenkohl & Politch, 1978). Subjecting pregnant rats to a stressor (handling and a saline injection) alters the capability of the HPA axis in the progeny to respond to a stressor as weanlings and as adults, as measured by plasma corticosterone and prolactin as well as brain content of norepinephrine and serotonin (Peters, 1982; Simmons, Miller, & Kellogg, 1984). Gorski, Harlan, and Christensen (1977) administered gonadal steroids to neonates, which resulted in altered patterns of gonadotrophin secretion and behavioral responses to ovarian hormones in adulthood. Levine (1967) handled rat pups for either 10 or 20 days after birth. He found that at weaning the handled pups had a diminished adrenocortical response to novel stimuli compared with nonhandled pups. In addition, nonhandled offspring of mothers who were handled in infancy also showed reduced adrenocortical response to novel stimuli compared with nonhandled offspring of nonhandled mothers. Meaney, Aitken, Van Berkel, Bhanagar, and Sapolsky (1988) reported that neonatal handling permanently increases the number of glucocorticoid receptors in the hippocampus, a region that has been linked to the negative-feedback inhibition of the HPA axis.

Thus, the evidence for the existence of critical periods of development in the immune system and endocrine systems is strong. Considering the observations presented in the previous section that fetuses and newborns are capable of immune cytokine production and that these cytokines, such as interleukins and interferons, can modulate the nervous and endocrine systems, it seems worthwhile now to investigate the role of antigenic stimulation at a critical period(s) in shaping the development of neuroendocrine systems.

DEVELOPMENTAL NEUROIMMUNOLOGY

Methodological Limitations in Developmental Studies

Most studies on the ontogeny of the human immune system have employed blood lymphocytes obtained from fetoscopy or after birth, from the placenta or newborn. Many conclusions about human B-lymphocyte function in the neonate have been based on analysis of immunoglobulins present in the serum of the newborn or infant. These studies are difficult to interpret both because of passively transmitted maternal IgG and its secondary interference with induction of in vivo antigenic challenge. Specific antigenic responses in vitro have proved difficult even in the adult human, with a few exceptions, for the most probable reason that peripheral blood is the only logistic source of lymphocytes. Since blood is available only in minimal quantities in neonates, such studies become impossible. Thus, there is a dearth of functional B-lymphocyte studies performed on the developing human, especially beyond the immediate neonatal period.

As already discussed, fetal and neonatal serum possesses high suppressor activity. Considering how interwoven the immune system is, one must pay attention as to the mechanism of action in an immune response before attributing results to one compartment of the immune system over another. For instance, immunoglobulin production in response to antigenic stimulation by Pokeweed mitogen (PWM) is low in neonates (Tosato, Magrath, Koski, Dooley, & Blaese, 1980). The interpretation of these data is complex, since PWM stimulates B cells but requires T-cell interaction to do so. Therefore, neonatal immunoglobulin response to PWM could be low due to immature B-cell function, high T-cell suppressor activity, T-helper cell immaturity, or some combination of the above. Most of our knowledge of the ontogeny of B-cell function is derived from results obtained from PWM and/or lipopolysaccharide (LPS), another T-dependent antigenic stimulus. If one wishes to study the development of B-cell function, one should also include an antigen such as Epstein-Barr virus (EBV), which stimulates B cells directly—that is, independent of T cells (Bird & Britton, 1979).

The bulk of immunological research has been and continues to be done in humans and mice. The maturational timetable is quite different in these two species: Humans are much more immunocompetent at birth than mice. In mice, proliferative responses exist at birth, but the maturation of the mitogenic response is not complete until the fourth week of life (Papiernik, Frydman, & Spira, 1980), unlike another type of immune response measure, the mixed lymphocyte reaction culture, which is mature at birth (Widmer & Cooper, 1979). There is a differential maturation pattern for the mixed lymphocyte reaction and cell-mediated lympholysis measures as well. These different patterns of maturation in the various immune compartments must be taken into account both in the design of experimental manipulations and, especially, in the interpretation of results. For instance, Ezine and Papiernik (1984) injected hydrocortisone hemisuccinate (HHC) into mice on the day of birth and studied cell-mediated immunity two weeks later. They observed a reduction in mitogen-induced proliferation, but no change in proliferation following allogeneic sensitization (MLR). Bearing in mind the maturation patterns discussed above, these observations make sense. In order to alter MLR response, one would probably need to perform prenatal HHC manipulations.

Thus, one must be careful to perform experimental manipulations at several different points in the maturational timetable and to measure several different types of immune responses before drawing conclusions. As our knowledge of the complexity of the interrelationship between the immune and neuroendocrine systems increases, we realize the difficulties inherent in ascribing causality to an immune system aberration or a neuroendocrine disturbance. In addition to regulating activity within their own respective systems, immune system products can influence neuroendocrine activity; conversely, neuroendocrine activity can affect immune development and function. Thus, the mechanism by which abnormalities occur can be complex indeed.

Suggested Areas for Future Research

The blood-brain barrier is not complete in newborns, which makes them particularly susceptible to CNS alterations caused by pathogens, endogenous immune system products, and so forth, that would not affect adults because of the inability of these products to cross the fully functioning blood-brain barrier in adults. Crnic and Pizer (1988) injected mice (s.c.) on one occasion with herpes simplex 1 virus on days 1, 2, 3, 4, or 5 after birth. When the animals were tested as adults, those that had been infected on days 1–4 were hyperactive compared with controls ($p < .01$) in an open field and in a Digiscan apparatus over a 24-hour testing period. Postmortem examination revealed that the virus had infected the vermal cortex and lateral hemispheres of the cerebellum. Thus, infection by a virus during the first 4 days after birth in mice can cause a behavioral change that is still present in 60-day-old animals.

We have investigated the effects of early viral infection or immune system products on subsequent neuroendocrine development. The male offspring of pregnant mice injected with TF5, a thymic extract, displayed enlarged adrenals as adults, whereas the female offspring had decreased ovary weights when tested as adults (O'Grady, Hall, & Goldstein, 1987). In another study of ours the virus used was Newcastle disease virus (NDV), which occurs naturally in avian species. Although NDV does not replicate well in mammals, it does induce IFN production. Rats were injected with NDV on Day 4 after birth and were left undisturbed until adulthood (3–4 months). At this time, it was observed that NDV-infected females had decreased adrenal gland weights and that NDV-infected males had a dose-dependent reduction in testes size compared with the appropriate controls (O'Grady & Hall, 1990). We are currently exploring these anatomical aberrations to determine whether there are physiological correlates.

In addition to studying the effects of NDV, we are also investigating the consequences of neonatal cytomegalovirus (CMV) infection in rats. Our preliminary findings suggest that CMV-infected animals experience a delay in CNS maturation as determined by attainment of certain developmental milestones such as negative geotaxis, acoustic startle response, and eye opening. Endocrine development seems to be altered as well, since onset of puberty is delayed in CMV-infected animals.

As discussed earlier, Gresser, Torey, Maury, and Chouroulinkov (1975) presented data on the effects of IFN in neonatal mice and rats. They noted a critical period for IFN's effect. If treatment began on the day of birth and continued for the first week of life, a wasting syndrome was induced, and the animals died. Gresser reports that IFN does not cross the placental barrier in mice. Given the lethal wasting syndrome he observed in neonates exposed to IFN, this seems to be a protection for the fetus. However, he raises the question that if a virus (such as rubella) crosses the placental barrier and induces IFN production in the fetus, are the embryotoxic effects that we have traditionally ascribed to the virus actually due to the virus or to IFN? Such questions need to be addressed.

Immunology is a relatively new field compared with other fields of science and is rapidly expanding as a result of advances in technology. But scientists have only recently started investigating the effects of an activated state of immunity and the subsequent release of immune system products on the nervous and endocrine systems. To date, the bulk of the work has utilized in vitro models; or, if in vivo models were used, they tended to be adult experimental subjects. Given the existence of critical periods in development and the known interactions between the immune and neuroendocrine systems, an area of research that greatly deserves attention is the effect of immune system products on the programming of neuroendocrine circuits if encountered at critical periods of development. With advances made in recombinant DNA technology, it is now feasible for scientists to obtain highly purified lymphokine and monokine preparations in sufficient quantities to carry out a sound experimental design. Availability of monoclonal antibodies against these immune system products allows us to discern the effects of these substances utilizing in vivo models.

ACKNOWLEDGMENTS

Some of the studies described in this chapter were supported by grants from the National Institutes of Health (DK41025 to Maureen P. O'Grady, Ph.D., and NS21210 to Nicholas R. S. Hall, Ph.D.). Nicholas R. S. Hall, was supported by an NIMH Research Scientist Development Award (MH00648). We greatly appreciate the excellent editorial assistance provided by Leah Schubach and Cecilia Figueredo during the preparation of this manuscript.

REFERENCES

Ader, R. (1981). *Psychoneuroimmunology*. New York: Academic Press.

Alford, R. H., Cartwright, B. B., & Sell, S. H. W. (1976). Ontogeny of human cell-mediated immunity: Age-related variation of an *in vitro* infantile lymphocyte transformation. *Infection and Immunity*, *13*, 1170.

Ambrose, C. T. (1964). Requirements for hydrocortisone in antibody-forming tissue cultivated in serum-free medium. *Journal of Experimental Medicine*, *119*, 1027.

Asklen, L. A., & Kalland, T. (1982). Adrenal gland influences on immune disturbances induced by neonatal exposure to diethylstilbestrol. *Cellular and Molecular Biology*, *28*(6), 587.

Bennich, H., & Johansson, M. (1971). IgE and immediate hypersensitivity. *Vox Sang*, *19*(1), 1.

Berkenbosch, F., Van Oers, J., del Rey, A., Tiders, F., & Besedovsky, H. (1987). Corticotropin-releasing factor-producing neurons in the rat activated by IL-1. *Science*, *238*, 524.

Bernton, E. W., Beach, J. E., Holaday, J. W., Smallridge, R. C., & Fein, H. G. (1987). Release of multiple hormones by a direct action of IL-1 on pituitary cells. *Science*, *238*, 519.

Besedovsky, H. O., del Rey, A., & Sorkin, E. (1981). Lymphokine-containing supernatants from Con A-stimulated cells increase corticosterone blood levels. *Journal of Immunology, 126*, 385.

Besedovsky, H. O., del Ray, A., Sorkin, E., Da Prada, M., Burri, R., & Honnegger, C. (1983). The immune response evokes changes in brain noradrenergic neurons. *Science, 221,* 564.

Besedovsky, H. O., Sorkin, E., Keller, M., and Muller, J. (1975). Changes in blood hormone levels during the immune response. *Proceedings of the Society for Experimental and Biological Medicine, 150*, 466.

Bird, A. G., Britton, S. (1979). Live human B cell activator operating in isolation of other influences. *Scandinavian Journal of Immunology, 9*(6), 507.

Blalock, J. E., & Stanton, J. D. (1980). Common pathways of interferon and hormonal action. *Nature, 283*, 406.

Carlsson, B., Gotheford, L., Ahlstedt, S., Hanson, L. A., & Winberg, J. (1976). Studies of *Eschenchui coli* O-antigen specific antibodies in human milk. *Acta Pediatrica Scandinavica, 65*(2), 216.

Chany, C., Mathieu, D., & Gregoire, A. (1980). Induction of Δ -4-3 ketosteroid synthesis by interferon in mouse adrenal tumor cell cultures. *Journal of General Virology, 50*, 447.

Crnic, L. S., & Pizer, L. I. (1988). Behavioral effects of neonatal herpes simplex type 1 infection of mice. *Neurotoxicology and Teratology, 10*, 381.

Dunn, A. J. (1988). Systemic interleukin-1 administration stimulates hypothalamic norepinephrine metabolism paralleling the increased plasma corticosterone. *Life Sciences, 43*, 429.

Ezine, S. & Papiernik, M. (1984). Abnormal T lymphocyte proliferation and cytotoxic responses induced by neonatal injection of hydrocortisone normalized by IL-2 addition. *International Journal of Immunology, 6*, 125.

Farah, J. M., Hall, N. R., Bishop, J. F., Goldstein, A. L., & O'Donohue, T. L. (1987). Thymosin fraction 5 stimulates secretion of immunoreactive beta-endorphin in mouse corticotrophic tumor cells. *Journal of Neuroscience Research, 18*, 140.

Felten, D. L., Felten, S. Y., Carlson, S. L., Olschowka, J. A., & Livnat, S. (1985). Noradrenergic and peptidergic innervation of lymphoid tissue. *Journal of Immunology, 135*(2), S755.

Fontana, A., Hengartner, H., de Tribolet, N., & Weber, E. (1984). Glioblastoma cells release IL-1 and factors inhibiting IL-2 mediated effects. *Journal of Immunology, 132*, 837.

Frohman, E. M., Vayuuegula, B., Van der Noort, S., & Gupta, S. (1988). Norepinephrine inhibits gamma-interferon-induced MHC class II (1a) antigen expression on cultured brain astrocytes. *Journal of Immunology, 17*, 89.

Gathings, W. E., Lawton, A. R., & Cooper, M. D. (1977). Immunofluorescent studies of development of pre-B cells, B lymphocytes and immunoglobulins isotype diversity in humans. *European Journal of Immunology, 7*(11), 804.

Giulian, D., & Lachman, L. B. (1985). IL-1 stimulating astroglial proliferation after brain injury. *Science, 228*, 497.

Goble, F. C., & Konopka, E. A. (1973). Sex as a factor in infectious disease. *Transactions of the New York Academy of Sciences, 35*(4), 325.

Gorski, R., Harlan, R. E., & Christensen, L. W. (1977). Perinatal hormonal exposure and development of neuroendocrine regulatory processes. *Journal of Toxicology and Environmental Health, 3*(1–2), 97.

Goya, R. G., Sosa, Y. E., & Meites, J. (1987). Half-life of plasma growth hormone in young and old conscious female rats. *Experimental Gerontology, 11*(1), 27–36.

Gresser, I. (1980). Usefulness of the results of studies on the anti-tumor effects of interferon in animals to interferon therapy of patients. *Recent Results in Cancer & Research, 75*, 226.

Gresser, I., Morimar, L., Verroust, P., Riviere, Y., & Guillon, J. C. (1978). Role of interferon in pathogenesis of virus diseases in mice as demonstrated by use of anti-interferon serum. 4. anti-interferon globulin inhibits development of glomerulonephritis in mice infected at birth with lymphocyte choriomeningitis virus. *Proceedings of the National Academy of Science, USA, 75*(7), 3413.

Gresser, I., Torey, M. G., Maury, C., & Chouroulinkov, I. (1975). Lethality of interferon preparation for mice. *Nature, 258*, 76.

Guillemin, R., Cohn, M., & Melnechuk, T. (1985). *Neural Modulation of Immunity*. New York: Raven Press.

Gunn, T., Reece, E., Metrakos, K., Colle, E. (1981). Depressed T cells following neonatal steroid treatment. *Pediatrics, 67*(1), 61.

Gunther, M. (1975). The neonate's immunity gap, breast feeding and cot death. *Lancet, 1*(7904), 441.

Hall, N. R., Goldstein, A. L. (1984). Endocrine regulation of host immunity: The role of steroids and thymosins. In R. L. Fenchiel & M. A. Chirigos (Eds.), *Immune modulation agents and their mechanisms* (p. 533). New York: Marcel Dekker.

Hall, N. R., McGillis, J. P., Spangelo, B. L., & Goldstein, A. L. (1985). Evidence that thymosins and other biologic response modifiers can function as neuroactive immunotransmitters. *Journal of Immunology, 135*(2), 806–811.

Hall, N. R., McGillis, J. P., Spangelo, B. L., Palaszynski, E., Moody, T. W., & Goldstein, A. L. (1982). Evidence for a neuroendocrine-thymus axis mediated by thymosin polypeptides. *Developmental Immunology, 17*, 653.

Herrenkohl, L. R. & Politch, J. A. (1978). Effects of prenatal stress on estrous cycle of female offspring as adults. *Experientia, 34*(9), 1240.

Ishizaka, K., & Ishizaka, T. (1967). Identification of gamma-antibodies as a carrier of reaginic activity. *Journal of Immunology, 99*, 1187.

Kauppila, A., Cantell, K., Janne, O., Kokko, E., & Vihko, R. (1982). Serum sex steroid and peptide hormone concentrations, and endometrial estrogen and progestin receptor levels during administration of human leukocyte interferon. *International Journal of Cancer, 29*(3), 291.

Klein, J. O., Remington, J. S., & Marcy, S. M. (1983). In J. S. Remington & J. O. Klein (Eds.), *Infectious diseases of the fetus and newborn infant* (p. 1), Philadelphia: Saunders.

Krueger, J., Dinarello, M., Wolff, M., Chedid, L., & Walter, J. (1984). Sleep promoting effects of endogenous pyrogen (IL-1). *American Journal of Physiology, 246*, R994.

Krzych, U., Strausser, H. R., Bressler, J. P., & Goldstein, A. L. (1979). Quantitative differences in immune responses during various stages of estrous cycle in female Balb-C mice. *Journal of Immunology, 121*(4), 1603.

Leslie, G. A., & Swate, T. E. (1972). Structure and biologic functions of human IgD. 1. Presence of IgD in human cord sera. *Journal of Immunology, 109*(1), 47.

Levine, S. (1967). Maternal and environmental influences on adrenocortical response to stress in weanling rats. *Science, 156*, 258.

Lu, C. Y., & Unanue, E. R. (1982). Ontogeny of murine macrophage functions related to antigen presentation. *Infection and Immunology, 36*(1), 169.

McGillis, J. P., Hall, N. R., Vahouny, G. V., & Goldstein, A. L. (1985). Thymosin fraction 5 causes increased serum corticosterone in rodents *in vivo*. *Journal of Immunology, 134*, 3952.

Meaney, M. J., Aitken, D. H., Van Berkel, C., Bhatnagar, S., & Sapolsky, R. M. (1988). Effect of neonatal handling on age-related impairments associated with the hippocampus. *Science, 239*, 766.

Monjan. A. A. (1981). Stress and immunological competence—studies in animals. In R. Ader (Ed.), *Psychoneuroimmunology* (p. 18). New York: Academic Press.

Ogra, S. S., & Ogra, P. L. (1978). Immunological aspects of human colostrum and milk. 1. Distribution characteristics, and concentrations of immunoglobulins at different times after onset of lactation. *Journal of Pediatrics, 92*(4), 546.

O'Grady, M. P., Hall, N. R. S., & Goldstein, A. L. (1987). Developmental consequences of prenatal exposure to thymosin: Long term changes in immune and endocrine parameters. *Society for Neuroscience, 13*, 1380.

O'Grady, M. P., & Hall, N. R. S. (1990). Postnatal exposure to Newcastle disease virus alters endocrine development. *Teratology, 41*, 623.

Olding, L. B., Benirsch, K., & Oldstone, M. B. (1974). Inhibition of mitosis of lymphocytes from human adults by lymphocytes from human newborns. *Clinical Immunology, 3*(1), 79.

Orava, M., Cantell, K., Kauppila, A., & Vihko, K. (1983). Interferon and serum thyroid hormones. *International Journal of Cancer, 31*(5), 671.

Orava, M., Cantell, K. & Vihko, R. (1985). Human leukocyte interferon inhibits human chorionic gonadotropin-stimulated testosterone produced by porcine Leydig cells in culture. *Biochemistry and Biophysics Research Communication, 127*(3), 809.

Osheroff, P. L. (1981). The effect of thymosin on glucocorticoid receptors in lymphoid cells. *Cellular Immunology, 60*(2), 376.

Papiernik, E., Frydman, R., & Spira, A. (1980). Effects of antenatal care and maternal nutrition on perinatal outcome. In S. Harel (Ed.), *At risk infant: A practical interdisciplinary approach to the prevention, discovery, assessment, and management of developmental disabilities* (pp. 167–171). Amsterdam: Excerpta Medica.

Peters, D. A. V. (1982). Prenatal stress effects on brain biogenic amine and plasma corticosterone levels. *Pharmacology, Biochemistry, and Behavior*, *17*(4), 721.

Pittard, W. B., Miller, K., & Sorensen, R. U. (1984). Normal lymphocyte responses to mitogens in term and premature neonates following normal and abnormal intrauterine growth. *Clinical Immunology and Immunopathology*, *30*, 178.

Pittard, W. B., Miller, K. M., & Sorensen, R. U. (1985). Perinatal influences on *in vitro* neonatal B lymphocyte differentiation in human neonates. *Pediatric Research*, *19*, 655.

Pittard, W. B., Schleich, D. M., Geddes, K. M., & Sorensen, R. U. (1989). Newborn lymphocyte subpopulations: The influence of labor. *American Journal of Obstetrics and Gynecology*, *160*, 151.

Rebar, R. W., Miyake, A., Low, T. L. K., & Goldstein, A. L. (1981). Thymosin stimulates secretion of LH-RF. *Science*, *214*, 669.

Ritter, M. A. (1977). Embryonic mouse thymocyte development: Enhancing effect of corticosterone at physiological levels. *Immunology*, *32*(2), 241.

Sapolsky, R., Rivier, C., Yamamoto, G., Plotsky, P. & Vale, W. (1987). Corticotropin RF-producing neurons in the rat activated by IL-1. *Science*, *238*, 522.

Schupf, N., Williams, C. A., Hugh, T. E., & Cox, J. (1983). Psychopharmacological activity of anaphylatoxin C3a in rat hypothalamus. *Journal of Neuroimmunology*, *5*, 305.

Shoham, S., Davenne, D., Cady, A. B., Dinarello, C. A., & Krueger, J. M. (1987). Recombinant TNF and IL-1 enhance slow-wave sleep. *American Journal of Physiology*, *253*(1), R142.

Simmons, R. D., Miller, R. K., & Kellogg, C. K. (1984). Prenatal exposure to diazepam alters central and peripheral responses to stress in adult rat offspring. *Brain Research*, *307*(1–2), 39.

Solomon, J. B. (1970). Unification of foetal and neonatal immunology. *Nature*, *227*, 895.

Spangelo, B. L., Judd, A. M., Ross, P. C., Hogin, I. S., Jarvis, W. D., Badamchian, M., Goldstein, A. L. & MacLeod, R. M. (1987). Thymosin fraction 5 stimulates prolactin and growth hormone release from anterior pituitary cells *in vitro*. *Endocrinology*, *121*, 2035.

Stiehm, E. R., & Fudenberg, H. H. (1966). Serum levels of immunoglobulins in health and disease—A survey. *Pediatrics*, *37*, 715.

Stites, D. P., Caldwell, J., Carr, M. C., & Fudenberg, H. H. (1975). Ontogeny of immunity in humans. *Clinical Immunology and Immunopathology*, *4*, 519.

Toivanen, P., Uksila, J., Leino, A., Lassila, O., Hirvonen, T. & Ruuskanen, O. (1981). Development of the mitogen responding T cells and natural killer cells in the human fetus. *Immunological Reviews*, *57*, 89.

Tosato, G., Magrath, I. T., Koski, I. R., Dooley, N. J., & Blaese, R. M. (1980). B cell differentiation and immunoregulatory T cell function in human cord blood lymphocytes. *Journal of Clinical Investigation*, *66*(2), 383.

Uehara, A., Gottschall, P. E., Dahl, R. R. & Arimura, A. (1987). Stimulation of ACTH release by human IL-1 beta, but not by IL-1 alpha, in conscious, freely-moving rats. *Biochemical and Biophysical Research Communication*, *146*(3), 1286.

Unander, A. M., & Olding, L. B. (1981). Ontogeny and postnatal persistence of a strong suppressor activity in man. *Journal of Immunology*, *127*(3), 1182.

Vilcek, J. (1964). Production of interferon by newborn and adult mice infected with Sindbis virus. *Virology*, *22*, 651.

Ward, I. L. (1972). Prenatal stress feminizes and demasculinizes behavior of males. *Science*, *75*(4017), 82.

Widmer, M. B., & Copper, E. L. (1979). Ontogeny of cell-mediated cytotoxicity: Induction of CTL in early postnatal thymocytes. *Journal of Immunology*, *122*(1), 291.

Willey, D. E., & Ushijima, R. N. (1981). The effect of thymosin on murine lymphocyte responses and corticosteroid levels during acute reovirus type 3 infection of neonatal mice. *Clinical Immunology and Immunopathology*, *19*, 35.

B.

NEURAL PLASTICITY

The authors of the chapters in this section address the issue of neural plasticity, but each from a different perspective. Each chapter describes strategies to discover the rules that regulate neural growth and plasticity during development. In each, the question of how experience alters brain development is key. The authors differ, however, in the model used to address this issue and on what they mean by "experience."

In Chapter 14, Hyson and Rubel use the well-defined auditory system in the chicken. The authors make the important point, too often overlooked in the critical initial steps of establishing research programs, that one's choice of model defines and often constrains the types of conceptual problems the scientist can address. Hyson and Rubel's "model system" was chosen for its simplicity, well-studied anatomy, and accessibility. The auditory system is anatomically defined, bilaterally symmetrical, and relatively homogeneous. The "experience" that the authors use is also defined and simplified. They have manipulated afferent input in several ways and measured the change in synaptic activity resulting from those manipulations. In their model system, they have exquisite control of both afferent input and the structure of the anatomical system. The questions that they ask, the role of changing experience at each level of processing, are possible by the nature of the model.

In Chapter 15, in contrast, Greenough and Sirevaag have studied a very different model system. Theirs is the study of the cerebral cortex, a complex and less constrained neural system. The "experience" used by Greenough and Sirevaag (environmental enrichment) is similarly less restrictive, and thus the components of the experience that contribute to the neuroanatomical changes they have discovered cannot be as clearly specified. The dependent measures in both chapters are similarly defined—changes that occur at the level of the cell. Greenough's work, while detailed to the level of the cell, is less reductionistic in the experience and in the anatomical system chosen for analysis. Both approaches are necessary, and both chapters detail the enormous strides that each group has made in understanding how experience is transduced into changes in neural structure.

In Chapter 16, Diaz describes methods by which the experience of the developing rat pup is reduced to a simple and standard level. His methods are elaborations of artificial rearing procedures originally defined by Messer and Thoman. Because the goal of his research is to specify how exposure to different nutrients and drugs affects brain development, he has eliminated possible confounds that might occur due to changes in the dam's behavior by eliminating the dam. In a sense, Diaz has removed his subjects from their environment, constructing his model system quite differently from that of Greenough and Sirevaag and Hyson and Rubel. The power of this method is that it allows the investigator to alter or replace specific components of the natural environment, such as odors or vestibular stimuli under complete experimental control. It is a particularly powerful technique to assess the developmental effects of nutrition, drugs, and other environmental stimuli so that one can more clearly separate the direct effects from indirect effects of treatment. In contrast to Chapters 14 and 15, Diaz has not specified particular neural systems for study. The strength of the model is that it can be applied to any number of "experiences" and any number of neural systems. His application of this model is to administer drugs through the diet or to alter the amount of composition of the diet, both of which bathe the brain. It is likely that he would see more general, less specified changes in particular neural systems.

Thus each of the authors has chosen a different level of analysis. Hyson and Rubel control specific afferent input and examine a delimited neural circuit. For Greenough and Sirevaag, experience is more general, and they have concentrated on a less restricted system, the cerebral cortex. Diaz, while restricting the pup's environment, has manipulated nutrition during development and has used more global indexes of changes in neuroanatomy. The types of questions asked, or that can be asked, are defined in large part by the model system that each has used. Often we become tied to our paradigms. It may be a model that we learned in graduate school, or one that we have successfully used for years, one we know well and are comfortable with. At times, the questions we must ask change or outgrow old models. It is up to the investigator to recognize that the information gathered is restricted by the model chosen. Or to put it another way, the questions that one wants to ask should in large part specify the paradigm that one chooses.

14

Methods for Studying Experiential Influences on Brain Development

RICHARD L. HYSON AND EDWIN W RUBEL

One fundamental goal of developmental psychobiology is understanding the biological mechanisms underlying experiential regulation of brain development. A host of examples show that sensory experience plays a crucial role in neural and behavioral development (see Globus, 1975; Gottlieb, 1976; Mistretta & Bradley, 1978; Movshon & van Sluyters; 1981; Sherman & Spear, 1982). Procedurally, such experiments usually involve some alteration of sensory information impinging on the organism during development. This is accomplished either by manipulation of the peripheral receptor system, by isolation of the subject from some type of sensory information, or by enhancement of the sensory environment as compared to control subjects. The effects of these manipulations are then assessed, following some intervening period, by various behavioral tasks or measures of neural structure or function. Published results point to a dramatic influence of sensory experience on the development of the brain and behavior. These same manipulations in adult animals, however, are often without effect or have a greatly reduced effect.

Along these lines, our laboratory has been examining the role of auditory experience in development of the brain stem auditory system of the chicken. Some of the basic questions we have addressed are: (1) When do peripheral and central components of the auditory system develop? (2) What are the roles of afferent input and experience on brain development? and (3) What are the biological mechanisms by which experience exerts its effect on the developing nervous system? In this chapter we show how these questions have guided an evolution of methodological approaches utilizing both in vivo and in vitro techniques.

METHODOLOGY

Choice of Organism

One of the first steps in developing a methodology is choosing the appropriate organism/system to study. The importance of this step cannot be overemphasized. Most of the fundamental advances in biological sciences in the past century can be tied, more or less directly, to the choice of preparations. Historically, one has only to

241

think of the discovery of the structure of DNA, unraveling the mechanisms of action potential generation, or examining cellular correlates of learning and memory. Choosing the best preparation is often referred to as a *model system* approach, meaning that a preparation (usually a particular species) has properties uniquely suited for addressing a particular biological issue (Kandel & Spencer, 1968). It is important to remember that a model system implicitly refers to a specific biological problem. A preparation useful for addressing one problem at one level of analysis may not be well suited for another problem or a different level of analysis. Thus, while it is extremely important to find a model system, it is equally important continually to reevaluate its usefulness as the biological problem or the level of analysis changes.

The model system we have chosen, the brain stem auditory system of the chicken, is particularly well suited for examining the role of auditory experience in development. This relatively simple, bilaterally symmetrical system is shown schematically in Figure 14.1. Signals from the cochlea are transmitted via the auditory nerve to second-order auditory neurons in nucleus magnocellularis (NM). The auditory nerve projection is strictly ipsilateral, and it is the only known excitatory input to NM. NM consists of a relatively homogeneous population of neurons. Each NM neuron receives input from two or three auditory nerve fibers and projects bilaterally to the third-order nucleus, nucleus laminaris (NL). NL consists of a layer of cells with symmetrical bipolar dendrites, extending dorsally and ventrally from the cell body layer. Axons

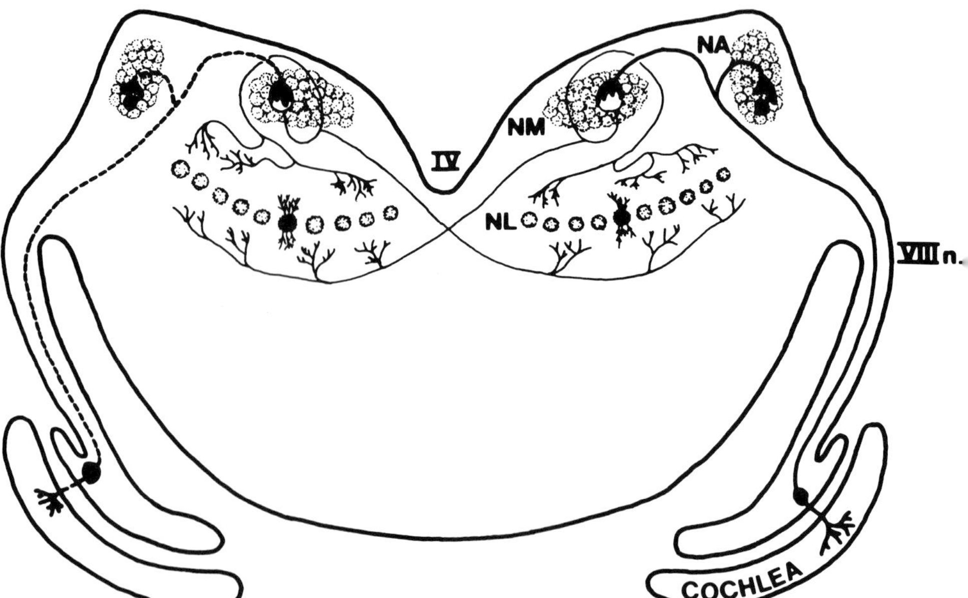

Figure 14.1. Schematic of a coronal section through the chick brain stem emphasizing the auditory pathways. The auditory nerve (VIII n.) projects to nucleus magnocellularis (NM). Each NM neuron projects bilaterally to nucleus laminaris (NL). (From "Organization and Development of the Avian Brain Stem Auditory System" by E. W Rubel and T. N. Parks, 1988. In J. Brugge, I. Hafter, and M. Merzenich (Eds.), *Functions of the Auditory System*, Figure 2, p. 10. New York: John Wiley. Reprinted by permission of Neuroscience Research Foundation, Inc.)

from NM synapse on the dorsal dendrites of the ipsilateral NL neurons and on the ventral dendrites of the contralateral NL.

This system offers several advantages for studying experiential influences on brain development: (1) The animal is relatively inexpensive and easily obtained. (2) All of the components (cochlea, NM, and NL) lie in a 1–2 mm thick slice taken in the coronal plane; as we will see, this is important for cellular analyses. (3) The normal organization and development of this system are known in considerable detail (Rubel & Parks, 1988). (4) Since NM consists of a relatively homogeneous population of cells, most manipulations will uniformly affect the neurons in this nucleus. (5) Finally, the bilaterally symmetrical organization of this system and the strictly unilateral projection of the auditory nerve allow one to make within-subject (and even within-tissue section) comparisons after various manipulations. This latter point needs further clarification. Since the ipsilateral auditory nerve is the only excitatory input to NM, unilateral manipulation of the auditory periphery will affect excitation of the ipsilateral NM but not of the contralateral nucleus. Thus, for most of the experiments we discuss, the appropriate control tissue is NM on the opposite side of the brain. This ability to make comparisons within the same tissue section is extremely advantageous when examining anatomical or biochemical properties, where absolute levels of a particular parameter are difficult or impossible to determine, but relative values are easily obtained.

ANALYSIS OF SENSORY DEVELOPMENT WITHOUT THE STIMULUS

A first and often difficult step in any research program dealing with experiential influences on brain development is to document the timetable of normal structural and functional ontogeny. One of the most interesting problems faced in these analyses is determining the functional connectivity of central nervous system pathways independent of receptor and transducer maturation. This issue becomes important for two reasons. First, to understand the mechanisms underlying developmental changes in the plasticity of a system (e.g., critical periods), it is necessary to understand thoroughly the developmental changes in functional connectivity. Second, to understand maturational patterns, it is often desirable to know if the periphery matures prior to more central regions, or vice versa. In fact, it is often claimed that the auditory system of birds and mammals develops in an orderly, sequential pattern with progressively more cephalad regions developing progressively later (Pujol, 1972). Unfortunately, studies that have attempted to support this hypothesis have used an intact organism and studied responses to sound at different levels of the system, ranging from the periphery to the cerebral cortex. This approach has the fatal flaw that responses at more cephalad locations *depend* on the processing at more peripheral sites. Any immaturity at the receptor level, for example, will necessarily be reflected at the brain stem, midbrain, thalamic, and cortical projection areas. In order to assess sequences of maturation at various levels of a system, each level must be examined independently of the other levels.

We developed an in vitro preparation of the entire chick embryo brain stem in order to address these issues. First, we wanted to know if the age at which neurons in NM become susceptible to modifications of their afferents corresponds to the time of functional synaptogenesis, as hypothesized by Levi-Montalcini (1949). Second, we wished to examine the relative timing of functional development in the central nervous system with that in the periphery. With an in vitro preparation we could examine the times at which presynaptic activity first results in postsynaptic action potentials, signaling the establishment of functional connections between the cochlea and each region of NM as well as between NM and NL (Jackson, Hackett, & Rubel, 1982).

Methodology

The in vitro brain stem preparation was adapted from various invertebrate preparations and the chick spinal cord preparations used by Landmesser and her colleagues (Landmesser, 1978). This preparation is described fully in earlier publications (Hackett, Jackson, & Rubel, 1982; Jackson, Hackett, & Rubel, 1982). Briefly, the chick is anesthetized by hypothermia; the egg is placed in a conventional freezer for 10–15 min. The egg is then opened, and the embryo's head is placed into ice cold, oxygenated avian Tyrode's solution for dissection (for this purpose petri dishes coated with approximately 1/4 in. of Sylgard [Dow Corning] are useful), and the head is pinned down. The brain stem is rapidly exposed by removing the dorsal cranium and the cerebellum. Then, the brain stem is carefully dissected free; each cranial nerve is cut, and the tissue is gently lifted free of the calvarium. The most important aspect of this dissection is to use sharp scissors and microsurgery scalpels to cut all nerves or fiber tracts as well as to cut off the forebrain and spinal cord. Stretching and torquing these structures seem to result in less physiologically viable preparations.

The brain stem is then transferred briefly to filter paper and glued with cyano-acrylic glue directly to the bottom of a recording chamber in which oxygenated Tyrode's solution is continually superfused over the tissue. It is allowed to come to room temperature but does not need to be heated. In fact, it is our opinion that embryonic tissue maintains its viability longer when kept at room temperature than when heated. The cellular physiological characteristics of the tissue may be less representative of a normal animal, but the tissue seems to be more robust, probably because of reduced proteolysis and better oxygenation. On the basis of field potentials, the preparation routinely remains viable for 8–12 h and often for 18 h in 9–17-day-old embryos. A few days before hatching it appears to become less viable, probably because it is larger and denser, thereby restricting adequate oxygenation. Since axoplasmic transport remains viable, we have used this preparation to great advantage for anatomical studies of cellular development (Young & Rubel, 1986), as well as for physiological investigations. A similar preparation has been used for comparable physiological and anatomical studies of cat fetuses by Shatz and her colleagues (Shatz & Kirkwood, 1984; Sretevan & Shatz, 1986).

In our physiological studies (Jackson et al., 1982) conventional recording techniques were used to characterize the ages at which extracellular postsynaptic potentials can be obtained. Postsynaptic responses in NM after eighth nerve stimulation were first recorded at about the same age that sound-evoked responses could be recorded in vivo (E11). Neurons in NL, however, could be driven by direct stimulation of NM prior to the age at which they could be driven by eighth nerve stimulation or sound. This suggests that the apparent immaturity of the higher-order NM-to-NL synapse after eighth nerve stimulation is actually due to the immaturity of synaptic connections between the eighth nerve and NM. Thus, for this aspect of neuronal development a simple peripheral-to-central sequence of maturation is not observed.

AFFERENT INFLUENCES

For the remainder of this chapter we discuss some of the methods we have used (and their rationale) in experiments that directly examine various cellular mechanisms underlying experiential regulation of brain development. Two conceptual issues must be addressed briefly before discussing our methodology or results. First, what do we mean by "experience"? Our view is that various forms of enrichment, deprivation, or modifications of sensory experience have in common that they represent relatively long-term (>seconds) modifications of the amount or pattern of afferent activity in the brain. If this view is accepted, the problem of understanding how experience regulates sensory system development becomes a subset of a broader question: How do chronic changes in the amount or pattern of input influence neuronal development? Thus, our research strategy becomes one of manipulating afferent input in a variety of ways and attempting to understand the regulation of neuronal ontogeny.

The second issue, derived from the first, is that we learned quite early in our investigations that it is impossible to predict accurately the direction, extent, or timing of *change* in synaptic activity in the central nervous system from the manipulation performed at the periphery. For example, if a manipulation led to reduced activation of an inhibitory circuit, then an unexpected increase in postsynaptic activity would occur. Additionally, delayed changes in activity may emerge through sprouting or transneuronal degeneration. Thus, we have found it important actually to measure the changes in pre- and postsynaptic activity at NM and NL as a function of each manipulation.

In Vivo Methods

An obvious first step in the analysis of experiential influences on brain development is to identify properties of the nervous system that are altered after some experiential manipulation. By using an extreme manipulation of sensory experience, one increases the likelihood of observing effects. It is then possible to use progressively more subtle manipulations and parcel out the subset of results regulated by each particpating factor. In the case of the auditory system, the most extreme manipulation is to deafen the animal. In the chicken, this is easily accomplished simply by removing the cochlea (Born & Rubel, 1985). This manipulation can be done at virtually any age, from early in embryogenesis through adulthood. After unilateral cochlea removal, we have allowed animals to survive for variable periods of time before examining the tissue for a variety of morphological and metabolic changes.

Early experiments (Born & Rubel, 1985; see Figure 14.2) revealed that within 48 hours after cochlea removal in hatchling chicks there is a decrease in the number and size of NM neurons on the side of the brain ipsilateral to cochlear ablation. This effect is age dependent. Removal of the otocyst in embryos is without effect until the age at which functional synaptic contacts have been made between the auditory nerve and NM neurons (Parks, 1979). After that time, decreased cell size and number are

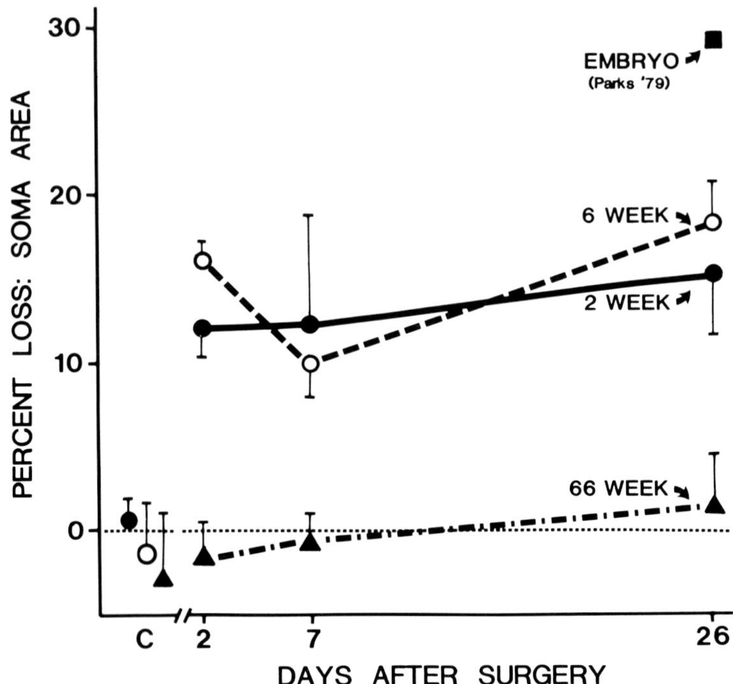

Figure 14.2. Percentage decrease in NM soma area as a function of survival period for chicks after unilateral cochlea removal at 2 weeks, 6 weeks, or 66 weeks posthatch. Mean percentage loss = 100× (mean area contralateral – mean area ipsilateral) / mean area contralateral. Thus, positive numbers indicate that cells ipsilateral to cochlea removal are smaller than those on the control side. Means from unoperated controls (C) of each age are shown (survival time pooled). Also plotted is a data point from Parks, 1979, in which the otocyst was removed on embryonic Day 2, and the chick survived until posthatch Day 26. Bars represent standard error of the mean. (From "Afferent Influences on Brain Stem Auditory Nuclei of the Chicken: Neuron Number and Size Following Cochlea Removal" by D. E. Born and E. W Rubel, 1985, *Journal of Comparative Neurology, 231,* Figure 5, p. 442. Reprinted by permission of Alan R. Liss, Inc.)

observed when the cochlea is removed in embryos and in 1- and 6-week-old hatchlings, but is not observed after cochlea ablation in fully adult animals (>1 year of age). This age dependence of the effect of cochlear ablation on NM neurons is particularly intriguing in that it does not simply follow the functional development of the auditory system. Auditory function is essentially mature within a few days after hatching, whereas deafferentation induces changes in NM even in 6-week-old birds. The biological basis for the change in sensitivity to deafferentation between 6 weeks and 1 year of age is at present unknown.

In addition to the changes in cell size and number, several other anatomical and metabolic properites of neurons in NM are senstitive to cochlea removal. These include changes in activity of Krebs cycle enzymes (Durham & Rubel, 1985), cytoskeletal proteins (Seftel, Deitch, & Rubel, 1986), and protein synthesis (Steward & Rubel, 1985). The changes in protein synthesis are particularly interesting. First, this change is observed after extremely short survival times. For example, within 1 h after cochlea removal, a 50% decrease in the incorporation of radioactive leucine into protein is observed in the ipsilateral NM. After approximately 6 h, two populations of cells appear to be distinguishable on the basis of protein synthesis. Approximately one third of the neurons in NM ipsilateral to cochlea removal appear to stop protein synthesis completely, while the other two thirds continue to make protein, although at a reduced level. The cells that stop protein synthesis altogether correspond to the population of cells that will die after longer survival times. Thus, it appears that protein synthesis can be used as an early marker for identifying which cells will live and which will die after more prolonged elimination of afferent input. These results suggest a possible cellular mechanism that contributes to the degeneration of deprived neurons. One early event in this process may be a disruption of the protein synthetic machinery within the cell. How deprivation causes the disruption of this machinery remains to be determined.

Defining "Experience": What Does the Cell See (Hear)?

One pitfall of many research programs is their precision in defining "experiential manipulation." During development, an animal's encoding of sensory information changes dramatically (Aslin, 1984; Rubel, Lippe, & Ryals, 1984; Hyson & Rudy, 1987). Thus, the presentation or withdrawal of a sensory stimulus must be defined in terms of the developing animal's current processing of that event. Likewise, as noted above, when studying neural properties, experience must be defined physiologically, not physically. That is, when analyzing the effect of an experiential manipulation on an individual neuron, one must be able to define the manipulation in terms of what that particular neuron experiences. The following example illustrates this point.

Given the dramatic changes in cell size and number observed after cochlea removal, Tucci and Rubel (1985) asked if more subtle changes in auditory experience might also affect the morphology of neurons in NM. Conductive hearing losses were produced by rupturing the tympanic membrane and removing the middle ear ossicle (columella) or placing silicone earplugs in one ear of young chicks. These manipulations result in nearly complete elimination of hearing, as judged by auditory-evoked potentials (ABR). However, no change in the size or number of NM neurons was observed after prolonged periods of hearing loss (see Figure 14.3).

Upon closer analysis of what these manipulations actually did to the "experience" of the NM neurons, this lack of effect is not surprising. The ABR threshold shift appears, at the surface, to be a good measure of the effects of this experiential manipulation. However, this synchronized response to sound is, in fact, not a good index of the overall change in the activity impinging on NM neurons. The high rate of spontaneous activity present in the auditory nerve maintained a high rate of activity in NM neurons. Electrophysiological recordings from NM reveal that although these manipulations produced a massive hearing loss, they did not alter the overall level of activity in NM in a profound manner. On the other hand, damaging the cochlea by puncture of the oval window as well as cochlea removal completely abolished excitatory activity in NM (Born & Rubel, 1984; Tucci, Born, & Rubel, 1987). These experiments strongly suggest that neural activity per se, and not the information provided about the environment, is the important factor for maintaining the NM cells. With these dependent variables, experience must be defined by the change in activity in the relevant area of the nervous system.

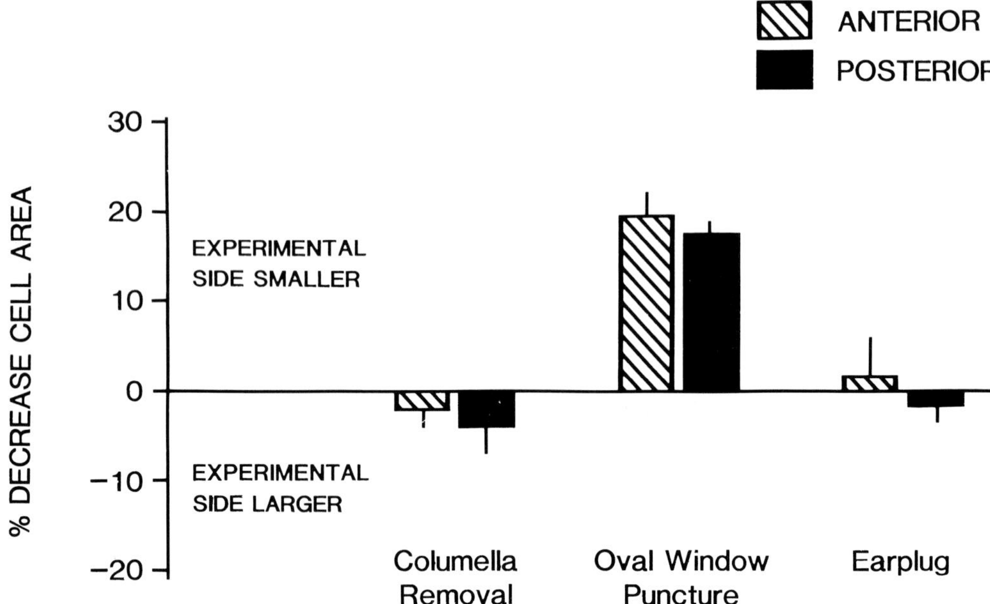

Figure 14.3. Percentage decrease in NM soma area 60 days after the various treatments. The ordinate is as described in Figure 14.2. Anterior and posterior refer to different portions of NM. Bars represent standard error of the mean. (From "Afferent Influences on Brain Stem Auditory Nuclei of the Chicken: Effects of Conductive and Sensorineural Hearing Loss on N. Magnocellularis" by D. L. Tucci and E. W Rubel, 1985, *Journal of Comparative Neurology, 238,* Figure 3, p. 377. Reprinted by permission of Alan R. Liss, Inc.)

A Reversible Ablation

There are several obvious limitations in the use of cochlea ablation for studying experiential influences on brain development. First, removing the cochlea damages the distal dendrites of the ganglion cells. The ganglion cells subsequently degenerate. Thus, one cannot be certain if an effect of cochlea ablation is due to a change in activity of the auditory nerve or to the chain of events resulting from mechanical damage and subsequent degeneration of the first-order neurons. A second obvious limitation is that cochlea ablation is permanent. One cannot begin to investigate the effects of recurring periods of deprivation or reversibility of the effects. Finally, since cochlea removal eliminates *all* eighth nerve and NM activity, it is not possible to assess changes to subtotal activity changes, that is, look at the "dose-response" function.

To get around some of these problems, Born and Rubel (1988) developed a means of blocking auditory nerve activity without causing permanent neural damage (see Dubin, Stark, & Archer, 1986, and Stryker & Harris, 1986, for use of a similar method in the visual system). We injected the drug tetrodotoxin (TTX, Sigma Chemical Company) into the perilymph of the inner ear without directly damaging the cochlea. TTX blocks the voltage-dependent sodium channels required for the generation of action potentials. Electrophysiological recording confirmed that the TTX injections completely eliminated action potentials in the auditory nerve and in NM neurons. After 2–3 h, the effects of the TTX injection wore off and activity returned.

Born and Rubel (1988) used this procedure to investigate the nature of the signals responsible for changes in protein synthesis, cell size, and cell number observed in NM after cochlea removal. Their data point to a role of afferent *activity* in the regulation of neuronal metabolic function. That is, changes similar to those observed after cochlea ablation were found when auditory nerve fibers were not damaged, but merely silenced for a few hours.

Since the effects of a single injection of TTX into the perilymph were relatively short acting, Born and Rubel had to make several injections over the course of their experiments. More recently, Pasic and Rubel (1989) have developed a means of extending the duration of action by embedding the TTX into a slow-release copolymer (Elvax, Dupont). In addition, they have extended this method to mammals. A pellet of this slow-release compound infiltrated with TTX is placed into the round window niche. The TTX then diffuses through the round window membrane to reversibly and noninvasively block eighth nerve action potentials. The duration of action can be adjusted by varying the proportions of TTX to Elvax (Langer, Brown, & Edelman, 1985). This preparation allows us to examine the effects of longer-term activity blockade and to study recovery of function after variable periods of blockade.

An in Vitro Approach to Examining Experiental Influences

The in vivo experiments described above, as well as those by other investigators, indicate that the activity (experience) of presynaptic elements regulates postsynaptic neuronal structure and function. Little is known, however, about the cellular mechanisms by which activity influences developing neural tissue. To address this question, we have utilized an in vitro slice preparation of the chick's brain stem auditory system. The key advantages of a brain slice preparation are ones of experimental control. First, one has greater control over the electrical activity of the neurons to be studied. For example, one can visually place stimulating and recording electrodes, thereby eliminating the inherent variability of blindly placing electrodes or using stereotaxic coordinates. Second, the in vitro preparation allows the investigator to control the external evironment of the neurons. This allows us to make controlled pharmacological manipulations, which are exceedingly difficult in vivo. The slice preparation is slightly more involved than the entire brain stem preparation described above; however, the smaller thickness of the slice allows (1) more rapid and complete changes of the extracellular environment when exchanging media and (2) better oxygenation of the tissue.

Procedure

The brain slice preparation used in our experiments is similar to that described by others (Martin, 1985; Oertel, 1983, 1985). Briefly, 5–15-day posthatch chicks are decapitated, and the brain stem is dissected free of the surrounding bone, taking care to leave as large an auditory nerve root as possible. The dissection is carried out in oxygenated room temperature artificial cerebrospinal fluid (ACSF). There are more published recipes for ACSF than there are for chocolate chip cookies (Crocker, 1969; DiScenna, 1987). But, like cookies, they all have generally the same ingredients. Everyone has a favorite. Ours contains 130 mM NaCl, 3 mM KCl, 2 mM $CaCl_2$, 2 mM

MgSO$_4$, 26 mM NaHCO$_3$, 1.75 mM NaH$_2$PO$_4$, and 10 mM d-glucose. Minor variations in this recipe do not appear to significantly affect our results.

After the brain stem is dissected free from the skull, 300-μm thick sections are cut using a tissue slicer (Frederick-Haer model OTS-3000). This slicing is done at room temperature in a reservoir of oxygenated ACSF.

Once the slice is obtained, it is transferred to the recording chamber. Our chamber is similar to the one used by Oertel (1985). The slice is completely submerged in 34°C oxygenated ACSF, which flows through the chamber at a rate of 3.5 mL/min. Slices maintained in this manner remain viable (as judged by extracellular field potentials) for 8–12 h.

Examining Experiential Influences Using a Brain Slice

The procedure for looking at the influence of experience using this brain slice preparation is straightforward. Recall from Figure 14.1 that the only excitatory input to NM comes from the ipsilateral auditory nerve, whose activity relies completely on the activity of the cochlea. Obviously, in this preparation both cochleas are removed. Thus, NM on both sides is "deprived." In order to mimic the case of unilateral deprivation, one simply has to provide an input into NM on one side of the brain by electrically stimulating the auditory nerve on that side. After a period of unilateral stimulation, one can then measure whatever cellular property one desires. It is important to note that this preparation preserves the ability to make within-subject (actually, within-tissue section) comparisons between stimulated and unstimulated NM neurons.

The initial results of these experiments are consistent with those that have been obtained in vivo. When we mimic the in vivo condition of unilateral cochlea ablation by unilaterally stimulating the auditory nerve, we find similar differences in protein synthesis between activated and inactive NM neurons. Within 1.5 h, stimulated NM neurons show greater levels of protein synthesis than unstimulated neurons.

Analysis of the Signals at a Cellular Level

The similarity between results of our in vivo and in vitro experiments is important; it indicates that the brain slice is an adequate preparation for examining cellular mechanisms underlying *the same* events we have examined in vivo. Having shown that changes in activity result in a transneuronal reduction in protein synthesis, the next questions involve further delineation of the signal by which afferent activity (experience) regulates protein synthesis in NM neurons. The possible signals for any activity-dependent modification of the nervous system can be broken down into two general categories: (1) The important signal could be released from the active presynaptic neurons. For example, the presynaptic terminal might release some "trophic substance" that binds to a postsynaptic receptor and activates a cascade of events influencing protein synthesis in the postsynaptic cell. Or (2) the signals could arise from a cascade of events resulting solely from changes in the rate of action potentials in the postsynaptic neuron, not requiring presynaptic activity or release of a trophic factor.

We have found that the increase in protein synthesis resulting from orthodromic activation of NM neurons appears to be dependent on synaptic release from the auditory nerve (Hyson & Rubel, 1989). When synaptic transmission is blocked by maintaining the slice in a low calcium and high magnesium ACSF, "stimulated" NM neurons do not make more protein than those on the unstimulated side. The question remained, however, whether electrical activity of the postsynaptic NM neuron alone is sufficient to up-regulate protein synthesis. In the slice preparation we can address this question by causing NM cells to discharge electrically in the absence of synaptic transmission. In one set of slices, we unilaterally antidromically activated NM neurons by electrically stimulating their axons near midline. We then assessed changes in protein synthesis as compared to similar slices in which the auditory nerve was stimulated. The results indicated that antidromic activation was not sufficient to increase protein synthesis. In fact, electrical stimulation of the postsynaptic neuron alone resulted in less protein synthesis than unstimulated neurons (see Figure 14.4).

Together, our results suggest that calcium-dependent release of some substance from the active auditory nerve is responsible for the transneuronal regulation of protein synthesis. The nature of this substance and the mechanism of its action on the postsynaptic neuron remain to be determined. This substance could be the neurotransmitter, but if this were the case, then, apparently, the mechanism of action is not merely turning on the cascade of events resulting from the generation of postsynaptic

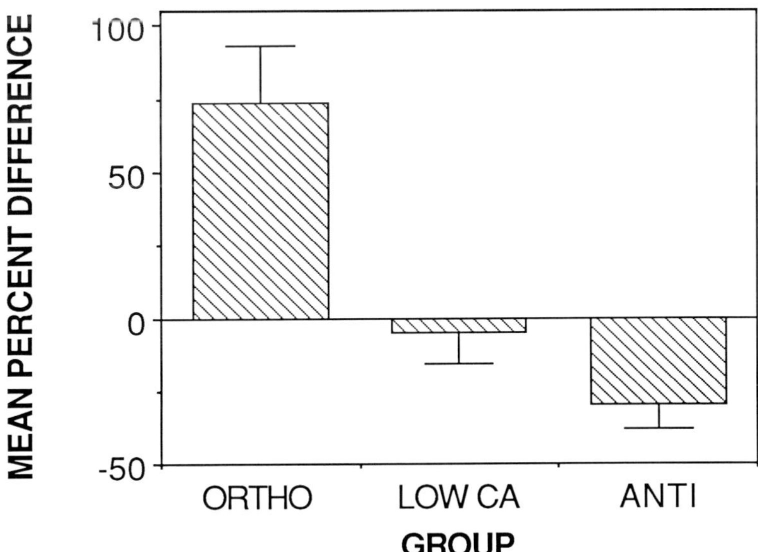

Figure 14.4. Mean percentage difference in protein synthesis for different groups of slices. Protein synthesis was assessed by video image analysis of 30–40 neurons on each side of the same tissue section. Mean percentage difference = 100 × (mean density for stimulated neurons- -mean density for unstimulate neurons) / mean unstimulated neurons. Thus, positive numbers indicate greater synthesis on the stimulated side of the brain. Ortho slices were unilaterally stimulated orthodromically for 1 h, Low Ca slices were stimulated orthodromically but in a low calcium/high magnesium medium; Anti slices were unilaterally stimulated antidromically for 1 h. Bars represent standard error of the mean.

action potentials. Its action could, however, be related to receptor-mediated events that occur prior to the generation of action potentials or to some event that is not directly related to action potentials. It is also possible that the important regulating substance is not the neurotransmitter per se but rather a substance that is coreleased from the active auditory nerve fibers. Future work will be directed at identifying the important regulating substance and the mechanism of its action.

PERSPECTIVES

In this chapter we have traced the evolution of methodological approaches we have used to address questions related to the role of experience in brain development. Some lessons have been learned from our experiences studying these issues. Most important, it is clear that a detailed definition of what is meant by "experience" is required. Experience must be defined in accordance with the level of analysis. In that we are studying the properties of neurons in nucleus magnocellularis of the chicken, our manipulations must be defined by what effect they have on the input to these particular neurons. As Tucci, Born, and Rubel (1987) discovered, a manipulation that logically should result in a dramatic experiential alteration may not cause a noticeable change in overall neuronal activity in regions that are high in spontaneous activity. Conversely, it is important to note that manipulations intended to be noninvasive and result in only minor and reversible effects on the periphery, when applied to young growing animals, can have lasting effects unintended by the experimenter. For example, in several species, chronic earplugs can cause middle ear disease, which can subsequently damage the inner ear. It is always necessary to determine directly whether a peripheral manipulation causes secondary changes in receptor structure or function.

In our search for biological mechanisms of experiential influences, we have redefined "experience" from a change in an organism's acoustic environment to a change in the neuronal activity impinging on a postsynaptic neuron. It now appears that a further redefinition is needed. A change in experience must now be defined by the effect it has on the local environment of the postsynaptic neuron. What ionic/chemical changes in the local milieu result from these manipulations?

Methodologies already exist to begin answering questions regarding the biological mechanisms that underlie experiential influences. The state of the art methods are well ahead of the questions that have been answered. One merely has to adopt the appropriate methods for addressing the question of interest. For example, in theory, to understand the changes in the local ionic/chemical environment as a result of an experiential manipulation, one could make use of ion-selective electrode or bioactive-dye techniques (DeWeer & Salzberg, 1986).

On the other hand, future methodological advances will be helpful for answering our questions. Perhaps the major drawback of our current methods is that we are able to analyze the effect of a given manipulation over time only by examining tissue at a given point in time—that is, by taking "snapshots" of dynamic processes. The development of methods for imaging and analyzing these dynamic cellular properties in real time is underway, but these methodologies are not yet readily usable (DeWeer & Salzberg, 1986; Inoue, 1986). Advances in this domain will provide exciting new techniques by which to explore developmental processes.

ACKNOWLEDGMENTS

Work reviewed in this chapter is currently funded by PHS grants DC 00393, DC 00395, and DC 00858.

REFERENCES

Aslin, R. N. (1984). Sensory and perceptual constraints on memory in human infants. In R. Kail & N. E. Spear (Eds.), *Comparative perspectives on the development of memory* (pp. 39–64). Hillsdale, NJ: Erlbaum.

Born, D. E., & Rubel, E. W (1984). Cochlea removal eliminates physiological activity in brain stem auditory nuclei of the chicken. *Society of Neuroscience Abstract, 10,* 843.

Born, D. E., & Rubel, E. W (1985). Afferent influences on brain stem auditory nuclei of the chicken: Neuron number and size following cochlea removal. *Journal of Comparative Neurology, 231,* 435–445.

Born, D. E., & Rubel, E. W (1988). Afferent influences on brain stem auditory nuclei of the chicken: Presynaptic action potentials regulate protein synthesis in nucleus magnocellularis neurons. *Journal of Neuroscience, 8,* 901–919.

Crocker, B. (1969). *Betty Crocker's cookbook.* New York: Bantam.

DeWeer, P., & Salzberg, B. M. (1986). *Optical methods in cell physiology.* New York: Wiley.

DiScenna, P. (1987). Method and myth in maintaining brain slices. In A. Schuur, T. J. Teyler, & M. T. Tseng (Eds.), *Brain slices: Fundamentals, applications and implications* (pp. 10–21). Basel: Kaiger.

Dubin, M. W., Stark, L. A., & Archer, S. M. (1986). A role for action-potential activity in the development of neuronal connections in the kitten retinogeniculate pathway. *Journal of Neuroscience, 6,* 1021–1036.

Durham, D., & Rubel, E. W (1985). Afferent influences on brain stem auditory nuclei on the chicken: Succinate dehydrogenase activity following cochlea removal. *Journal of Comparative Neurology, 231,* 446–456.

Globus, A. (1975). Brain morphology as a function of presynaptic morphology and activity. In A. H. Riesen (Ed.), *The developmental neuropsychology of sensory deprivation* (pp. 9–91). New York: Academic Press.

Gottlieb, G. (1976). The roles of experience in the development of behavior and the nervous system. In G. Gottlieb (Ed.), *Studies on the development of behavior and the nervous system* (Vol. 3, pp. 25–54). New York: Academic Press.

Hackett, J. T., Jackson, H., & Rubel, E. W (1982). Synaptic excitation of the second and third order auditory neurons in the avian brain stem. *Neuroscience, 7,* 1455–1469.

Hyson, R. L., and Rubel, E. W (1988). Transneuronal regulation of protein synthesis in the brain stem auditory system of the chick requires synaptic activation. *Journal of Neuroscience, 9,* 2835–2845.

Hyson, R. L., & Rudy, J. W. (1987). Ontogenetic change in the analysis of sound frequency in the infant rat. *Developmental Pyschobiology, 20,* 189–207.

Inoue, S. (1986). *Video microscopy.* New York: Plenum.

Jackson, H., Hackett, J. T., & Rubel, E. W (1982). Organization and development of brain stem auditory nuclei in the chick: Ontogeny of postsynaptic responses. *Journal of Comparative Neurology, 210,* 80–86.

Kandel, E. R., & Spencer, W. A. (1968). Cellular neurophysiological approaches in the study of learning. *Physiology Review, 48,* 65–134.

Landmesser, L. (1978). The development of motor projection patterns in the chick hind limb. *Journal of Physiology, 284,* 391–414.

Langer, R., Brown, L., & Edelman, E. (1985). Controlled release and magnetically modulated release systems for macromolecules. *Methods in Enzymology, 142,* 399–422.

Levi-Montalcini, R. (1949). The development of the acoustico-vestibular centers in the chick embryo in the absence of the afferent root fibers and of descending fiber tracts. *Journal of Comparative Neurology, 91,* 209–242.

Martin, M. R. (1985). Excitatory amino acid pharmacology of the auditory nerve and nucleus magnocellularis of the chicken. *Hearing Research, 17,* 153–160.

Mistretta, C. M., & Bradley, K. M. (1978). Effects of early experience on brain and behavioral development. In G. Gottlieb (Ed.), *Studies on the development of behavior and the nervous system* (Vol. 4, pp. 215–247). New York: Academic Press.

Movshon, J. A., & van Sluyters, R. C. (1981). Visual neural development. *Annual Review of Psychology, 32,* 477–522.

Oertel, D. (1983). Synaptic responses and electrical properties of cells in brain slices of the mouse anteroventral cochlear nucleus. *Journal of Neuroscience, 3,* 2043–2053.

Oertel, D. (1985). Use of brain slices in the study of the auditory system: Spatial and temporal summation of synaptic inputs in cells in the anteroventral cochlear nucleus of the mouse. *Journal of the Acoustical Society of America, 78,* 328–333.

Parks, T. N. (1979). Afferent influences on the development of the brain stem auditory nuclei of the chicken: Otocyst ablation. *Journal of Comparative Neurology, 183,* 665–678.

Pasic, T. R., & Rubel, E. W (1989). Rapid changes in cochlear nucleus cell size following blockade of auditory nerve electrical activity in gerbils. *Journal of Comparative Neurology, 283,* 474–480.

Pujol, R. (1972). Development of tone-burst responses along the auditory pathway in the cat. *Acta Oto-laryngolica, 74,* 383–391.

Rubel, E. W, Lippe, W. R., & Ryals, B. M. (1984). Development of the place principle. *Annals of Otology, Rhinology, and Laryngology, 93,* 609–615.

Rubel, E. W, & Parks, T. N. (1988). Organization and development of the avian brain stem auditory system. In J. Brugge, I. Hafter, & M. Merzenich (Eds.), *Functions of the auditory system* pp. 3–92. New York: Wiley.

Seftel, M. A., Deitch, J. S., & Rubel, E. W (1986). Immunocytochemical analysis of cytoskeletal proteins in normal and deafferented nucleus magnocellularis neurons. *Abstracts of the Association for Research in Otolaryngology, 9:* 8–9.

Shatz, C. J., & Kirkwood, P. A. (1984). Prenatal development of functional connections in the cat's retinogeniculate pathway. *Journal of Neuroscience, 4,* 1378–1397.

Sherman, S. M., & Spear, P. D. (1982). Organization of visual pathways in normal and visually deprived cats. *Physiological Review, 62,* 738–885.

Sretevan, D. W., & Shatz, C. J. (1986). Prenatal development of retinal ganglion cell axons: Segregation into eye-specific layers within the cat's lateral geniculate nucleus. *Journal of Neuroscience, 6,* 234–251.

Steward, O., & Rubel, E. W (1985). Afferent influences on brain stem auditory nuclei of the chicken: Cessation of amino acid incorporation as an antecedent to age-dependent transneuronal degeneration. *Journal of Comparative Neurology, 231,* 385–395.

Stryker, M. P., & Harris, W. A. (1986). Binocular impulse blockage prevents the formation of ocular dominance columns in cat visual cortex. *Journal of Neuroscience, 6,* 2117–2133.

Tucci, D. L., Born, D. E., & Rubel, E. W (1987). Changes in spontaneous activity and CNS morphology associated with conductive hearing loss in chickens. *Annals of Otology, Rhinology, and Laryngology, 96,* 343–350.

Tucci, D. L., & Rubel, E. W (1985). Afferent influences on brain stem auditory nuclei of the chicken: Effects of conductive and sensorineural hearing loss on n. magnocellularis. *Journal of Comparative Neurology, 238,* 371–381.

Young, S. R., & Rubel, E. W (1986). Embryogenesis of arborization pattern and topography of individual axons in nucleus laminaris of the chicken brain stem. *Journal of Comparative Neurology, 254,* 425–459.

15

A Neuroanatomical Approach
to Substrates of Behavioral Plasticity

WILLIAM T. GREENOUGH AND ANITA M. SIREVAAG

STATE OF THE FIELD

Questions to Be Addressed

The key issues in this research area, neural plasticity of the development process, have to do with the mechanisms by which the developing brain and the mature brain store information. Ample indications of critical or sensitive periods suggest that there may be different mechanisms in development and adulthood, although common mechanisms may also exist. At the time our laboratory began its work in this area in the late 1960s, a very few reports had indicated that the structure of nerve cells might be sensitive to experience in development (e.g., Coleman & Riesen, 1968; Cragg, 1967; Holloway, 1966; Valverde, 1967). Thus, the initial questions, most of which have yet to be entirely answered, were:

1. Do neurons show morphological responsiveness to behavioral experiences such as dark-rearing and differential environmental complexity?
2. If so, are these morphological responses involved in the storage of extrinsically originating information?
3. Is there more than one cellular mechanism of information storage?
4. Do the mechanisms change with lifespan developmental stage?

We have added a few other questions along the line, notably, What changes other than neurons or synapses, why and how, and who does what to whom in initiating, stabilizing, and maintaining the process? but the basic questions above still establish the boundaries of the research. As it turns out, question (1) was fairly easy to answer, (3) and (4) appear tentatively to be answered in the affirmative, and (2) has consumed an enormous amount of energy. We leave it to the reader of this and our continuing work to judge the adequacy of our answer.

Brief Background

We addressed these questions within a paradigm in which rats are reared in environmental complexity (EC), social cages (SC), or individual cages (IC), which is described in more detail in the following pages. Many colleagues in the "harder" areas of neuroscience are troubled by our choice of the complex (termed by others "enriched") environment paradigm for much of our research on neural plasticity. We selected it for two primary reasons: (1) We wanted a paradigm in which input from behavioral experience mediated by sensory and memory systems was paramount in terms of the effects on brain—that is, one in which known endocrine or metabolic effects were minimal. This does not appear in general to be true for visual deprivation involving dark-rearing, for example, in which pronounced maternal endocrine effects may be transmitted to the infant subjects (e.g., Mos, 1976; Eayrs & Ireland, 1950). We are suggesting that we think the EC-IC paradigm is free of extraneous factors; every change in the environment, no matter how seemingly specific, has consequences beyond those one may desire to study. We do believe that the various metabolic effects of our EC, SC, and IC environments, that can be measured in the periphery are, by and large, not paralleled in the brain (because we do not have space to cover that work here, the interested reader is referred to Black, Sirevaag, Wallace, Savin, & Greenough, 1989). (2) We wanted an experiential sledgehammer—that is, something that had a sufficiently profound effect upon brain development that we could detect brain changes that might be fairly subtle. As an undergraduate student preparing for a seminar taught by Donald Stein in 1964, the senior author first began to realize that the "enriched environment" research then being described at Berkeley by Rosenzweig, Bennett, Krech, and Diamond might fill that need.

There was evidence that the complexity of the rearing environment had substantial behavioral effects. Hebb (1949) reported that home-reared pet rats were superior to laboratory rats on learning tasks. A more typical laboratory study by Forgays and Forgays (1952) reported that maze performance of rats improved with the increasing complexity of the environment in which they were reared. The work of the Berkeley group presented similar evidence for meaningful parameters of brain (e.g., Bennett, Diamond, Krech, & Rosenzweig, 1964; Rosenzweig, Bennett, & Diamond, 1972). While many of their early cellular and vascular measurements were inconsistent or not repeated by others (e.g., Diamond, Krech, & Rosenzweig, 1964, vs. Diamond et al., 1966, regarding glial cells; Mollgaard, Diamond, Bennett, Rosenzweig, & Lindner, 1971, vs. Diamond, Lindner, Johnson, Bennett, & Rosenzweig, 1975, regarding synapse size; Diamond et al., 1964, vs. Black, Sirevaag, & Greenough, 1987, regarding capillary density), the demonstration that the weight and thickness of the cerebral cortex could be reliably altered by experience provided a solid foundation for a detailed quantitative neuroanatomical approach. Measurements of specific brain areas have indicated that the cerebellum, brain stem (subcortical areas), and olfactory bulbs did not greatly contribute to the differences in whole-brain weights. Rather, the total cortex was heavier in ECs than in ICs (e.g., Rosenzweig et al., 1972). A standardized sample from occipital cortex of ECs was typically 6–10% heavier than in ICs (e.g., Rosenzweig et al., 1972; Rosenzweig & Bennett, 1978; Diamond et al., 1972). Of the 175 EC-IC littermate comparisons made between 1960 and 1969, EC cortices averaged 6.4% heavier, and the cortex of the EC littermate was heavier in 133 comparisons.

Limitations of Available Methodology

A conceptual issue that requires brief mention is the interpretation of results from the complex environment paradigm. Almost certainly the complex environment we describe in our methods section falls short of providing the amount of stimulation and information available to the average feral rat. In this sense, use of the term "enriched" may be misleading, and our laboratory dropped it in 1973 after it was criticized on these grounds by an anonymous reviewer. What the paradigm *does* provide is three different levels of environmental stimulation and available information associated with substantial neuroanatomical differences. There is good reason to believe that these differences result from information processing and not from endocrine or metabolic consequences of the differential housing conditions (Rosenzweig et al., 1972; Black et al., 1989). Thus, the paradigm has proved very valuable for investigating the integrated response of the visual cortex to information storage demands. More specific paradigms can be used to evaluate the roles of these changes with regard to behavior (e.g., Chang & Greenough, 1982; Black et al., 1987).

The optimal methodology for resolving issues regarding the hypertropy and genesis of specific cell types and cellular components such as synapses would involve "magic bullets," highly specific histological markers of the type only dreamed of in 1970 and just becoming available in some cases now. For example, if one wanted to see whether neurons were forming new synapses, the ideal methodology would be one in which entire neurons could be viewed and their synapses visualized such that they could be counted. We have something like magic bullets for *astrocytes* today, which we have been using in parallel studies (see below). However, the dominant methods for visualizing *neurons and their synapses* involve two separate techniques: the *Golgi techniques for visualizing dendritic fields* and *transmission electron microscopy for visualizing synapses*.

In principle, quantitative analysis of dendritic fields of neurons stained by the Golgi procedures can provide a quite accurate assessment of the amount of postsynaptic space for synapses. In practice, several problems must be addressed.

1. First, there have been reports that Golgi procedures may not impregnate the entire dendritic field of neurons, so that estimates would be unreliable or artificially low. There have been reports and unpublished suggestions that dye-injection studies reveal larger dendritic fields than are seen in Golgi studies. Our examination of dye-injected neurons from other laboratories has indicated that the differences stem largely from the superior ability to reconstruct through serial sections with dyes (see below) and from the fact that larger somata are more likely to be penetrated by electrodes. In general, across a range of injected cells, our examinations of others' tissue indicate that dye injection is far more likely to yield incomplete impregnation of dendrites and spines than is properly conducted Golgi staining (see Greenough, 1984)

2. The various Golgi procedures (rapid, Cox, and Kopsch are the primary variants) stain only a small fraction of neurons in most tissues. If they stained all neurons, they would be useless, of course, because the ability to isolate and follow the

processes of individual neurons would be lost. However, the capriciousness of the staining raises the possibility that the stained neurons are not a representative sample. It is even possible that different populations of neurons might be stained in different experimental groups.

3. Only a portion of the dendritic field can be studied with Golgi procedures, because distal dendrites are typically truncated by the cut surface of the tissue section. The comparatively high density of stained processes makes reconstruction of dendritic fields across adjacent sections very difficult and impractical for studies in which large numbers of subjects and neurons are to be quantified. We select neurons that are centered in the thickness of the tissue section in order to minimize this problem, and we have avoided some very large neuron types that greatly exceed section thickness. Nonetheless, this criticism is valid with regard to the accuracy of estimates of group differences in dendritic field size, which will be underestimated by incomplete data collection. As the primary outcome of the EC-IC studies has been the greater dendritic field size in EC animals, and the pruning effect is greater on average for larger dendritic fields, the attenuation will tend to *underestimate the size of EC fields* more than of ICs and hence to *underestimate the magnitude of EC-IC differences*. Given this source of attenuation and the problems discussed in the section below on quantitative analysis of dendritic fields, the relative agreement between our early light microscopic estimates of an EC-IC difference of about 20% in dendritic field size in upper visual cortex (Volkmar & Greenough, 1972; Greenough & Volkmar, 1973) and our light-electron microscope estimates of an EC-IC difference of about 25% in the number of synapses per neuron (Turner & Greenough, 1985; Sirevaag & Greenough, 1987) may be somewhat surprising. Equally surprising is the similarity of the magnitude (for excitatory synapses) of the group difference in a study that used cats rather than rats (Beaulieu & Colonnier, 1987).

4. A final and quite valid criticism of the use of Golgi staining to estimate synapse differences is that Golgi impregnation reveals only the amount of postsynaptic surface, and there is no way to be sure that innervation density is comparable across different experimental conditions. While Globus, Rosenzweig, Bennett, and Diamond (1973) found spine frequency on layer V visual cortex pyramidal neurons of EC rats to be equal to or higher than that of IC rats, this does not resolve the issue for nonspiny neurons, nor is there evidence that conclusively relates spine density to innervation density.

The net effect of these methodological limitations—particularly (2) and (4)—is that converging evidence using other procedures is necessary to corroborate the Golgi findings. Our approach, the use of combined light and electron microscopic density estimates to estimate the average number of synapses per neuron, gives up any hope of the "magic bullet" approach because we relinquish knowledge of which synapses are on which neurons, but all neurons and synapses are visualized, thus obviating most criticisms of the Golgi technique. The primary criticisms of older synapse density estimates in EC and IC rats (e.g., Diamond et al., 1966; Mollgaard et al., 1971) is that the estimates were not corrected for differences in the size of synapses and neuronal nuclei. The somewhat more modern methods we used (Turner & Greenough, 1985) corrected for size differences, but were based upon certain geometric approximations to the shape of a neuronal nucleus or a synaptic contact. The newest shape-independent methods avoid this pitfall, although the electron microscopic approach still can assess only averages. Thus, for example, the simultaneous loss and gain of synapses by neurons inferred from Golgi studies (e.g., Juraska, 1982; Greenough & Chang, 1988) would not be detected by these procedures.

METHODS USED

Behavioral Methods: The Complex Environment Paradigm

Complex environment studies typically place same-sex (typically male) animals into three environments: complex (EC), social condition (SC), or individual cage (IC), where they remain until behavioral testing or sacrifice for morphological studies. For SC and IC we use standard transparent cages. For EC, we fill a 1-m³ cage both horizontally and vertically with toys so that the rats have as complex a setting as possible; some other investigators use many fewer toys (e.g., Rosenzweig et al., 1972; Bhide & Bedi, 1984a, 1984b). Our pool of toys numbers in the hundreds and is regularly replenished by visits to local garage sales, for example. As our published studies indicate (e.g., Greenough & Volkmar, 1973; Sirevaag & Greenough, 1985), we change the toys in the cage each day while the animals are placed in a 1.2-m² play arena, also equipped with a daily change of toys. We take great pains to observe the animals to make certain that they are interacting with each other and the environment. In one case a few years ago, we noted that the animals were not exhibiting their usual exuberant exploratory and play behavior; investigation revealed that cats had been moved into a colony room separated from the rats' housing by four metal doors so that the cats' meowing was not noticeable to the human observer. Nonetheless, after the cats were moved, normal play behavior resumed.

The complex environment offers a multisensory experience. No single sensory modality appears to account for all of the morphological differences between EC and IC rats, even within cortical tissue of that modality. Blinded rats show gross morphological differences in visual cortex similar to sighted ones; thus, visual input is not necessary for morphological differences, even though differences occur in the visual cortex (Rosenzweig et al., 1969). Nonvisual cortex as well as noncortical brain regions are also affected (Greenough, Volkmar, & Juraska, 1973; Floeter & Greenough, 1979; Juraska, Fitch, Henderson, & Rivers, 1985; Rosenzweig et al., 1972). These data suggest that what is being studied in this paradigm is qualitatively or functionally different from what is being studied in a typical developmental sensory deprivation study, in which the effects are typically limited to the modality under study and occur only at specific ages.

We typically use a split-litter design to minimize intersubject variability. Pre-experimental variance is reduced to partial genetic heterogeneity (rats are multiple ovulators) and individually unique incidents in the common environment that the rats have shared since conception.

While both brain and behavioral effects of EC versus IC housing may vary with age (see Greenough, 1986, and Black & Greenough, 1986, for reviews of studies involving age and possible sensitive period effects), substantial effects of environmental complexity occur in adults (Uylings, Kuypers, Diamond, & Veltman, 1978; Juraska, Greenough, Elliott, Mack, & Berkowitz, 1980; Hwang & Greenough, 1986). Thus, the functional plasticity affected by complex environment housing may differ from that tied to sensitive periods of development (e.g., effects of dark-rearing on visual cortex; Cragg, 1975; Borges & Berry, 1976). There may, however, be a period around the age of weaning when, because of the natural activity level of the pups, environmental effects are manifest. There is a significant history of reports of increased locomotor activity in rat pups between about 28 and 35 days of age (Baenninger, 1967; Calhoun, 1962; Panksepp, 1981; Spear & Brake, 1983; our unpublished observations). While we have not varied onset age systematically, we have found that small but detectable increases in visual cortex dendritic branching are very rapidly manifest at this age, within as little as 4 days (Kilman, Wallace, Withers, & Greenough, 1988), although additional effects continue to be manifest with increasing exposure duration, at least up to the 30-day duration of most experiments (Volkmar & Greenough, 1972; Greenough & Volkmar, 1973; Juraska, 1984). Similarly, Ferchmin and Eterovic (1986), using a somewhat different EC condition, detected quite rapid effects at about this age. We are still in the process of determining the age dependence of some other effects of these rearing conditions on the brain (see the section below on capillaries). We assume that dendritic and synaptic changes would continue to accumulate with continuing lifetime demands for information storage.

Quantitative Golgi Analysis of Dendritic Fields

We began using different Golgi staining but settled upon the Glaser and Van der Loos (1981) method because of its reliability and because it appears to impregnate a representative population of nerve cells (Pasternak & Woolsey, 1975). A detailed protocol from a recent Society for Neuroscience Short Course is available upon request from the author. In early studies, camera lucida projection drawings were made (e.g., Greenough & Volkmar, 1973). For a radial neuron, collapsing the three-dimensional dendritic field to a two-dimensional drawing attenuates true branch length by about 30% (the sine of the average 45° deviation from horizontal is .71). The relative attenuation is the same regardless of dendritic field size, so this does not affect the *magnitude* of the difference between experimental groups. However, on average, compression to two-dimensional projection drawings will *increase the variance* in length of branches because it affects individual branches to a greater or lesser extent, depending upon their orientation. Thus, projection drawing with a camera lucida *works against detecting a group difference statistically.* (We are indebted to Janice Juraska for having demonstrated this to us.) In practice, with the sample sizes that we use and with statistics based upon litters or subjects with nerve cells treated as repeated within-subject measures, the effect of projection drawing on statistical power is probably relatively small. We now use a computer-aided Zeiss microscope with stepping-motor-driven specimen stage controls that allow accurate three-dimensional dendritic field quantification (Wann, Woolsey, Dierker, & Cowan, 1973; DeVoogd, Chang, Floeter, Jencius, & Greenough, 1981). This system certainly adds to the absolute accuracy of our data. We have no evidence, however, that it has provided significantly more group discriminative power than two-dimensional projection drawings. Commercially available dendritic tracking systems (e.g., Eutectics Neuron Tracking System, Eutectic Electronics, Inc.) have many of the same basic features as our system and should prove satisfactory for most purposes.

While we have used other methods of dendritic field quantification, such as two-

and three-dimensional grids (the latter never published; Greenough, Carter, Steerman, & DeVoogd, 1977), we commonly use the methods adopted from Sholl (1956), the concentric ring (or sphere) intersection analysis, and from Coleman and Riesen (1968), the quantification of numbers of branches at each order of bifurcation away from the soma or apical dendrite and their lengths. More elaborate methods of dendritic field analysis have been proposed, but none seems to have been sufficiently compelling to have caught on (Pysh & Weiss, 1979; Uylings, Smit, & Veltman, 1975), and the analyses we use are by far the most common. These are detailed in many published sources (e.g., DeVoogd et al., 1981; Camel, Withers, & Greenough, 1986), as well as in the Society for Neuroscience Short Course writeup mentioned previously (Greenough, 1984).

The details of the branching analysis have evolved considerably over the years. The idea that terminating and bifurcating branches had different distributions and should be analyzed separately arose from the work of Uylings et al. (1978), who detected differences in nonbifurcating branch length that we had previously ignored, in their report that dendritic field differences occurred in adult rats placed in EC and IC environments. Our work subsequent to this also separated out branches that were truncated by the section plane (Juraska et al., 1980). The original studies with Volkmar (Greenough & Volkmar, 1973; Volkmar & Greenough, 1972) had included cut-off, terminating, and bifurcating branches in the same-length analyses. It is not surprising that those studies, which detected differences in the overall amount of dendrite, led us to focus on changes in branch number and to ignore changes we might have detected in branch length! Separate analysis of basilar and apical dendritic fields and the treatment of apical dendrites as sources of primary branches throughout their length (rather than incrementing apical shaft order each time an oblique branch leaves the shaft), were arbitrary decisions later validated by detection of differences in basilar but not apical dendrites (Juraska et al., 1980) or in particular regions of apical dendrites (Greenough, Juraska, & Volkmar, 1979; Chang & Greenough, 1982). Modified analysis methods are also necessary for certain cell types. For example, in Purkinje cells of the cerebellar cortex, one type of excitatory afferent, the climbing fiber from the inferior olive, terminates on the main branches. Another type, the parallel fiber axons that conduct mossy fiber input from other brain stem nuclei, terminates on the spiney branchlets. These two parts of the field must be analyzed separately if differential responses with regard to these afferents are to be detected.

Light-Electron Microscopic Estimates of the Number of Synapses per Neuron

Golgi studies have shown that ECs have larger dendritic fields in the occipital cortex than ICs, but as noted above, they cannot discern new synapse formation directly. To assess synapse number change, it is necessary to estimate synaptic density and neuronal density to calculate numbers of synapses per neuron. Earlier estimates of neuronal density among environmental groups were made by counting the number of nuclei in a micrograph. However, assuming equal density, larger nuclei or synapses are more likely to be sampled in a tissue section and hence will inflate density estimates. One can use stereological procedures to correct the two-dimensional information for size differences. (*Stereology* refers to the science or technology of accurately describing the properties of volumes on the basis of samples of the volumes; for example, deriving accurate, or "unbiased," estimates of the three-dimensional characteristics of a volume of brain tissue from two-dimensional tissue sections requires stereological procedures.) Our studies of this issue in the EC-IC paradigm used older numerical density (Nv) methods that involved assumptions regarding the regularity of shape of synapses and neurons. More recent stereological methods no longer require measuring the diameter of nuclei and are independent of nuclear shape (Braendgaard & Gundersen, 1986). While the methods we used have been superseded, there is little reason to believe that the newer methods will alter the outcome significantly. However, we will not detail those older methods, referring those interested to the newer ones (Braendgaard & Gundersen, 1986).

Results of a study with the older methods appear in Figure 15.1. A very important point is that the *density of synapses alone* may provide little useful information regarding the formation or loss of synapses. This is because both the neural components of the new synapses (axon, dendrite, increased neural soma size) and changes in nonneuronal tissue components such as the glial and vascular changes described below expand the total volume, pushing neurons and synapses apart. Thus, in the upper visual cortex of EC and IC rats, synaptic density is nearly the same. Neuronal density is more diluted by these tissue components in the EC rats, and the number of synapses per neuron, the best estimate of the net addition of synapses, is greater in the EC rat.

Development of New Methodology for Astrocytic Quantification

Figure 15.1 indicates substantial dilution of neurons by other tissues in EC rats—more than could be accounted for by increased dendritic and synaptic volume alone (Sirevaag & Greenough, 1987). Work by the Berkeley group and others often suggested that differences in glial cells could be involved. Diamond et al. (1964) initially reported that glial number and the glia to neuron ratio (G/N) did not differ between EC and IC rats. Later, however, they reported that ECs had 14% more glia per columnar sample of cortex and a greater G/N than ICs (Diamond et al., 1966). Based on thymidine incorporation, higher gliogenesis rates were also reported in EC than IC rats by Altman and Das (1964). Szeligo and Leblond (1977) reported a nonsignificant trend for EC rats to have more astrocytes than ICs. ECs had significantly more oligodendrocytes than SCs or ICs as well as a greater G/N. None of these reports applied stereological methods. Pooling astrocytes and oligodendrocytes in a stereologically corrected study, Bhide and Bedi (1984a, 1984b) reported the G/N to be significantly greater in ECs than in ICs. However, a different pattern was observed when astrocytes and oligodendrocytes were considered separately. Clearly, a coherent view had yet to emerge.

We began work on glial cells using conventional methodology (Sirevaag & Greenough, 1987) to report that the volume fraction, Vv (the percentage of cortex occupied), of nuclei was significantly greater in ECs for both astrocytes and oligodendrocytes. The greater oligodendrocyte Vv was probably due to the small numerical density differences among the groups, which were not themselves statistically reliable. The greater astrocytic Vv was due, at least in part, to the ECs' significantly greater nuclear volume, since the density of astrocytes did not differ among the groups. The larger EC astrocytic nuclear size suggested that EC astrocytes may have larger trees consisting of more processes. Finally, in the late 1980s, we had a question for which a

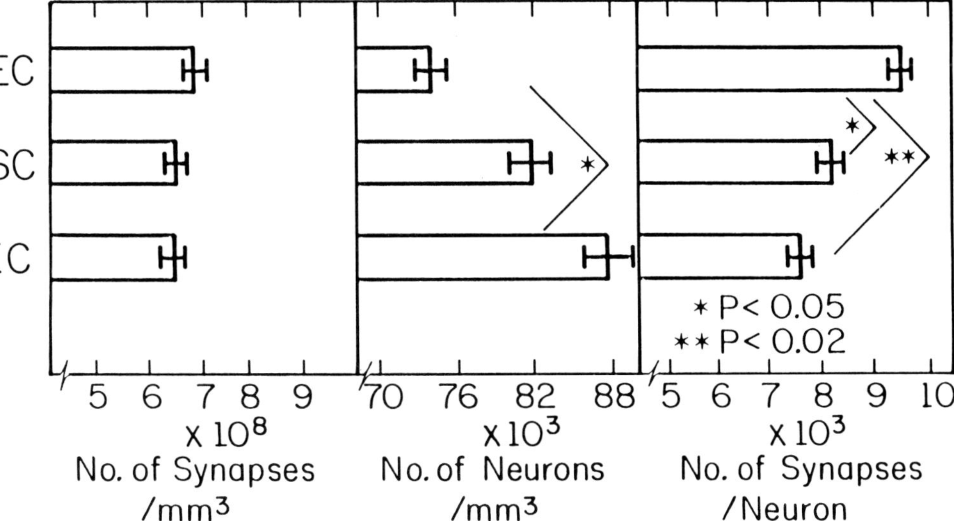

Figure 15.1. Synaptic and neuronal density and synapses per neuron, corrected for group differences in size, in upper visual cortex of rats reared for 30 days after weaning in environmental complexity (EC), social cages (SC), or individual cages (IC). Lower density of neurons in EC and SC rats reflects increases in neural processes and associated tissue elements. (After A. M. Turner and W. T. Greenough, 1985, "Differential Rearing Effects on Rat Visual Cortex Synapses. I. Synaptic and Neuronal Density and Synapses per Neuron," *Brain Research, 329*, pp. 195–203. Copyright 1984, Elsevier Science Publishers.)

"magic bullet," or special label, was available. There were antibodies to *glial fibrillary acidic protein (GFAP)*, a constituent specific to astrocyte processes.

To investigate possible differences in astrocytic size, we used monoclonal GFAP antibodies to label astrocytic processes (Sirevaag, Bell, & Greenough, 1989). With highly sensitive and specific antibodies and the greater sensitivity of recently available avidinbiotin peroxidase techniques, we impregnated the processes of astrocytes almost as intensely and seemingly as thoroughly as the Golgi procedures impregnate dendrites. Because many known or proposed functions of astrocytes involve transport across the membrane, the amount of astrocyte membrane per unit tissue volume, or *surface density (Sv)*, was chosen as the appropriate functional measure. Because of the relative ease of identification and visualization of astrocytes we could accurately quantify the astrocytic processes present in the tissue. Sv was obtained by counting the intersections of astrocytic processes with a superimposed cycloid are grid (Baddeley, Gundersen, & Cruz-Orive, 1986). A cycloid grid ensures that an unbiased sample of astrocytic processes (a requirement for three-dimensional stereological estimates) is obtained from the precisely oriented, nonrandom coronal sections used in this and most neuroanatomical measurements. EC rats had a greater Sv of labeled processes than their IC littermates. Using the same tissue, we were able to estimate the numerical density (Nv) of astrocyte nuclei. With both Sv and Nv, we could estimate the mean surface area of processes per astrocyte, and we found that the differences were in astrocytic size rather than in numbers of astrocytes. We have also recently found astrocyte surface density changes in the dentate gyrus molecular layer of the hippocampal formation following the induction of long-term potentiation, an electrical stimulation-based model of the adult memory process (Isaacs, Marks, Sirevaag, Chang, & Greenough, 1989). Thus, these methods have shown a dynamic property of astrocytes across two different neural plasticity paradigms and brain areas.

Analyses of Capillaries

An issue of considerable interest is whether the vascular system keeps up in some way with all of these volumetric changes. If the capillary system stayed static, it would be diluted by about 25% in the upper visual cortex—a substantial reduction in perfusion with oxygen and other metabolites. The conventional belief was that capillary formation in the rat visual cortex had ceased by the age that weanling rats are introduced to the complex environment (Rowan & Maxwell, 1981). Early studies of capillaries and complex environments were complicated by methodological errors (see Sirevaag, Black, Shafron, & Greenough, 1988, for a more complete discussion). One study reported no difference in the frequency of capillary profiles (Walsh, Budtz-Olsen, Penny, & Cummins, 1969), and another reported that ECs had slightly *fewer* capillaries than ICs (Diamond et al., 1964), which supported the dilution view. When we eliminated some methodological difficulties by using relatively thin (0.5-μm) epoxy-embedded tissue sections and adapting both standard stereological approaches and direct measurements of projection drawings (Black et al. 1987), we found that ECs had a *greater* volume fraction (Vv) and average capillary diameter than SCs or ICs. Capillary frequency can be used to estimate the average distance between a blood vessel and a random point in the cortex (Homer, 1984). ECs had a shorter average distance and so a higher density of capillaries than SCs or ICs. This implied capillary sprouting in EC rats. A direct test of that implication by counts of india ink-filled blood vessels showed that EC rats had more capillary branches than ICs and thus that, under conditions of sufficient demand, capillaries continue to sprout beyond the age at which this process was reported to terminate (Sirevaag et al., 1988).

There is also an interesting contrast in the capillary data. Just as the finding that dendritic growth occurred in the adult and middle-aged EC rat visual cortex reversed beliefs that this sort of neuronal plasticity was limited to development, vascular measures in older animals have shown that not all components of juvenile plasticity survive maturation and aging. Whereas capillary density *overcompensates* for synaptic number and tissue volume increases in periadolescent EC rat visual cortex, capillary density just compensates for tissue volume increases in young adults and progressively fails to keep up in older rats. This raises the curious possibility that the brain retains synaptic plasticity but loses the metablic plasticity necessary to support it. It is not likely, however, that a feral animal would live the sheltered existence of a laboratory rat until adulthood, then to be plunged into a complex, real-life situation. It is probably the case that existing vasculature, conditioned by species-typical early experience, could handle the smaller localized changes that would mediate aspects of day-to-day learning and memory.

PERSPECTIVES

In the late 1960s it seems to have been generally assumed that, once the pattern of synapses had been laid down during development, little could be done, aside, perhaps, from synaptic efficacy change, to alter that pattern. Demonstrations that experience manipulations altered synapse formation or structure in early development were considered quite radical and exciting (and like many demonstrations of plastic neural phenomena, widely disbelieved when first reported). It seems safe to say that few neuroscientists would have predicted the generality of morphological plasticity in postweaning brain, much less in that of the adult brain which has since been uncovered.

What do we know now that we did not know then? Plasticity of synapse formation, or at least of dendritic structural alteration, appears to be nearly a lifelong process, with perhaps only the most elderly exhibiting significant reductions in this capacity (Connor, Melone, Yuen, & Diamond, 1981; Green, Greenough, & Schlumpf, 1983; Greenough, McDonald, Parnisari, & Camel, 1986; Juraska et al., 1980; Uylings et al., 1978). Quantification of synaptic numbers per neuron has confirmed interpretation of dendritic field analyses as reflecting net synaptogenesis (Turner & Greenough, 1985; Sirevaag & Greenough, 1987; Hwang & Greenough, 1986). Studies using within-subject designs have indicated specificity of these changes to input via normal channels to the brain, as opposed to metablic or endocrine effects (e.g., Chang & Greenough, 1982), and comparisons of endocrine/metabolic-sensitive peripheral measures with brain measures appear to indicate a remarkable independence of brain effects from peripheral metabolic fluctuations (Black et al., 1989) Peripheral dendritic fields show similar surprising plasticity (Purves, Hadley, & Voyvodic, 1986), although forces that influence this plasticity remain to be ascertained. Possibly related plastic phenomena such as axonal sprouting have added to the credibility of these reports, if not to the understanding of the purpose of synaptogenesis in adulthood (Cotman & Nieto-Sampedro, 1984).

With regard to synaptogenesis in very early development, a revolution has occurred. In the late 1960s the questions were how chemospecification of nerve cell connections worked and how experience could act through neural activity modulation to influence this process. We have not really answered those questions, but our perspective has changed dramatically with the knowledge that in many, if not all, developing mammalian nervous system components production of extra synapses and selective elimination or preservation of functionally appropriate subsets of them is a key aspect of the establishment of the initial wiring diagram. Social acceptability of this view was initiated by Changeaux and Danchin (1977), who proposed analogies between supernumerary synapse elimination at the developing neuromuscular junction and the development of functional systems in the central nervous system. The jury is still out on whether some similar process is involved in the accumulation of synapses with experience in adulthood. We have argued that the initiation of synaptogenesis in adults is determined or influenced by experience (e.g., Black & Greenough, 1986; Greenough, 1984), but leave open the possibility that a selective elimination/preservation process operates upon experience-dependent synapses.

Both astrocytic processes and microvasculature increase dramatically in the EC cortex, suggesting, at the very least, greatly increased metabolic demands from the EC brain. The glial cell differences may indicate a qualitatively or merely a quantitatively different functional role in neuronal and synaptic processing. Glia, especially astrocytes, regulate extracellular $H+$, $K+$, and $Cl-$ and are also involved in uptake and possibly storage of at least some neurotransmitters or their precursors (Kuffler, 1967; Newman, Frombach, & Odette, 1984; Kimelberg, 1983). Ionic and neurotransmitter concentrations are primary factors in the regulation of synaptic firing; therefore, glial differences among the environmental groups could reflect either differences in the use of synapses by these groups or differences in the regulation of synaptic efficacy. There is also some indication that glial processes may be involved in initiating synaptogenesis (Tweedle & Hatton, 1984; Meshul, Seil, & Herndon, 1987), so determining how many of these potential glial roles are important in the "turned-on brain" of the EC rat should prove quite interesting.

Sirevaag and Greenough (1987) reported more peak capillary volume per synapse, neuron, and glial cell in ECs than in SCs or ICs. This implies that EC rats require greater access to nutrients, carbohydrates, and oxygen than SC or IC rats. Regional blood flow studies support this conclusion (Berman, Goldman, & Altman, 1985). Young, middle-aged, and old rats exposed to complex environments had significantly higher blood flow in several brain areas, including the occipital cortex, than did SCs (ICs were not studied). These results indicate another cortical component that pushes neurons farther apart and explain the thicker EC cortex and lower neuronal density. They also provide evidence for an EC-IC difference in cortical metabolism that suggests the EC cortex may be capable of more efficient and/or protracted synaptic communication.

It is clear that the plasticity of capillaries decreases vastly more rapidly with age than does that of synapses. One can imagine an oxygen-starved brain region that had failed to develop until late in life in an otherwise healthy brain—or a deprived whole brain, never entirely able to come up to speed despite proper experience and a good bit of adult information storage. Whether this sort of thing might be common in human development is an open question at this point. A picture that is emerging from these studies is one of a coordinated set of processes involved in developmental plasticity underlying brain information storage. The processes that support neuronal plasticity from a metabolic perspective appear to be more temporally restricted in development than is the neuronal plasticity itself. Whether this represents a compromise between the ideal needs of the brain and the capacity of vasculature to remain plastic or whether there is some benefit gained by this early restriction is unclear.

In addition to these conceptual advances, there has been an integrative methodological contribution from the general body of research, of which that described here is but a small part. Twenty years ago, the term "quantitative neuroanatomy" was nearly an oxymoron. Anatomy in general was a descriptive science, and anonymous reviews of grant proposals might argue that if detailed stereological and statistical analyses were necessary to demonstrate an effect, then there probably really was not a very important one present (if any). Now, while the special skills of the trained histologist remain important, the necessity of precise, quanitative description in anatomy has been recognized. Within neuroscience, much of the credit for this turnaround, particularly with regard to use of proper statistics, goes to those investigators trained as psychologists, for whom statistics has long been a fundamental element of their discipline's tool kit.

The contribution of this work to *developmental theory in psychology*, to the broad set of ideas concerning human developoment, has not yet been of major consequence. We have elsewhere elaborated some theoretical perspectives that developmental theorists might consider (Black & Greenough, 1986; Greenough, Black, & Wallace, 1987). These include:

1. The fact that experiential and intrinsically specified development patterning *both* work through regulation of the expression of an organism's genome and of its gene products and probably often involve many genes in common.

2. The facts that experiential patterning of the organization of the nervous system is a *major* component of the development process in mammals and that this is probably particularly so in humans.

3. The theoretical views that there appear to be two mechanisms for the incorporation of environmentally originating information, depending upon whether the type of experience involved is of a type common to all species members and hence incorporated in evolution into "experience-expectant" developmental processes or unique to individuals and hence acquired through "experience-dependent" developmental processes, with different synapse-generation rules.

4. The proposal that in the experience-dependent case it is the experience, and not an intrinsic unfolding, that is largely responsible for driving the developmental process (although organisms such as mammals appear predisposed to seek such experience).

5. The likelihood that the effects we see magnified by the relative extremes of our rearing environments parallel the processes of *normal* development.

These views are, at this point, merely theoretical. A reasonable number of findings are compatible with these views, but certainly the case for all of these points has not been made "beyond reasonable doubt." In particular, the demonstration of *specific functional roles* of the changes in connectivity that we have described still eludes us. Although a host of other research challenges could be cited, this is probably the biggest one currently facing us. As we begin to define the roles these changes play in brain function and behavior, rather than merely speculating upon them, there may be a basis for developmental theory at the behavioral level and at the cellular and brain levels to begin to find some common ground.

ACKNOWLEDGMENT

Preparation of this paper and research not otherwise reported was supported by the Office of Naval Research and by NIMH 35321, 43830, and 40631, NIH training grant No. HD 07333, A Center for Learning and Memory award from NSF (BNS 88-21219), and the Epilepsy Foundation of America.

REFERENCES

Altman, J., & Das, G. D. (1964). Autoradiographic examination of the effects of enriched environment on the rate of glial multiplication in the adult rat brain. *Nature, 204,* 1161–1163.

Baddeley, A. J., Gundersen, H. J. G., & Cruz-Orive, L.-M. (1986). Estimation of surface area from vertical sections. *Journal of Microscopy, 142,* 259–276.

Baenninger, L. P. (1967). Comparison of behavioural development in socially isolated and grouped rats. *Animal Behaviour, 15,* 312–323.

Beaulieu, C., & Colonnier, M. (1987). Effects of the richness of the environment on the cat visual cortex. *Journal of Comparative Neurology, 266,* 478–494.

Bennett, E. L., Diamond, M. C., Krech, D., & Rosenzweig, M. R. (1964). Chemical and anatomical plasticity of brain. *Science, 146,* 610–619.

Berman, R. F., Goldman, H., & Altman, H. J. (1985). Decreased cerebral blood flow and impaired learning in rats associated with aging. *Society for Neuroscience Abstracts, 11,* 727.

Bhide, P. G., & Bedi, D. S. (1984a). The effects of a lengthy period of environmental diversity on well fed and previously undernourished rats. I. Neurons and glial cells. *Journal of Comparative Neurology, 227*, 296–304.

Bhide, P. G., & Bedi, D. S. (1984b). The effects of a lengthy period of environmental diversity on well fed and previously undernourished rats. II. Synapse to neuron ratios. *Journal of Comparative Neurology, 227*, 305–310.

Black, J. E., & Greenough, W. T. (1986). Induction of pattern in neural structure by experience: Implications for cognitive developoment. In M. E. Lamb, A. L. Brown, & B. Rogoff (Eds.), *Advances in developmental psychology* (Vol. 4). Hillsdale, NJ: Erlbaum.

Black, J. E., Jones, A. L., Anderson, B. J., Isaacs, K. R., Alcantara, A. A., & Greenough, W. T. (1987). Cerebellar plasticity: Preliminary evidence that learning, rather than repetitive motor exercise, alters cerebellar cortex thickness in middle-aged rats. *Society for Neuroscience Abstracts, 13*, 1596.

Black, J., Sirevaag, A. M., & Greenough, W. T. (1987). Complex experience promotes capillary formation in young visual cortex. *Neuroscience Letters, 83*, 351–355.

Black, J. E., Sirevaag, A. M., Wallace, C. S., Savin, M. H., & Greenough, W. T. (1989). Effects of complex experience on somatic growth and organ development in rats. *Developmental Psychobiology, 22*, 727–752.

Borges, S., & Berry, M. (1976). Preferential orientation of stellate cell dendrites in the visual cortex of the dark-reared rat. *Brtain Research, 112*, 141–147.

Braendgaard, H., & Gundersen, H. J. G. (1986). The impact of recent stereologic advances on quantitative studies of the nervous system. *Journal of Neuroscience Methods, 18*, 19–38.

Calhoun, J. C. (1962). *The ecology and sociology of the Norway rat*. Bethesda, MD: U.S. Department of Health, Education and Welfare.

Camel, J. E., Withers, G. S., & Greenough, W. T. (1986). Persistence of visual cortex dendritic alterations induced by postweaning exposure to a "superenriched" environment in rats. *Behavioral Neuroscience, 100*, 810–813.

Chang, F.-L., & Greenough, W. T. (1982). Lateralized effects of monocular training on dendritic branching in adult split-brain rats. *Brain Research, 232*, 283–292.

Changeaux, J.-P., & Danchin, A. (1977). Biochemical models for the selective stabilization of developing synapses. In G. A. Cottrell & P. M. Usherwood (Eds.), *Synapses*. New York: Academic Press.

Coleman, P. D., & Riesen, A. H. (1968). Environmental effects on cortical dendritic fields. I. Rearing in the dark. *Journal of Anatomy (London), 102*, 363–374.

Conner, J. R., Melone, J. H., Yuen, A. R., & Diamond, M. C. (1981). Dendritic length in aged rats' occipital cortex: An environmentally induced response. *Experimental Neurology, 73*, 827–830.

Cotman, C. W., & Nieto-Sampedro, M. (1984). Cell biology of synaptic plasticity. *Science, 225*, 1287–1294.

Cragg, B. G. (1967). Changes in visual cortex on first exposure of rats to light: Effect on synaptic dimensions. *Nature, 215*, 251–253.

Cragg, B. G. (1975). The development of synapses in kitten visual cortex during visual deprivation. *Experimental Neurology, 46*, 445–451.

DeVoogd, T. J., Chang, F.-L., Floeter, M. K., Jencius, M. J., & Greenough, W. T. (1981). Distortions induced in neuronal quantification by camera lucida analysis: Comparisons using a semiautomated data acquisition system. *Journal of Neuroscience Metyods, 3*, 285–294.

Diamond, M. C. (1967). Extensive cortical depth measurements and neuron size increases in the cortex of environmentally enriched rats. *Journal of Comparative Neurology, 131*, 357–364.

Diamond, M. C., Krech, D., & Rosenzweig, M. R. (1964). The effects of an enriched environment on the histology of the rat cerebral cortex. *Journal of Comparative Neurology, 123*, 111–120.

Diamond, M. C., Law, F., Rhodes, H., Lindner, B., Rosenzweig, M. R., Krech, D., & Bennett, E. L. (1966). Increases in cortical depth and glia numbers in rats subjected to enriched environments. *Journal of Comparative Neurology, 128*, 117–125.

Diamond, M. C., Lindner, B., Johnson, R., Bennett, E. L., & Rosenzweig, M. R. (1975). Differences in occipital cortical synapses from environmentally enriched, impoverished, and standard colony rats. *Journal of Neuroscience Research, 1*, 109–119.

Diamond, M. C., Rosenzweig, M. R., Bennett, E. L., Lindner, B., Lyon, B., & Lyon, L. (1972). Effects of environmental enrichment and impoverishment on rat cerebral cortex. *Journal of Neurobiology, 3*, 47–64.

Eayrs, J. T., & Ireland, K. F. (1950). The effect of total darkness on the growth of the newborn albino rat. *Journal of Endocrinology, 6*, 386–397.

Ferchmin, P. A., & Eterovic, V. A. (1986). Forty minutes of experience increase the weight and RNA content of cerebral cortex in periadolescent rats. *Developmental Psychobiology, 19*, 511–519.

Floeter, M. K., & Greenough, W. T. (1979). Cerebellar plasticity: Modification of Purkinje cell structure by differential rearing in monkeys. *Science, 20*, 227–229.

Forgays, D. G., & Forgays, J. W. (1952). The nature of the effect of free-environments experience in the rat. *Journal of Comparative Physiology and Psychology, 45*:322–328.

Glaser, E. M., & Van der Loos, H. (1981). Analysis of thick brain sections by obverse-reverse computer microscopy: Application of a new, high clarity Golgi-Nissl stain. *Journal of Neuroscience Methods, 4*, 117–125.

Globus, A., Rosenzweig, M. R., Bennett, E. L., & Diamond, M. C. (1973). Effects of differential experience on dendritic spine counts in rat cerebral cortex. *Journal of Comparative Physiology and Psychology, 82*, 175–181.

Green, E. J., Greenough, W. T., & Schlumpf, B. E. (1983). Effects of complex or isolated environments on cortical dendrites of middle-aged rats. *Brain Research, 264*, 233–240.

Greenough, W. T. (1984a). Structural correlates of information storage in the mammalian brain: A review and hypothesis. *Trends in Neuroscience, 7*:229–233.

Greenough, W. T. (1984b). Anatomical substrates of behavioral plasticity. In C. W. Cotman & R. F. Thompson (Eds.), *Plasticity in neurobiology: Cell to Behavior* (1984 Society for Neuroscience Short Course Syllabus), pp. 42–53. Washington, DC: Society for Neuroscience.

Greenough, W. T. (1986). Enduring brain effects of differential experience and training. In M. R. Roseenzweig & E. L. Bennett (Eds.), *Neural Mechanisms of Learning and Memory*, pp. 255–278. Cambridge, MA: MIT Press.

Greenough, W. T., Black, J. E., & Wallace, C. S. (1987). Experience and brain development. *Child Development, 58*, 539–559.

Greenough, W. T., Carter, C. S., Steerman, C., & DeVoogd, T. J. (1977). Sex differences in dendritic patterns in hamster preoptic area. *Brain Research, 126*, 63–72.

Greenough, W. T., & Chang, F.-L. (1988). Plasticity of synapse structure and pattern in the cerebral cortex. In E. G. Jones & A. Peters (Eds.), *Cerebral cortex* (Vol. 7, pp. 391–440). New York: Plenum.

Greenough, W. T., Juraska, J. M., & Volkmar, F. R. (1979). Maze training effects on dendritic branching in occipital cortex of adult rats. *Behavioral and Neural Biology, 26*, 287–297.

Greenough, W. T., McDonald, J. W., Parnisari, R. M., & Camel, J. E. (1986). Environmental conditions modulate degeneration and new dendrite growth in cerebellum of senescent rats. *Brain Research, 380*, 136–143.

Greenough, W. T., & Volkmar, F. R. (1973). Pattern of dendritic branching in occipital cortex of rats reared in complex environments. *Experimental Neurology, 40*, 491–504.

Greenough, W. T., Volkmar, F. R., & Juraska, J. M. (1973). Effects of rearing complexity on dendritic branching in frontolateral and temporal cortex of the rat. *Experimental Neurology, 41*, 371–378.

Hebb, D. O. (1949). *Organization of behavior.* New York: Wiley.

Holloway, R. L. (1966). Dendritic branching: Some preliminary results of training and complexity in rat visual cortex. *Brain Research, 2*, 393–396.

Homer, L. D. (1984). Moments of distributions of distances to nearest capillary. *Microvascular Research, 27*, 114–116.

Hwang, H.-M., & Greenough, W. T. (1986). Synaptic plasticity in adult rat occipital cortex following short-term, long-term, and reversal of differential housing environment complexity. *Society of Neuroscience Abstracts, 12*, 1284.

Isaacs, K. R., Marks, A., Sirevaag, A. M., Chang, F.-L., & Greenough, W. T. (1989). Long-term potentiation alters glial processes in the dentate gyrus. *Society of Neuroscience Abstracts, 15*.

Juraska, J. M. (1982). The development of pyramidal neurons after eye opening in the visual cortex of hooded rats: A quantitative study. *Journal of Comparative Neurology, 212*, 208–213.

Juraska, J. M. (1984). Sex differences in dendritic response to differential experience in the rat visual cortex. *Brain Research, 295*, 27–34.

Juraska, J. M., Fitch, J. M., Henderson, C., & Rivers, N. (1985). Sex differences in dendritic branching of dentate granule cells following differential experience. *Brain Research, 333*, 73–80.

Juraska, J. M., Greenough, W. T., Elliott, C., Mack, K., & Berkowitz, R. (1980). Plasticity in adult rat visual cortex: An examination of several cell populations after differential rearing. *Behavioral and Neural Biology, 2*, 157–167.

Kilman, V. L., Wallace, Withers, G. S., & Greenough, W. T. (1988). 4 days of differential housing alters dendritic morphology of weanling rats. *Society of Neuroscience Abstracts, 14*, 1135.

Kimelberg, H. K. (1983). Primary astrocytes in culture—a key to astrocyte function. *Cellular and Molecular Neurobiology, 3*, 1–16.

Kuffler, S. W. (1967). Neuroglial cells; Physiological properties and a potassium mediated effect of neuronal activity on the glial membrane potential. *Proceedings of the Royal Society of Britain, 168*, 1.

Meshul, C. K., Seil, F. J., & Herndon, R. M. (1987). Astrocytes play a role in regulation of synaptic density. *Brain Research, 402*, 139–145.

Mollgaard, K., Diamond, M. C., Bennett, E. L., Rosenzweig, M. R., & Lindner, B. (1971). Quantitative synaptic changes with differential experience in rat brain. *International Journal of Neuroscience, 2*, 113–128.

Mos, L. P. (1976). Light rearing effects on factors of mouse emotionality and endocrine organ weight. *Physiology and Psychology, 4*, 503–510.

Newman, E. A., Frambach, D. A., & Odettte, L. L. (1984). Control of extracellular potassium levels by retinal glial cell K+ siphoning. *Science, 225*, 1174–1175.

Panksepp, J. (1981). The ontogeny of play in rats. *Developmental Psychobiology, 14*, 327–332.

Pasternak, J. F., & Woolsey, T. A. (1975). On the 'selectivity' of the Golgi-Cox method. *Journal of Comparative Neurology, 160*, 307–312.

Pysh, J. J., & Weiss, M. (1979). Exercise during development induces an increase in Purkinje cell dendritic tree size. *Science, 206*, 230–232.

Purves, D. R., Hadley, D., & Voyvodic, J. (1986). Dynamic changes in the dendritic geometry of iandividual neurons visualized over periods of up to three months in the superior cervical ganglion of living mice. *Journal of Neuroscience, 6*, 1051–1060.

Rosenzweig, M. R., & Bennett, E. L. (1978). Experimental influences on brain anatomy and brain chemistry in rodents. In G. Gottlieb (Ed.), *Studies on the development of behavior and the nervous system: Vol. 4. Early influences.* New York: Academic Press.

Rosenzweig, M. R., Bennett, E. L., & Diamond, M. C. (1972). Chemical and anatomical plasticity of brain: Replications and extensions. In J. Gaito (Ed.), *Macromolecules and behavior* (2nd ed.). New York: Appleton-Century-Crofts.

Rosenzweig, M. R., Bennett, E. L., Diamond, M. C., Wu, S.-Y., Slagle, R. W., & Saffran, E. (1969). Influences of environmental complexity and visual stimulation on development of occipital cortex in rat. *Brain Research, 14*, 427–445.

Rowan, R. A., & Maxwell. D. S. (1981). Patterns of vascular sprouting in the postnatal development of the cerebral cortex of the rat. *American Journal of Anatomy, 160*, 247–255.

Sholl, D. A. (1956). *Organization of the cerebral cortex.* London: Methuen.

Sirevaag, A. M., Bell, O., & Greenough, W. T. (1989). Rats in complex, social and individual environments exhibit astrocytic plasticity after 30 and 60 days of environmental exposure. *Abstracts of the Society for Neuroscience, 15*:610.

Sirevaag, A. M., Black, J., Shafron, D., & Greenough, W. T. (1988). Direct evidence that complex experience increases capillary branching and surface area in visual cortex of young rats. *Developments in Brain Research, 43*, 299–304.

Sirevaag, A. M., & Greenough, W. T. (1985). Differential rearing effects on rat visual cortex synapses. II. Synaptic morphometry. *Developments in Brain Research, 19*, 215–226.

Sirevaag, A. M., & Greenough, W. T. (1987). Differential rearing effects on rat visual cortex synapses. III. Neuronal and glial nuclei, boutons, dendrites, and capillaries. *Brain Research, 424*, 320–332.

Spear, L. P., & Brake, S. C. (1983). Periadolescence: Age-dependent behavior and psychopharmacological responsivity in rats. *Developmental Psychobiology, 16*, 83–109.

Szeligo, F., & Leblond, C. P. (1977). Research response of the three main types of glial cells of cortex and corpus callosum in rats handled or exposed to enriched, control and impoverished environments following weaning. *Journal of Comparative Neurology, 172,* 247–264.

Turner, A. M., & Greenough, W. T. (1985). Differential rearing effects on rat visual cortex synapses. I. Synaptic and neuronal density and synapses per neuron. *Brain Research, 329,* 195–203.

Tweedle, C. D., & Hatton, G. I. (1984). Synapse formation and disappearance in adult rat supraoptic nucleus during different hydration states. *Brain Research, 309,* 373–376.

Uylings, H. B. M., Kuypers, K., Diamond, M. C., & Veltman, W. A. M. (1978). Effects of differential environments on plasticity of dendrites of cortical pyramidal neurons in adult rats. *Experimental Neurology, 62,* 658–677.

Uylings, H. B. M., Smith, G. J., & Veltman, W. A. M. (1975). Ordering methods in quantitative analysis of branching structures of dendritic trees. In G. W. Kreutzberg (Ed.), *Advances in neurology* (Vol. 12). New York: Raven Press.

Valverde, F. (1967). Apical dendritic spines of the visual cortex and light deprivation in the mouse. *Experimental Brain Research, 3,* 337–352.

Volkmar, F. R., & Greenough, W. T. (1972). Rearing complexity affects branching of dendrites in the visual cortex of the rat. *Science, 176,* 1445–1447.

Walsh, R. N., Budtz-Olsen, O. E.,Penny, J. E., & Cummins, R. A. (1969). The effects of environmental complexity on the histology of the rat hippocampus. *Journal of Comparative Neurology, 137,* 361–366.

Wann, D. F., Woolsey, T. A., Dierker, M. L., & Cowan, W. M. (1973). An on-line digital-computer system for the semiautomatic analysis of Golgi-impregnated neurons. *IEEE Transactions on Biomedical Engineering, 20,* 233–247.

16

Experimental Rearing of Rat Pups
Using Chronic Gastric Fistulas

JAIME DIAZ

STATE OF THE FIELD

One of the fundamental issues for the discipline of developmental neuroscience is determining the conditions necessary for normal brain growth. Two aspects of early life that have been closely examined are nutrient intake and environmental stimulation. Experiments examining nutrient intake during early brain growth have typically restricted mother's diet or have manipulated litter size.

Brain growth in rat pups has been shown to be adversely affected by low maternal dietary levels of such specific components as iron (Weinberg, Dallman, & Levine, 1980) and tryptophan (Ashley & Curzon, 1981), to mention a few. Reducing the general levels of protein in the diet of lactating mothers may lead to permanent brain growth retardation (see Morgane et al., 1978, for a review). One of the major problems with experiments that seek to effect changes in rat pups via mother's milk is that the presence of maternal nutrient reserves makes it difficult to assess accurately the extent of the nutritional insult to the pup (Zamenhof, 1981). In addition, a confounding effect of such dietary manipulations may be to alter dynamic interactions between mother and pup that may be critical to normal development.

Many such studies have relied upon manipulations of litter size in rodents to vary the presumed amount of nutrients available during infancy (Bernstein, 1978; Faust, Johnson, & Hirsch, 1980; Winick & Noble, 1967; Wurtman & Miller, 1976a). The rationale is that when litter size is small, individual pups will get more milk, and when litter size is large, individual pups will get less. Rearing rat pups in very large litters does lead to reduced adiposity in adulthood as well as to decreased adipocyte number (Faust et al., 1980; Johnson, Stern, Greenwood, Zucker, & Hirsch, 1973; Knittle & Hirsch, 1968). Furthermore, some studies have reported that when rat pups are reared in small litters, there is a permanent increase of adult body weight (Johnson et al., 1973; Winick & Noble, 1967) and adiposity (Johnson et al., 1973). However, other studies have failed to replicate either of these findings (Bernstein, 1978; Wurtman & Miller, 1976a).

The manipulations of environmental stimulation early in life have provided provocative results. Environmental enrichment or increased levels of stimulation have been shown to alter brain processes, and environmental deprivation or isolation have been shown to have detrimental effects on brain processes (see Uphouse, 1980, and Greenough, chap. 15, this volume, for reviews). These environmental manipulations have been typically imposed after the animals have been weaned because of the difficulty of having a sustained increased level of stimulation in a nursing litter of rats. There is a particular stage of brain development that occurs in all mammals called the *brain growth spurt*. Since the brain growth spurt occurs during the nursing period of the rat, postweanling experimental manipulations will not affect this phase of development.

The brain growth spurt is the period of time when the brain is growing at its fastest rate (Dobbing, 1970). For humans, the brain growth spurt runs approximately from midgestation through the fourth postnatal year (Dobbing & Sands, 1973). In the rat, the brain growth spurt occurs entirely postnatally, from day 4 through day 25. Three critical events occur during this time: (1) multiplication of glial cells during the first part of the brain growth spurt; (2) rapid myelination (Dobbing, 1974); and (3) proliferation of synapses and dendritic arborization (Davidson, 1977). Furthermore, it appears that the developing brain during this time may be more readily affected by adverse influences than at other times in development and that such insults during the brain growth spurt may be permanent (Dobbing, 1970).

Given the nature of these organizational events in brain development that occur in the brain growth spurt, it seems that disrupting this phase of brain development may have significant consequences on the development of behavior. The general theme of the issues I addressed initially was the possibility that drugs may disrupt these events in brain development. In particular, I wanted to examine those drugs that neonates and infants are likely to experience either through pediatric medications, like phenobarbital, or prenatally, like alcohol.

The methodological difficulty of examining the brain growth spurt in rat pups is essentially the problem of disrupting normal nursing patterns, especially drug-induced reductions in nutrient intake, and of altering maternal behavior. The solution to this dilemma is to have a rat pup preparation that does not have to rely on mother for nutrients or warmth.

The procedure of rearing rat pups in the absence of mother, so-called artificial rearing, was first introduced by Michael Messer (Messer, Thoman, Terrasa, & Dallman, 1969) and later resurrected and significantly modified by Ted Hall (Hall, 1975). We have further modified the protocol to address our initial pharmacological questions.

THE METHODOLOGY OF ARTIFICIAL REARING

Basically, artificial rearing involves (1) surgically implanting a chornic gastric cannula through which the rat pup receives all its nutrients; (2) keeping the animal warm; and (3) incrementing the amount of milk delivered to approximate normal growth. Since we initially wanted to model periods in human brain growth, we began our experiments at 4 days of age. However, rat pups can be encannulated as early as 24 h old.

Intragastric Cannula Construction and Installation

Cannula construction involves three steps: (1) measure 20-cm lengths of PE-10 tubing (Clay-Adams); (2) form a small flange on one end with a heat source; and (3) attach a small, soft plastic disk to one end. The actual length of the cannula is a matter of convenience. We have found, given our syringe-delivery system (see below) and the size of our cups, that 20-cm lengths are optimal. The flange should be formed with care so as not to make it too large (the outer diameter of PE-10 is 0.6 mm, and the flange should be approximately 1–1.2 mm). All that is really necessary is to slightly increase the diameter of the end so as to retain the plastic disk. An actual flame may be difficult to control, so we typically use a match end that has just been heated on an alcohol lamp. Disks, made with a standard hole punch and ordinary plastic sandwich bags, are approximately 0.7 cm in diameter. A fine hole is punched in the center of each disk with a 23-gauge needle. The unflanged end of the cannula is forced through the center hole, and the disk is then gently pushed down the cannula to rest against the flange.

A unique feature of the implantation procedure is use of the oral route. After being lightly anesthetized with halothane, the rat pup is placed on its back and its mouth gently opened. A fine wire (stainless steel, 0.25 mm in diameter) sheathed with soft Silastic tubing (0.3 mm i.d., 0.6 mm o.d., Dow Corning, VWR Scienfitic) is carefully and gently slid down the pup's esophagus and into the stomach. There are two possible outcomes to this part of the procedure: The wire will go down the esophagus or down the trachea. The more dorsal hole is the esophagus, so if you keep the wire on the roof of the animal's mouth and on the dorsal surface of the throat, the wire will be properly placed. If the wire does go down the animal's trachea, you will feel a spongy resistance as it enters the bronchial tubes. It is extremely important during this part of the procedure to stop if you feel a spongy resistance and start over again, even if it means reanesthetizing the animal. Once the wire is in the stomach, the pup is shifted onto its right side. This maneuver renders the stomach visible as a prominent whitened area beneath the skin and enables locating and positioning the end of the coated wire. When the tip of the coated wire is in position against the side wall of the stomach (which can be visualized through the very thin skin), the Silastic tubing is withdrawn slightly at the mouth end, exposing the cutting tip of the wire. The wire is then pushed through the stomach, peritoneal wall, and skin until 2 or 3 cm emerges from the pup's flank. The wire is securely held at the pup's flank while the Silastic tubing is completely removed at the mouth end. The unflanged end of the cannula is then friction-fitted over the end of the wire, which is still protruding from the pup's mouth. If the cannula is not securely fitted onto the wire, it will slip off when you are pulling the wire/cannula assembly through the stomach and peritoneal walls, and you will have to start all over again. The wire, followed by the cannula, is then pulled down the esophagus and out the abdomen. As the polyethylene disk passes down the esophagus it collapses and reopens in the stomach. The cannula is pulled through until the disk is flush against the stomach wall. It is important to have the cannula exist in a lateral position (see Figure 16.1). The closer the cannula is to midline, the more likely the animal is to get its foot caught in the cannula and pull it out or dislodge it.

Using the wire as a needle, the cannula is threaded through a fold of skin on the pup's back, forming a slight loop between the abdominal exit of the cannula and the

Figure 16.1. An anesthetized 4-day-old rat pup with an intragastric cannula. Note the lateral placement over the pup's leg.

pup's back, (not illustrated). The cannula is secured on either side of this fold of skin by a plastic disk fashioned from Tygon tubing (Tygon size R-3603; Fisher Scientific). Just as with the plastic disks on the ends of the cannulas, a standard hole punch will give you an adequate size disk, the center of which you lance with a 23-gauge needle. This provides a small amount of slack so that the cannula cannot easily be pulled out of the stomach. The entire cannulation procedure takes between 4 and 6 min and requires no incisions or sutures.

Incubator

The incubator consists of a large, partly filled, temperature-controlled water bath (VWR 1240, Model 13309-704, is a good general-purpose water bath for this) in which can be floated several plastic cups, each weighted so as to prevent tipping over. A large washer taped to the center of the floor of the cup seems to be adequate. Following cannula implantation, the pups are housed individually in individual plastic cups (11 cm in diameter and 7.5 cm deep), which fit into the counter-weighted cups. Each animal cup contains a small amount of bedding material and is covered with a perforated plastic lid. When the animals are 4–12 days old, the bedding is changed every third day. Older animals need daily changes of bedding. The water temperature is initially set at 40 °C when the pups are cannulated (typically at 4 days of age). When the animals are approximately 13 days old, the water temperature is gradually reduced (about a degree a day), so that when the rats are 18 days of age the bath is at room temperature. The intragastic cannulas emerge from the holes in the lids of the cups and are connected to nearby syringes containing a milk formula. The pups live in the cups and receive gentle tactile and vestibular stimulation as the cups float in the bath. It is important to allow the cups to float as freely as possible so that they can rotate in the bath as the pups move, preventing twisted cannulas.

Infusion Schedule

Syringes containing the formula diet are mounted on infusion pumps (Harvard Apparatus Syringe Pumps, Series #35). The pumps are programmed to infuse the diet for 5 min every half hour, 24 h a day, resulting in approximately 45 feeds per day (allowing time each day to clean the syringes and associated tubing and to weigh the animals and clean their cups). The programming consists of solid state circuitry to time the desired interval of on time for the infusion pumps. Our particular circuit consists of (1) a time base generator, (2) CMOS logic to turn on a transistor 5 min every 30 min, and (3) a relay, closed by the transistor, that actually switches on power to the infusion pumps. A variety of commercially available devices will do this kind of task. It should be noted that in our previous experiments (Diaz, Schain, & Bailey, 1977; Diaz & Schain, 1978) the formula diet was infused continuously without any problems. However, we have now adopted this intermittent feeding schedule because it more closely approximates the rate at which maternally reared pups receive milk. The amount of diet infused typically starts at 3 ml/day and is incremented daily by approximately 0.6 ml.

The formula used in the initial experiments is a modification of the milk-based diet of Messer et al. (1969). The modifications include addition of extra protein (casein) to the formula (Diaz, Stamper, Moore, Schacher, & Petracca, 1982) and use of different oils as a source of fat. The precise recipe is given in Table 16.1 (ingredients available from Sigma Chemical Co.). The formula is prepared under clean conditions in batches of 500 ml, and daily aliquots are frozen until needed. The syringes are cleaned and fresh milk formula used each day.

At the time of cleaning, the pups are removed from the cups, weighed, have their cannulas flushed with demineralized water (distilled water may not be necessary, but at least have minerals removed from this water), and have the loops formed on their backs by the cannulas adjusted when necessary to accommodate the animals' growth. At this time, the animals have their anogenital area lightly stroked with tissue to stimulate urination and defecation. We have used cotton applicators, paint brushes (stiff and soft, to more closely resemble mother's tongue), toilet tissue, and Kimwipes, and there does not seem to be an appreciable difference among them. There is ample evidence in the cup bedding that the animals are capable of urinating and defecating without this procedure even at 5 days of age. We maintain this procedure for several

Table 16.1 Milk Formula and Mixing Protocol

1. Mineral Solution
   Measure:
Iron gluconate:	.0207 g
Copper gluconate:	.0105 g
Zinc gluconate:	.0160 g

   Add this to 10 ml of distilled water.
2. Water Solution
   Measure:
   Deoxycholic acid (sodium salt): 0.200 g
   Add this to 200 ml of distilled water.
3. Weigh 30 g of corn oil and put directly into a clean blender.
4. Add 5 ml of Polyvisol vitamins.
5. Add 5 ml of the mineral solution from step 1.
6. Add the following amounts:
Methionine:	0.200 mg
Tryptophan:	0.250 mg
Riboflavin:	0.005 mg
7. Add 15 g of casein hydrolysate.
8. Add 375 ml of evaporated milk.
9. Add 85 ml of the water solution from step 2.
10. Mix the ingredients at a moderately high speed for approximately 90 seconds.
11. Pour the formula into jars and store in the freezer.

reasons: First, and more important, maternal stimulation is a critical part of the pup's early life, and the stroking procedure keeps some tactile stimulation in the animal's experience. Second, this stroking procedure allows us to monitor the general state of the animal. Third, this type of handling may hinder the development of bloating (see below).

Common Problems

Rat pups are surprisingly sturdy and tolerate the surgical procedure quite nicely. However, maintaining the animals for long periods of time (10–14 days) can be difficult. The following are some common pitfalls.

Bloating

Abdominal distension is perhaps the most common problem with the artificial rearing procedure. The animals initially show signs of gastric retention, and if allowed to progress, this abdominal distension will impede breathing and be fatal to the pup. As mentioned above, stroking the animal may help delay bloating and perhaps ameliorate a slight bloating condition. However, we have found that the bile salt deoxycholic acid can prevent and in some cases reverse the bloating process (Diaz et al., 1980). Our formula has sodium deoxycholate added as a prophylactic measure, and in cases of bloating we flush the animal's cannula with approximately 0.1 ml 100 mg/% of sodium deoxycholate. When this does not work to alleviate the bloating, we dilute that day's amount of formula by as much as 30% with distilled water. While diluting helps the bloating condition, the animal will lose weight the next day. Because of the weight loss, we do not dilute more than once for a pup in an experiment. Subsequent bloating will result in dropping that animal from the study and returning it to a lactating mother.

Pulled and Disconnected Cannulas

As the animals get older, they are able to dislodge or pull the cannula from their stomachs and to disconnect their cannulas from the syringe tubing. In the case of a pulled cannula, the animal may be reencannulated by essentially the same surgical procedure with the exception that older animals are considerably more difficult to encannulate. For one thing, there is no longer the transparency of the 4-day-old pup. You must rely more on "feel" than sight. More important, the skin is tougher, which means that the wire must be firmer, and the cannula is more likely to slip off the wire as you pull the wire through the skin. In the case of disconnected cannulas, the connection between the PE 10 tubing of the cannula and the PE 50 of the syringe tubing should be made tighter. This can be done by applying heat to the animal's cannula and causing a *slight* flange.

In either case, having solved the immediate problem still leaves the experimenter with a significant tactical problem: how much formula to give the animal? If the animal pulled or disconnected its cannula so that the period of time it was not receiving milk is relatively short and it has not lost weight, then these animals are not treated any differently than the others. What is usually the case is that pulled or disconnected animals will lose considerable weight and stop growing. These animals should not receive as much formula the next day. Just how much to cut back depends on the amount of weight loss and can be estimated on the basis of percentage of body weight.

Bath Temperature

As rat pups' fur comes in (approximately Day 13), they become less poikilothermic and begin to regulate temperature themselves. The temperature of the bath must be gradually reduced at this age (about 1°C a day) to accommodate this maturational process. Condensation on the inside of the cups is a good indicator that the bath temperature should be reduced. Bedding in the cups should be changed daily at this time as well.

HOW THIS PROCEDURE HELPS ANSWER THE INITIAL QUESTIONS

The clinical use of drugs that act on the central nervous system is widespread in infant patient populations, in particular, phenobarbital. These medications are given during critical brain growth periods. Because previous studies showed that early phenobarbital administration hindered normal nursing in rats (Schain & Watanabe, 1976), we used the artificial rearing procedure to examine this question of chronic barbiturate exposure in infants. Nutrient intake was identical for both barbiturate and vehicle groups, and their body weight curves were the same. However, those neonatal rats that received phenobarbital experienced a marked interruption of early brain growth, which was reflected in brain levels of DNA, RNA, protein, and cholesterol as well as in gross brain weight (Diaz et al., 1977; Diaz & Schain, 1978). These phenobarbital-treated rats also exhibited behavior patterns similar to those of children on barbiturate medication, namely, a hyperresponsivity to stimulation (Diaz & Schain, 1978).

The artificial rearing procedure was also used to develop an animal model of the fetal alcohol syndrome (FAS). As noted above, the brain development that occurs during the third trimester in the human occurs during postnatal days 1–8 in the rat. As with the barbiturates, the artificial rearing procedure enabled drug exposure in rat pups without the characteristic complication of decreased nutrient intake in the drug group. In this case, the effects of brief ethanol exposure early in brain development could be determined. We observed a brain weight reduction of 19% coupled with a 31% reduction of brain volume when rat pups were exposed to ethanol for only 4 days during postnatal days 5 through 9. This brain growth deficit resulted even though the ethanol animals had overall body weights equal to their vehicle-treated siblings (Diaz & Samson, 1980). This was the first animal model of FAS to show a definite true microcephaly (reduced brain to body weight ratio) independent of body weight gain.

ARTIFICIAL REARING APPLIED TO THE PROBLEM OF EARLY ONSET OBESITY

Most recently, we have been pursuing the artificial rearing procedure as a model for infantile obesity. The influence of early dietary history upon the etiology of adult obesity in humans remains an important but unresolved issue. Most commonly used animal models of neonatal overnutrition have significant methodological problems. Many such studies have relied on manipulations of litter size in rodents to vary the presumed amount of nutrients available during infancy (see review above).

There are fundamental problems of interpretation when litter size is varied. For one, self-regulation of food intake develops prior to weaning in rats (Hall & Rosenblatt, 1977), so that reducing litter size does not necessarily induce uniform overfeeding (if at all) throughout the weaning period. Furthermore, the stimulus for milk production by the dam is reduced in small litters, and individual pups may in fact get normal (or close to normal) nutrition (Russell, 1980). Finally varying litter size will change the amount of maternal and sibling stimulation for the individual pup (Grota & Ader, 1977). The effect of the increased or decreased maternal stimulation upon adult feeding and adiposity is unknown.

Other experiments have used supplementary gavage to overfeed pups (Czajka-Narins & Hirsch, 1974; Wurtman & Miller, 1976b). However, such manipulations cannot control for induced effects on suckling behavior; that is, pups may compensate and suckle less milk. Moreover, only a relatively small volume can be supplemented conveniently. Consistent with these shortcomings, these studies have noted a small effect of gavage on body weight and adiposity at weaning but no long-term effect on adult weight.

A more effective model would allow complete experimental control of the total nutrient intake, and specific nutrient content, of the food given to pups during a critical period. Variation and control of the nutrient content of the diet may be relevant to human adiposity, since many human infants are bottle fed. Bottle feeding per se, as well as the caloric density of the specific formulas used, can have a striking effect upon weight gain in human babies (Fomon, Filer, Thomas, Rogers, & Proksch, 1969). Therefore, a rat model in which the pups are given a controlled milk formula as their only source of nutrition is desirable.

In our initial study (West, Diaz, & Woods, 1982), female rat pups were fed a milk formula (see Table 16.1) by intragastric infusion so that their growth rate matched that of their maternally reared siblings or else were overfed the same diet so as to exceed the growth of their siblings. They were gastrostomy reared from Day 4 through Day 18 postpartum. By Day 18 the overfed pups weighed significantly more than both maternally reared pups and pups fed intragastrically but maintained at the same growth rate as maternally reared pups. On Day 18 the catheters were removed from gastrostomized pups, and maternally reared pups were removed from their dams. Thereafter, all groups were maintained ad libitum on a standard diet (rat chow pellets). The rats that have been overfed as pups remained significantly heavier than either control group throughout the experiment (around 250 days). At sacrifice, they were found to have significantly larger fat pads than control rats; that is, they were obese.

In a second experiment (West, Diaz, Roddy, & Woods, 1987), male pups were fed a milk formula by intragastric infusion. One group received the same formula used in the previous experiment in amounts calculated to match the growth rate of their maternally reared siblings. A second group received the same diet but with a supplement of fat to achieve overnutrition. The latter group had accelerated growth and remained significantly heavier than either control group as adults. As in the first experiment, they were also more obese.

Early overnutrition by either method resulted in adult obesity characterized by significantly larger fat depots. We found that this was due to an increase of both fat cell size and fat cell number. There was also an increase of lipoprotein lipase (LPL) activity in rats overnutritionalized as pups. LPL is the rate-limiting enzyme for the storage of fat and thus might be a casual factor in the rats' obesity. We also found that gastrostomy rearing per se, without the necessity of overfeeding, caused changes in the adioposity of adults. Rats given the milk formula diet we use from Day 4 through Day 18 of life weighed the same as maternally reared siblings as adults, but had significantly more fat due to more adipocytes. This is presumably related to some aspect of the formula composition relative to maternal milk. Collectively, these findings suggest that both the quantity and quality of food received during periods of early growth and development (and/or the early rearing environment) affect long-term weight and adiposity of rats.

In summary, these first two experiments have demonstrated that (1) the procedure of chronic gastrostomy feeding during the nursing period can render both male and female rat pups obese; (2) this obesity persists throughout life; (3) while the gastrostomy-rearing-induced obesity is expressed frankly in the overfed animals, those animals that were gastrostomy reared but weight matched to their maternally reared siblings had relatively large fat depots as adults; and (4) experimental adult obesity can be induced either by feeding rat pups a greater quantity of formula or by feeding a formula with a higher fat content. This is the first animal model which unambiguously demonstrates that adult adiposity can be influenced to a such a large degree by early deitary manipulations.

ADVANTAGES AND LIMITATIONS OF ARTIFICIAL REARING

Experimental protocols developed to overcome certain methodological problems will usually incur the price of a new set of constraints that the experimenter must address.

The artificial rearing protocol offers both distinct advantages and substantial new disadvantages that must be weighed by the experimenter.

The principle advantages of using this artificial rearing procedure is to control any experimentally induced differential intake of nutrients. For example, administering a central nervous system depressant to nursing rats may result in sedation, significantly interfering with the animal's ability to nurse normally. This, in turn, may lead to a change in the pup's body weight and subsequent changes in maternal and sibling behavior directed to that animal. The artificial rearing procedure obviates these problems. In fact, *any* experimental procedure, not only pharmacological ones, that may disrupt normal ongoing nursing behavior carries that risk. Within the artificial rearing protocol, all the animals, regardless of experimental manipulation, receive the same amount of nutrients. It is important to remember that because of the unique nature of the artificial rearing environment, both experimental and control animals must be artificially reared so that comparisons are made between these groups and not to animals that have not experienced this peculiar early environment.

The principle disadvantage of the artificial rearing procedure is, of course, the rearing of these animals in an environment that is very foreign to the animal's normal situation. Normally, rat pups have constant contact with their siblings in the nest and show a particular dynamic behavior of readjusting their positions to maximize warmth. Moreover, their nursing is characterized by searching and competing for nipple attachment, followed by shifting to another nipple when they have stripped one dry. In the artificial rearing procedure one removes the only two really robust behaviors these animals exhibit at this stage of life. The artificially reared animals do not have to work for warmth, since they are housed in a temperature-controlled and moist water bath, and they do not have to work for their nursing; in fact, they do not even have to swallow. The animal has been disengaged from its environment.

Appreciate that artificial rearing does not impose sensory isolation. For example, these animals undergo unique vestibular and tactile stimulation. Rather, the artificial rearing procedure is a severe social deprivation situation. The removal of sibling and maternal interactions during critical periods of brain growth may be a significant negative factor in the development of the animals. In fact, one of the common findings of artificially reared experiments (Messer et al., 1969; Diaz et al., 1982; Diaz, Moore, Petracca, & Stamper, 1983) is that the artificially reared animals usually have smaller brains that the mother-reared weight-matched siblings. The interaction of environmental and nutritional variables that are likely to mediate this reduced brain size remain to be described.

PERSPECTIVES

As with many new experimental procedures, the artificial rearing procedure has opened several unique data bases that address familiar issues in a new light. The fundamental question of what is actually necessary for normal brain development to occur can be examined with artificial rearing. By reintroducing various components of the animal's natural environment, both social (namely, sibling and maternal interactions) and nutritional (namely, how much rat milk is included in the formula), one can reexamine a significant and recurring theme in developmental neuroscience.

The artificial rearing procedure has provided not only a powerful new tool in the assessment of phramacological agents administered early in life but also a different tool in the study of development in general. For drug assessment, the field is better able to ask more precise questions concerning drugs and development. Which systems are disrupted more than others, and what are the time courses for these changes? For the developmental scientist, the artificial rearing procedure has caused a reexamination of fundamental, and previously untestable, issues concerning the importance of nutritional history and social interactions early in life.

Future directions for the artificial rearing procedure will probably take at least two distinct routes. One will be improving the technique, especially the formula. This procedure is predicated on knowing what nutrients are necessary and sufficient to ensure normal growth, making the formula a critical part of the artificial rearing procedure. The formulas that have been used for artificial rearing, while expedient, are not similar to the make-up of mother's milk and may be inadequate for proper growth, especially brain growth. John Edmond and his group at the Mental Retardation Research Center and the Department of Biological Chemistry at the University of California at Los Angeles are making significant improvements in correcting the glaring dissimilarities in the macronutrients between the currently used formulas and mother's milk.

Another direction in improving and changing the technique will involve the route of milk delivery. The introduction of a device on which rat pups would attach by Hoshiba (1986) and the introduction of the "palate" cannula by Susan Henning (Blake, Lau, & Henning, 1988) represents technological change that has interesting implications. Both of these new methodologies require the animal to actively participate in the acquisition of nutrients. This participation may have significant consequences for neural development.

With the Hoshiba device, pups learn to suckle on artificial nipples connected to a formula source. Although there is not precise control of individual pup nutrient intake, this type of device allows for sibling interactions, and thus this method of artificial rearing is more similar to the environment of a mother-reared nest of pups. We have built a chamber according to Hoshiba's methods but have not been successful in getting rat pups to thrive. It will be interesting to see how other laboratories replicate and modify Hoshiba's device.

The palate cannula involves a chronic cannula that is positioned through the animal's palate so that the animal has to swallow the formula. As with the artificial rearing procedure discussed above, the palate cannula has the main advantages of total control of nutrient intake as well as control of the animal's environment. In addition, an alternate placement of the palate cannula in a forward position allows the animal to reject formula delivery by letting it spill out its mouth. This opens the possibility for a variety of experiments on taste preference and conditioned avoidance. Moreover, the direct consequence on brain growth of the pup's active role in nutrient intake may be the reason the brain weights of the palate cannula animals are so similar to their sibling mother-reared controls (Blake, Lau, & Henning, 1988). However, this procedure seems to work best on animals 13 days or older. When we tried palate cannulas on 4-day-old pups, we found that they did not thrive, and we did not find normal brain weights compared to mother-reared sibling controls.

The other route of future artificial rearing experiments will involve the application of the procedure to different developmental questions, in particular nutritional and behavioral questions. The initial applications of the artificial rearing procedure have typically been "broad stroke" data bases that essentially have demonstrated the utility of the procedure. More precise use of artificial rearing will address better questions concerning the description of true critical periods in development. For example, artificial rearing for relatively brief periods of time (3–4 days) during different phases of brain growth may have different effects on adult behavior patterns. Another example would be in overnutrition and undernutrition during brief times early in life and its impact on peripheral organ growth as well as on behavior. The use of the artificial rearing procedure by Drs. Kirby and Robillart at the Cardiovascular Center, University of Iowa, to examine the effects of salt in hypertensive-prone strains of rats is another example of a new direction for the procedure. The use of the procedure by Drs. Myron Hofer and Harry Shair and their group at Columbia University to examine the development of attachment behavior and isolation stress is a good example of the behavioral directions for the artificial rearing procedure (Hofer, Shair, & Murochick, 1989).

Determining the impact and the permanency of disrupting the earlier stages of development is a matter of vital concern. Whether the insult is pharmacological, nutritional, or environmental in nature, research efforts should be directed at determining the parameters of potential insults to human development.

REFERENCES

Ashley, D., & Curzon, G. (1981). Effects of long-term low dietary intake on determinants of 5-hydroxytryptamine metabolism. *Journal of Neurochemistry, 37,* 1385–1393.

Bernstein, I. (1978). Re-evaluation of the effect of early nutritional experience on body weight. *Physiology and Behavior, 21,* 821–823.

Blake, H., Lau, C., & Henning, S. (1988). A new method for the artificial raising of infant rats: The palate cannula. *Physiology and Behavior, 42,* 495–498.

Czajka-Narins, D., & Hirsch, J. (1974). Supplementary feeding during the preweaning period. *Biology of the Neonate, 25,* 176–185.

Davidson, A. (1977). *Biochemical correlates of brain structure and function.* New York: Academic Press.

Diaz, J., Moore, P., Petracca, F., Schacher, J., & Stamper, C. (1981). Artificial rearing of preweanling rats: The effectiveness of direct intragastric feeding. *Physiology and Behavior, 27,* 1103–1105.

Diaz, J., Moore, E., Petracca, F., & Stamper, C. (1983). Somatic and central nervous system growth in artificially reared rat pups. *Brain Research Bulletin, 11,* 643.

Diaz, J., & Samson, H. (1980). Impaired brain growth in neonatal rats exposed to ethanol. *Science, 208,* 751–753.

Diaz, J. Samson, H., Keesler, D., Stamper, C., Moore, E., & Robisch, E. (1980). Experimental necrotizing enterocolitis: The possible role of bile salts in its etiology and treatment. *Pediatric Research, 14,* 595.

Diaz, J., & Schain, R. (1978). Chronic phenobarbital administration: Effects upon behavior and brain of artificially reared rats. *Science, 199,* 90–91.

Diaz, J., Schain, R., & Bailey, B. (1977). Phenobarbital-induced brain growth retardation in artificially reared rat pups. *Biology of the Neonate, 32,* 77–82.

Diaz, J., Stamper, C., Moore, E., Schacher, J., & Petracca, F. (1982). Artificial rearing of rat pups with a protein-enriched formula. *Journal of Nutrition, 112,* 841–847.

Dobbing, J. (1970). Undernutrition and the developing brain. In W. Himwich (Ed.), *Developmental neurobiology* (pp. 241–261). Springfield, IL: Charles C. Thomas.

Dobbing, J. (1974). The later development of the brain and its vulnerability. In J. Davis (Ed.), *Scientific foundations of paediatrics* (pp. 565–577). London: Heinemann.

Dobbing, J., & Sands, J. (1973). Quantitative growth and development of human brain. *Archives of Disease in Childhood, 48,* 757–767.

Faust, I., Johnson, P., & Hirsch, J. (1980). Long-term effects of early nutritional experience on the development of obesity in the rat. *Journal of Nutrition, 110*, 2027–2034.

Fomon, S., Filer, L., Jr., Thomas, L., Rogers, R., & Proksch, A. (1969). Relationship between formula concentration and rate of growth of normal infants. *Journal of Nutrition, 98*, 241–254.

Grota, L., & Ader, R. (1977). Continuous recording of maternal behavior in the rat. *Animal Behaviour, 17*, 722–729.

Hall, W. (1975). Weaning and growth of artificially reared rats. *Science, 190*, 1313–1316.

Hall, W., & Rosenblatt, J. (1977). Suckling behavior and intake control in the developing rat pup. *Journal of Comparative Physiology and Psychology, 91*, 1232–1247.

Hofer, M., Shair, H., & Murochick, E. (1989). Isolation distress and maternal comfort responses of two-week-old rat pups reared in social isolation. *Developmental Psychobiology, 22*, 553–566.

Hoshiba, J. (1986). An automatic feeder for infant rats. *Laboratory Animal Science, 36*, 682–685.

Johnson, P., Stern, J., Greenwood, M., Zucker, L., & Hirsch, J. (1973). Effect of early nutrition on adipose cellularity and pancreatic insulin release in the zucker rat. *Journal of Nutrition, 103*, 738–743.

Knittle, J., & Hirsch, J. (1968). Effect of early nutrition on the development of rat epididymal fat pads: Cellularity and metabolism. *Journal of Clinical Investigation, 47*, 2091–2098.

Messer, M., Thoman, E., Terrasa, A., & Dallman, P. (1969). Artificial feeding of infant rats by continuous gastric infusion. *Journal of Nutrition, 98*, 404–410.

Morgane, P., Miller, M., Kemper, T., Stern, W., Forbes, W., Hall, R., Bronzino, J., Kissane, J., Hawrylewicz, E., & Resnick, O. (1978). The effects of protein malnutrition on the developing central nervous system in the rat. *Neuroscience and Biobehavioral Review, 2*, 137–230.

Russell, J., (1980). Milk yield, suckling behavior and milk ejection in the lactating rat nursing litters of different sizes. *Journal of Physiology, 303*, 403–415.

Schain, R., & Watanabe, K. (1976). Origin of brain growth retardation in young rats treated with phenobarbital. *Experimental Neurology, 50*, 806–809.

Uphouse, L. (1980). Re-evaluation of mechanisms that mediate brain differences between enriched and impoverished animals. *Psychology Bulletin, 88*, 215–232.

Weinberg, J., Dallman, P., & Levine, S. (1980). Iron deficiency during early development in the rat: Behavioral and physiological consequences. *Pharmacology, Biochemistry and Behavior, 12*, 493–502.

West, D., Diaz, J., Roddy, S., & Woods, S. (1987). Long-term effects on adiposity following preweanling nutritional manipulations in the gastrostomy-reared rat. *Journal of Nutrition, 117*, 1259–1264.

West, D., Diaz, J., & Woods, S. (1982). Infant gastrostomy and chronic formula infusion as a technique to overfeed and accelerate weight gain of neonatal rats. *Journal of Nutrition, 112*, 1339–1343.

Winick, M., & Noble, A. (1967). Cellular response with increased feeding in neonatal rats. *Journal of Nutrition, 91*, 179–182.

Wurtman, J., & Miller, S. (1976a). Effect of litter size on weight gain in rats. *Journal of Nutrition, 106*, 697–701.

Wurtman, J., & Miller, S. (1976b). The effect on growth of feeding supplemental milk or baby food to the suckling rat. *Life Sciences, 19*, 769–776.

Zamenhof, S. (1981). Maternal nutrient storage and efficiency in production of fetal brain tissue in rats. *Biology of the Neonate, 39*, 45–51.

C.

NEUROPHARMACOLOGY

The three chapters in this section address the questions of how specific neurotransmitters in defined circuits act to regulate behavior, and how changes that occur during the development of these neural substrates relate to the changing behavioral repertoire of the maturing animal. There are common themes and assumptions implicit in each of these diverse chapters. Among them, that there are specific neural changes which can be associated with behavioral and physiological development; that important neural changes include the maturation of specific neurochemicals located in anatomically defined pathways and nuclei; and that anatomical as well as biochemical methods are required to assess the association between neural development and behavioral maturation.

Although the general developmental issue each chapter addresses contains common themes, the methodologies that each chapter brings to bear on these issues are diverse and provide examples of the wide range of tools the neuroscientist can use to attack developmental questions. In Chapter 17, Wilson, Sullivan, and Leon utilize olfactory learning as a model system to study changing brain-behavior relationships during development. Olfactory learning is a robust phenomenon that allows precise experimental control outside the litter and that has an essential role in the survival of the young. These authors have chosen an approach that measures the change in glucose utilization following olfactory conditioning. The advantages of the method are that it is noninvasive, can be studied in the awake and behaving animal, and maps the consequence of learning in the entire system in an economical and complete manner. The authors have begun to define the neurochemical bases of early olfactory learning by first assessing the neurochemical inputs to the bulb and then determining how the development of these inputs relates temporally to the behavioral phenomenon. For example, the ontogeny of the cholinergic system is out of synchrony with the development of the behavior and thus is not a likely candidate to mediate changes in olfactory function associated with learning. By combining metabolic mapping with an understanding of the development of specific neural circuits, intelligent first approximations can be made with regard to the role of various neurotransmitters. Electrophysiological techniques, guided by the 2-deoxyglucose results, further specify the electrical changes induced by early olfactory learning.

The approach taken by Moran and Coyle, in Chapter 18, is quite different from that of Wilson et al. The problem that they faced was how to destroy a single class of neurons selectively to assess the effects that the loss of that population of cells would have on neural and behavioral development. Although the use of toxins to destroy neurons that use specific neurotransmitters is common, lesions of other classes of neurons are rare. Moran and Coyle chose a mitotic inhibitor that would disrupt cells with a common birthday. The advantage of this unique approach is the ability to eliminate classes of neurons rather specifically, although it is clear that MAM treatment leads to a subsequent cascade of alterations in neural development. What they describe in their chapter are the biochemical, anatomical, and behavioral sequelae of neural development in the absence of those cells. This approach is reminiscent of studies that have used mice with mutations that result in the abnormal development of specific cell types (e.g., reeler mouse). These mutation studies have been essential to our understanding of the role of those cells in normal developmental processes, such as migration within the cerebellum. Conceptually, the experimental elimination of classes of cells parallels that approach and has the further advantage of flexibility in the selection of cells to be eliminated. Although this approach is in its infancy (no pun intended), the richness of the method is apparent.

Finally, in the last chaper of this section, Barr has taken classic methods from behavioral neuroscience, such as chemical and electrical stimulation, and applied them to issues of development. The conceptual framework for which Barr modified these methods is similar for both adult and infant studies: The brain is organized anatomically, and methods to study the function of the brain, either in developing or mature animals, must tap into that anatomic specificity.

The theme of anatomical specificity is constant in the three chapters of this section, despite three uniquely different approaches. The conceptual biases that direct the authors' methods are searches for ways to assess specific effects of defined anatomical pathways during development. In a sense, these methods mirror changes that are occurring in the field of neuropharmacology as studied in adult animals. Some scientists (e.g., pharmacologists) have classically viewed the brain as sort of a chicken soup to which one could add drugs as flavoring to change its function. Others (nuroanatomists) have historically viewed the brain as a wiring circuit in which the important information is made up of connections between two or more brain regions. What is the modern approach? It combines both views, and these three chapters provide excellent examples of the modern approach applied to development. Wilson et al. are studying the functional anatomy of well-defined circuits in the brain, and they hope to specify which aspects of that circuitry are involved in early olfactory learning. Moran and Coyle have used a developmental toxin to eliminate a specific set of cortical neurons and have described the consequences of development in the absence of those cells. Barr has defined methods to stimulate distinct neural structures to assess changes in behavior as a function of age. But each chapter also emphasizes the neurochemical and functional aspects of these circuits. These types of studies might allow for an understanding of how neural systems are assembled during development to express or inhibit certain behaviors.

17

A Search for the Neural Mechanisms of Olfactory Learning in Young Rats

DONALD A. WILSON, REGINA M. SULLIVAN,
AND MICHAEL A. LEON

Despite recent advances in understanding the neurobiology of learning and memory in mature animals, little is known about the neural mechanisms of early learning. In the mature central nervous system (CNS), a variety of structures, pathways, neurotransmitter systems, and synaptic events have been implicated in learning and memory (for reviews, see Lynch, McGaugh, & Weinberger, 1984, and Squire, 1987). The majority of theories regarding learning and memory in the mature CNS center around the synapse, with modification of either synaptic efficacy or synaptic number as the mechanism of information storage (e.g., Hebb, 1949; Lynch, 1986; Teyler & DiScenna, 1984).

Storage of information in the developing CNS, however, presents a unique problem for these synapse-based theories of learning and memory. Specifically, how can information stored at a synapse during development be retrieved after extensive synaptogenesis, neurogenesis, and myelination have occurred between learning (training) and remembering (testing)? That is, early learning requires that memories be stored within a constantly changing, incomplete system, a requirement not made of memory mechanisms in the mature CNS. Thus, the possibility arises that the mechanisms of learning in the immature CNS are not identical to those in the mature CNS.

STRATEGIES FOR THE SEARCH

In the search for the neural mechanisms of early learning, at least two directions can be taken. One approach is to identify the kinds of neural plasticity that occur in the developing CNS and compare these types with those in the mature CNS. Thus, the question might be asked, What tools are available to the developing brain to store information, and how do they compare with those of the mature brain? Research of this type might, therefore, examine responses to injury (Cotman, Nieto-Sampedro, & Harris, 1981), altered sensory environments (Weisel & Hubel, 1965; Diamond, Rosenzweig, Bennett, Lindner, & Lyon, 1972; Greenough, Volkmar, & Juraska, 1973), or electrical stimulation (Wilson, 1984) during development and compare these responses to those of the mature CNS. For example, Wilson originally began searching for neural mechanisms of early learning by examining activity-dependent synaptic facilitation (long-term potentiation) in the rat forebrain during postnatal development (Wilson & Racine, 1983; Wilson, 1984; Wilson, Willner, Kurz, & Nadel, 1986). However, once the types and characteristics of neural plasticity during development have been identified, the question remains concerning how and whether they are involved in early learning. That is, the ability to demonstrate an event in an experimental situation does not imply that the same event actually occurs in real life.

The work of Sullivan and of Leon, however, has been guided by a second approach, which involves teaching the newborn a specific task and then identifying neural correlates of that learning (Leon et al., 1987; Sullivan & Hall, 1988). While this approach obviously gets more directly at the question of how the developing brain stores information, it is not without problems. First, a task has to be selected that newborns readily learn and are behaviorally competent to perform. Second, even a developing brain is a big place, and techniques have to be developed to limit the possible critical sites or pathways for learning and memory before detailed analyses of neural mechanisms can begin. Finally, after neural *correlates* of early learning have been identified, they have to be further examined in order to select the necessary and sufficient neural *mechanisms* of early learning.

In addition to selecting a general research plan for investigating the neural mechanism of early learning, the technical details for studying neonates must be developed. The rest of this chapter focuses on how we selected a behavioral learning task for neonates and how we modified existing psychobiological techniques to examine neural plasticity in the developing nervous system.

EARLY OLFACTORY LEARNING

It has been extensively documented that the mother-infant interaction in altricial mammals is heavily dependent on olfactory cues (for reviews, see Alberts, 1976, Leon, 1983, and Rosenblatt, 1976). Infant rats appear to learn about important odors at or even before birth, and these odors subsequently modulate critical aspects of their behavior.

The first demonstrations of experience-based olfactory modulation of pups' behavior were done in a naturalistic manner, with little control over odor presentation (Leon, 1975; Galef & Kaner, 1980). Not until pups were removed from the nest and the temporal relationship between the odor and reinforcer were manipulated did it become evident that pup odor-preference development could be modeled by classical conditioning (Brake, 1981; Johanson & Hall, 1982; Johanson & Teicher, 1980). Moreover, the same temporal constraints as those that govern adult classical conditioning appear to control pup olfactory classical conditioning (Johanson & Hall, 1982; Sullivan & Hall, 1988). However, within the context of classical conditioning, specific differences between infant and mature rats emerge. Most notably, infant rats exhibit increased motor activity during reinforcement presentations, such as during milk delivery (Johanson & Hall, 1982; also see Kehoe & Blass, 1986, for descriptions of another possible type of infant learning). Although this behavioral activation is not necessary for a stimulus to function as a reward (Kucharski, Johanson, & Hall, 1986), those stimuli that induce behavioral activation are most likely to function as a reward (Sullivan, Brake, Hofer, & Williams, 1986; Sullivan, Hofer, & Brake, 1986). For example, infant rats will learn to prefer an odor provided that the odor has been paired (forward or simultaneous) with *optimally arousing* stimulation (Do, Sullivan, & Leon, 1988; Pedersen, Williams, & Blass, 1982; Sullivan, Brake, Hofer, & Williams, 1986; Sullivan, Hofer, & Brake, 1986; Sullivan, McGaugh, & Leon, in preparation). The *optimally arousing* stimulation may be any of a myriad of different stimuli, including milk (Johanson & Hall, 1982; Sullivan & Hall, 1988), tactile stimulation produced by stroking or tail pinch (Sullivan, Brake, Hofer, & Williams, 1986; Sullivan, Hofer, & Brake, 1986), maternal odors (Sullivan, Hofer & Brake, 1986) activation of the noradrenergic system (Pedersen et al., 1982; Sullivan, Wilson, & Leon, 1989), and electrical brain stimulation (Moran, Lew, & Blass, 1981).

The classical conditioning paradigm we use to produce olfactory learning in rat pups relies on the pairing of a novel odor and tactile stimulation produced by stroking the pup's body vigorously with a small brush (Pedersen et al., 1982; Sullivan, Brake, Hofer, & Williams, 1986; Sullivan, Hofer, & Brake, 1986). Although only a one-trial, 10-min conditioning session is necessary for acquisition of an odor preference (Sullivan et al., 1986a, 1986b; Sullivan & Leon, 1987), we use 18 daily, 10-min trials, starting on postnatal Day 1, to more approximate the pups' natural olfactory learning situation while allowing maximal experimental control. The pups are removed from the nest for the training sessions and placed in individual glass training chambers. Their only experience with the conditioned odor, therefore, is during the 10-min training sessions. On the day of testing, postnatal Day 19, the pups are again removed from the nest for testing of conditioned behaviors.

Thus, this detailed behavioral analysis provides us with a very specific learning situation with newborns, in which we have experimental control of the significant external variables. While most studies of the neural correlates of learning employ such a model system (Cohen, 1984; Kandel, 1976; Thompson, 1986; Weinberger & Diamond, 1987; Woody, 1982), our task has the advantage of mimicking a learning situation that actually does, and must, occur during normal development of the pup. Early olfactory learning, therefore, appeared to be an ideal behavioral model to search for the neural mechanisms of early learning.

Having defined a learning task, the next step was to look for changes in the CNS associated with this learning. Where to look? Electrophysiological recordings or neuroanatomical analyses from random locations in the brain might eventually have detected some differences between control and conditioned brains, but more than likely the process would have taken many years without some initial indication of where to focus attention. An alternative was to map the general activity of whole brain regions in response to the conditioned stimulus simultaneously, in order to identify specific locations that are modified with learning. Electroencephalographic (EEG) recordings are commonly used for this purpose in mature animals (e.g., Freeman & Schneider, 1982); however, given the small and delicate nature of the immature rat skull, prolonged attachment of EEG electrodes on moving pups was impractical.

In contrast, radiolabeled metabolic markers, such as ^{14}C-2-deoxyglucose (2-DG), can be injected systemically with minimal discomfort to the pup and provide fairly detailed (50–100-μm resolution) maps of CNS metabolic activity. This technique has been used to assess olfactory bulb responses to both novel (Astic & Saucier, 1982) and familiar (Coopersmith & Leon, 1984) odors in rat pups. Thus, it was possible to inject a pup with 2-DG and expose it to the conditioned stimulus and then examine the brain for areas of differential activity.

The 2-DG technique was initially developed for use with mature animals and required modification for use in neonates (discussed below). As we have already described, in the 2-DG technique demonstrated modified olfactory bulb metabolic activity in response to conditioned odors in neonates (Sullivan & Leon, 1986). This modified activity was not generalized but, instead, was localized in odor-specific regions of the bulb (Coopersmith, Henderson, & Leon, 1986). Thus, with the information provided by this 2-DG mapping, we were able to focus our detailed functional analysis of neural activity (single-unit recordings) on a region we knew was modified by early learning. Using single-unit recording techniques, again modified for neonates (see below), we found specific changes in neural response patterns to conditioned odors (Wilson, Sullivan, & Leon, 1987). Importantly, our subsequent neurophysiological work suggests that had we recorded randomly in the olfactory bulb, rather than in a localized region, no learning-associated changes would have been detected (Wilson & Leon, 1988).

METHODS

General survey of activity

The major source of energy for both mature and developing neurons is glucose, and, in general, the more active a neuron, the greater its demand for glucose. The 2-DG technique uses this situation to "trick" active cells into labeling themselves. 2-DG is a form of glucose that can be taken up, but not readily metabolized, by active neurons. Thus, when 2-DG is introduced via systemic injections, active neurons will take up 2-DG along with glucose. The gradient of metabolically trapped, intracellular 2-DG across neurons then provides an indication of differences in activity during the uptake period. This intracellular 2-DG can be most easily observed and quantified by using a radiolabel (e.g., ^{14}C-2-DG) and standard autoradiography. Autoradiography involves exposing thin brain sections (20-μm) to X-ray film. Differences in ^{14}C uptake will produce differences in beta particle emission that is revealed as differentially exposed X-ray film. Relative variations in exposure can then be ascertained via analysis of relative optical densities in the autoradiograph of brain sections. Optical density can be easily quantified for comparisons between brain regions or between experimental conditions (see below for a description of our quantification technique).

Since several factors influence 2-DG uptake (e.g., glycogen incorporation and the hexose monophosphate shunt), it may not solely reflect neural activity. However, 2-DG uptake and neural activity are highly positively correlated in many systems (Gallistel et al., 1983; Mata et al., 1980; Sokoloff, 1981; Yarowski & Ingvar, 1981). For example, patterns of 2-DG uptake have been shown to correspond to differential single-unit activity in several sensory systems (Gonzalez-Lima & Scheich, 1984; Maier & Scheich, 1983; Schoppman & Stryker; 1981; Wilson & Leon, 1988). In the olfactory system, focal 2-DG uptake in olfactory bulb glomeruli is directly related to the intensity of electrical stimulation of the olfactory nerve (Greer, Stewart, Kauer, & Shepherd, 1981). Furthermore, as described below, spatial patterns of glomerular layer 2-DG uptake in response to learned odors correspond to spatial patterns of mitral/tufted cell single-unit activity (Wilson, Sullivan, & Leon, 1985, 1987; Wilson & Leon, 1988). Thus, whatever nonneuronal elements affect 2-DG uptake patterns in the olfactory bulb, differential 2-DG uptake probably reflects differential neural activity in this system. This independent confirmation of the 2-DG/neural activity link is a critical step in interpreting 2-DG autoradiography.

The 2-DG technique is clearly a useful tool in assessment of differential brain activity. However, proper interpretation of 2-DG autoradiographs requires an understanding of the limitations of the technique. Briefly, these limitations are:

1. Activity measured by the 2-DG technique is summated over a long period of time (generally 45 min) to ensure that unused 2-DG is not measured in the autoradiography. Thus, observing brief, momentary changes in neural activity is not possible.

2. Spatial resolution is such that individual cells are not discriminable, at least with the ^{14}C technique described here. The exception is use of a high-resolution technique, such as ^{3}H-2-DG, that sacrifices the recognition of spatial pattern within the brain for such resolution (Lancet, Greer, Kauer, & Shepherd, 1982).

3. Changes in 2-DG uptake may reflect either excitatory or inhibitory synaptic/neural activity and may be indicative of activity in intrinsic neurons or in efferents to the region of interest.

4. The relative contribution of glial and neuronal tissue, and of unrelated metabolic events, to 2-DG uptake is not certain.

The majority of these limitations, however, can be overcome with a combination of techniques, as described in this chapter.

The basic procedures for 2-DG autoradiography in mature animals can be used with neonatal animals with a few minor changes. Detailed descriptions of ^{14}C-2-DG autoradiography in adult rodents have been published elsewhere (Sokoloff, 1981; Gallistel et al., 1983). Our procedure for neonates is as follows.

^{14}C-2-DG can be obtained either mixed with ethanol or with saline. If obtained in ethanol, the ethanol has to be evaporated and then replaced with saline. Sigma Chemical Company supplies ^{14}C-2-DG in saline (100 μmCi/ml), which is our preferred solution. On receipt, we aliquot the stock solution into 0.2 ml units and store them in microfuge tubes at 0° C. Fifteen to 20 min before an animal is to be injected, the appropriate number of units are thawed.

The primary difference between neonates and mature animals is the route of 2-DG administration. The most economical method of 2-DG injection in mature animals is intravenous because intraperitoneal or subcutaneous injection requires a prohibitively large quantity of expensive 2-DG injected into each subject. The young animal is small enough, however, so that the 2-DG may be injected subcutaneously without much expense. Thus, in addition to expense, time is saved because catherization prior to testing is unnecessary, and the subject is left unstressed by the surgery and the cannula. No differences in 2-DG autoradiography are seen with either route of administration (Gallistel et al., 1983). Using the subcutaneous injection route, we use a dose of 20 μCi/100 g body weight.

Immediately after injection, the pup is placed in a test chamber, which is a glass cylinder with an airtight cap in which a sensitive pressure transducer is imbedded. Pups are then exposed to the test stimulus (an odor that has previously been experienced during classical conditioning training). The stimulus exposure lasts 45 min, for reasons discussed above. In order to prevent receptor adaptation during this exposure, the odor is refreshed for 1 min before each of eight 4.75-min periods during the test. Also during this test, respiration rate is monitored during the eight periods with a Columbus Instruments respiration monitor interfaced to an Apple II computer. This allows determination of the role of modified respiration rate in the expression of modified 2-DG uptake responses. The use of the pressure transducer allows monitoring of respiration without stressing or restricting the pup.

During the 2-DG test, the animal is exposed to a controlled stimulus and/or performs a controlled task. Given that the 2-DG is meant to detect changes in brain metabolism that are specific to the stimulus or task in question, control of the test environment is particularly important. Our animals are placed in a glass test container that is surrounded by gray foam rubber. This foam rubber sheath helps to limit visual stimulation and attenuate extraneous noises. Testing is always done in a quiet room.

Our stimuli for 2-DG testing are odors, which are controlled with flow-dilution olfactometers. These olfactometers present the odor as a dilution of saturated odor vapor. The diluted odor is presented to the animal via Tygon tubing (Fisher Scientific). Given that Tygon tubing will absorb some odor, the tubing and flow meters are always changed when the stimulus odor is changed.

Following the 45-min test, the pup is quickly decapitated and the olfactory bulbs dissected over ice and frozen in 2-methylbutane (-45°C). Since the 2-DG is water soluble, the delay from decapitation to freezing should be no more than 5 min to prevent nonspecific spread of the label. The immature state of the pup skull facilitates the dissection. The skull can be grossly removed using ordinary surgical scissors and fine dissection around the bulbs performed with microscissors. The dissection involves removing the skull overlying the cortex with a central cut between the hemispheres (easily observed through the thin neonatal skull) and then peeling away the skull to either side. Forceps and microscissors (Fine Science Tools) are then used to remove the remaining bone from around the bulbs (both on the top and sides). The olfactory nerves are severed by carefully cutting below each bulb, starting at the rostral pole. At this point, the bulbs should be free from connective tissues, but still attached to the olfactory peduncles and the rest of the brain. The brain is then removed from the skull, starting at the caudal end and gently lifting with a small spatula or periosteal elevator. If the skull is placed upside down during this final step, the brain and olfactory bulbs will come out easily. The brain should then be sectioned with a scalpel or razor blade, coronally, about 2–3 mm anterior to the cerebellum, and mounted on a freezing stage with Histomount (Fisher Scientific). Care should be taken to orient the bulbs at this point to standardize the eventual angle at which the bulbs will be sectioned. The freezing stage is then slowly lowered into a bath of 2-methylbutane kept at $-45°C$ with dry ice and left 30–60 s until all bubbling stops. Rapid lowering of the tissue will produce cracks in it. The bulbs can then be sectioned immediately or stored at $-70°C$ for several months.

The bulbs are sectioned at 20 μm in a cyrostat (Reichert Scientific) at $-17°C$, and every third section is picked up on a 20-mm square coverslip (thickness 1 mm). To prevent the water-soluble 2-DG from spreading as the section thaws, the coverslips are rapidly dried on a slide warmer for at least 5 min. The use of thin coverslips, rather than thick slides, facilitates the drying. Once the entire area of interest within the brain has been cut, the coverslips are attached to cardboard with either double-sided tape or epoxy. The cardboard is precut to match the size of the X-ray film (Kodak SB-5; suggested vendor: Merry X-Ray). Also attached to the cardboard are ^{14}C standards (ARC) to assist in the quantitative analysis of the autoradiographs. The cardboard with attached sections and film is then placed in an X-ray exposure cassette (Du Pont; suggested vendor: Merry X-Ray) , without a window or intensifying screen but with foam in the interior to ensure close apposition of the film with the brain sections, for 8–10 days. The film is then removed and processed using standard X-ray development procedures. Autoradiographs are analyzed with an image-analysis system (Imaging Research, Inc.). (For a theoretical discussion of image-analysis systems, see Ramm, Kulick, Stryker, & Frost, 1984; Ramm & Kulick, 1985).

Our method of quantification of optical density relies on the fact that 2-DG uptake in the periventricular core of the olfactory bulb is consistently low, regardless of odor used, and is not affected by learning (Coopersmith & Leon, 1984; Sullivan & Leon, 1986). Thus, measurements of focal density in the glomerular layer (which is affected by learning) can be expressed as a ratio of glomerular layer uptake to periventricular core uptake. Use of this ratio minimizes interpretation problems stemming from differences in background uptake between sections and differences in section thickness. Furthermore, it eliminates the need for repeated sampling of blood levels of ^{14}C-2-DG that is required in determination of exact levels of 2-DG uptake (Sokoloff, 1981).

Single-Unit analysis

Since we were interested in determining the neural mechanisms of early olfactory learning, greater cellular resolution than that provided by 2-DG autoradiography was essential. Thus, we began an electrophysiological study of olfactory bulb responses to learned odor cues, recording extracellularly from single olfactory bulb output neurons, the mitral and tufted cells. Mitral/tufted cells receive direct input from the olfactory receptors via synapses in the glomerular layer. Rather than recording from random locations throughout the bulb, however, as has been standard procedure in olfactory bulb electrophysiological studies (e.g., Harrison & Scott, 1986; Chaput & Lankheet, 1987), we recorded from neurons in relatively precise locations in the bulb. The spatial patterns of focal 2-DG uptake that were previously determined provided an excellent map to guide our electrode placements toward regions most likely to demonstrate modified neural activity following early learning.

Since electrophysiological localization of these areas in mobile pups is difficult (although desirable), the pups must be immobilized and anesthetized. The most commonly used anesthetics for neurophysiology are pentobarbital (Nembutol, Abbott Laboratories) or urethane (Sigma Chemical Company). The surgical dosage of barbiturates varies greatly with age. Immature animals are much more sensitive, both behaviorally (Mirkin, 1970) and physiologically (Wilson & Racine, 1985), to barbiturates than mature animals, and care must be taken to avoid overdose. An effective surgical dose of Nembutol for a 7-day-old (20-g) rat is 20–30 mg/kg, compared to 65 mg/kg for a mature rat. Sensitivity to urethane does not appear to vary with age (Wilson & Racine, 1985); however, a surgical dose of urethane (1.5 g/kg) will incapacitate animals of all ages for 24 h or more, a duration that may be intolerable if survival of the neonate is desired. Furthermore, with any anesthetic, body temperature must be maintained during the recording session, particularly in neonates (34–36 °C). We use a Hamilton Aquamatic heating pad, which circulates thermistatically controlled warm water through a pad placed below the animal (Baxter Hospital Supply). This warms the animal without introducing a source of electrical noise into the recording chamber. For prolonged recording sessions (more than 5 h), normal saline (1–5 ml, i.p. or s.c.) may be provided to raise hydration levels and improve the general condition and survivability of the animal.

Once the neonate is anesthetized, the next step is to implant the electrodes, generally using a stereotaxic apparatus (Narishige stereotaxic SN-1, Medical Systems, Inc.). The olfactory bulb is fortunately isolated from the rest of the brain and can actually be seen through the thin skull of the neonatal rat; thus, a special neonatal sterotaxic atlas is not required for our work. We have recorded successfully from rat olfactory bulb as young as 3 days old. The pups are placed in the stereotaxic apparatus using standard ear bars. The ear bars, however, are gently inserted below the ears, at the back of the jaw, so that the skin is dimpled and the skull is resting on the points of the ear bars. A standard incisor bar is inserted in the rat's mouth to further stabilize the head.

The scalp is then resected with a scalpel. If the animal is to recover, this incision should be kept as small as possible and still allow access to the skull. The skin edges of the incision are held out of the way with small alligator clips or seraphin clips. The skull of the neonate, as already stated, is very delicate and paper thin. However, attempting to insert the electrodes directly through the skull will invariably indent the skull prior to penetration, resulting in CNS physiological depression, contusions, and potential internal bleeding and edema. The preferred method of electrode implantation is to drill a hole over the structure of interest using a dental or hand-held drill. This hole can be either just slightly larger than the diameter of the electrode or may involve removal of a large (2–3-mm ^{2}) window. If, for example, the intention is to make single-unit recordings from several areas of a single structure, the larger hole is preferred. The use of a larger hole also reduces the chance of damaging the micro-electrode on bone as it passes through the skull. The primary disadvantage of opening a large hole and exposing more CNS is the increased likelihood of tissue damage from drying of the CNS surface. Thus, the area of exposed CNS must be kept moist with either warmed (34°C) mineral oil or physiological saline. The dura may either be punctured after the skull has been removed or left intact. We routinely leave it intact and have no problem inserting glass microelectrodes for extracellular recording through it.

Electrodes used for recording and stimulating in the mature CNS can be used in neonates. For single-unit recording, we use microfilament capillary glass (1-mm O.D., A-M Systems) pulled on a Narishige horizontal electrode puller. The electrodes are filled with 0.5 M KCl, providing electrode resistances in normal saline of 5–15 mohms. The use of capillary glass with a microfilament increases electrolyte filling time and decreases the need to remove bubbles from within the electrode. Stimulating electrodes consist of Teflon-coated stainless steel wire (0.28 mm, A-M Systems). Additional details of our recording and stimulating procedures in the neonatal rat CNS are presented in Wilson et al. (1987), Wilson and Leon (1988) and Wilson (1984).

As with 2-DG testing, odors are presented to the pup from flow-dilution olfactometers. Since each cell is tested with several odors, the Tygon tubing from each olfactometer meet at a single, multiple-branch glass connector, one branch of which presents the odors to the animal. The glass connector is used to prevent odor contamination of the final stimulus-presenting tube. Clean air is presented between stimuli to wash out the stimulus lines.

When recording evoked potentials or single-unit activity in neonatal animals, three facts are immediately apparent in comparison with mature animals: (1)Evoked potentials are smaller in amplitude; (2) there is generally a longer latency from the stimulus to the onset of the response; (3) and spontaneous activity of single neurons is generally slower in neonates. The small amplitude and low spontaneous activity is generally attributed to immature levels of synaptic contact and activity, while the longer response latencies are due to incomplete myelination of central pathways. These factors combine to make detection and analysis of neural activity in the immature CNS more difficult. For example, single-unit recording in the olfactory bulb involves advancing the microelectrode slowly until the electrode approaches a neuron. The appearance of a neuron is signaled by its spontaneous activity. If, however, the majority of neurons are silent for extended periods, many cells will be missed. In the olfactory bulb, spontaneous activity levels of mitral cells at 1 week is less than half that of cells at 3 weeks of age (Wilson & Leon, 1986).

We dealt with the problem of slow spontaneous activity in the bulb by stimulating antidromically or orthodromically at a set rate while slowly advancing the microelectrode. When searching for mitral/tufted cells, antidromic activation of their axons by stimulation of the lateral olfactory tract will induce an antidromic spike that may be observed when the recording electrode is near the cell. Thus, the presence of a nearby, silent mitral/tufted cell can be detected by recording evoked action potentials.

A point to mention here is the extreme importance of having a stable, healthy preparation when attempting to describe the characteristics of the immature CNS. It is critical, for example, to know if a slow spontaneous activity rate in a particular system is characteristic of the system at that age or is simply due to an increased sensitivity of the immature system to the experimental procedures used to examine it. Short of replication by several laboratories under a variety of conditions, the best way to avoid problems of interpretation is to do everything possible to maintain a healthy preparation. In most cases this involves controlling body temperature, anesthetic level, and hydration and may involve monitoring heart rate and respiration to detect deterioration in the condition of the pup.

APPLICATION

The primary experimental design used throughout our work on neural correlates of learning has involved whole litters receiving the same training condition. Such a design avoids problems of odor contamination between pups within a litter. For testing, 2 pups (1 male and 1 female) from a single litter were used for each of the tests: behavioral testing, 2-DG autoradiography, and single-unit recording. This split-litter testing allowed many pups from a single litter to be used without invalidating statistical analysis, while maintaining low testing variability. It was also required to prevent pups from receiving differential stimulus exposure during the tests. For example, a pup that had been used in the behavior test would be exposed to the test odor more or less time depending on its relative preference for the odor. Subsequent testing of these pups with 2-DG or single-unit recording, therefore, would be confounded by the differential odor exposure received during the behavior test.

Using 2-DG autoradiography, we found that the olfactory-learning paradigm that produces a behavioral odor preference is associated with a modified neural response of the olfactory bulb to subsequent presentations of that odor (Coopersmith & Leon, 1984; Sullivan & Leon, 1986). This modified response is characterized by an enhancement of 2-DG uptake in odor-specific regions of the bulb glomerular layer. Was this the engram, and, therefore, the end of our search? To answer this question, the first task was to strengthen the link between the neural change and the behavioral change. That is, we needed to know if the modified neural response was the cause of, or was caused by, the change in behavior. Thus, each observed neural correlate or hypothesized mechanism of early learning was rigorously examined to determine its necessary and/or sufficient relationship to the change in behavior.

A good example was a series of experiments designed to determine the role of modified respiration in the expression of modified olfactory bulb responses. In mature rabbits, olfactory learning produces conditioned sniffing to the learned odor (Freeman, Viana Di Prisco, Davis, & Whitney, 1983). If conditioned sniffing occurs in neonates as well, the modified neural response we observed may simply be due to a modification in stimulus sampling (changed behavior) by the subject. The enhanced response would, in this case, represent a normal response to increased receptor stimulation rather than a real change in the response of the olfactory system, which could represent encoding of learned information.

Two strategies were employed to examine the role of modified respiration in the expression of enhanced olfactory bulb responses. First, respiration was monitored and quantified during the 2-DG test (Coopersmith & Leon, 1984; Sullivan & Leon, 1986; Sullivan, Wilson, Kim, & Leon, 1988). When this method failed to detect changes with learning, the possibility remained that subtle changes in respiration rate or volume had occurred but were not detectable with the variety of methods we had used.

The second strategy involved experimentally controlling respiration (Sullivan et al., 1988). Trained animals were anesthetized with urethane, as described earlier, and tracheotomized. (The details of the tracheotomy and artificial sniff procedure are described in Sullivan et al., 1988.) This technique allowed precise control over both the volume and rate of respiration during the 2-DG test. The results suggested that even under conditions of identical respiration and, thus, equal stimulus exposure during the test, pups had an enhanced olfactory bulb response to learned odors.

In addition to examining the role of primary afferent and centrifugal olfactory bulb inputs in the modified responses to learned odors, we have extended our search for mechanisms to the single neuron level. As described above, we used the spatial information obtained with 2-DG autoradiography to guide our microelectrode placements to known regions of change. Single-unit recordings could help uncover what the enhanced metabolic response to learned odors meant in terms of neural activity. That is, 2-DG autoradiography works very well for delineating regions of increased activity, but cannot differentiate between activity of excitatory and inhibitory connections. Since we wanted to understand the neural mechanisms of early learning, it was necessary that we know precisely how the neural response was changed by learning. Once we have completely described the characteristics of the "learned response," we can go on to discover how and why the response was changed by learning.

Single-unit recordings from olfactory bulb output cells, mitral/tufted cells, revealed that in regions of enhanced glomerular layer 2-DG uptake, there was an increase in suppressive responses and a decrease in excitatory responses to the learned odor (Wilson et al., 1985, 1987; Wilson & Leon, 1988). These results were basically the opposite of what we might have predicted given the increased 2-DG response in the area of primary afferent input to the mitral/tufted cells. A common assumption about the mechanisms of learning is that learning involves a strengthening of excitatory connections and increased output of a neural circuit (e.g., Hebb, 1949). Thus, the enhanced 2-DG response should have represented an increase in excitation of mitral/tufted cells. Instead, early olfactory learning actually reduces olfactory bulb output by enhancing inhibition of mitral/tufted cells.

Furthermore, mitral/tufted cells not associated with focal regions of enhanced 2-DG uptake were not modified by learning (Wilson & Leon, 1988). Thus, had we not done the initial 2-DG mapping and used that information to guide our electrode placements, we most likely would not have observed any changes in single-unit responses. The combination of the 2-DG and single-unit recording techniques, therefore, has been integral to our success to this point in two ways. First, the 2-DG mapping was necessary to allow us to find changes in single-unit activity. Second, the single-unit recording was necessary to allow us to interpret the enhanced 2-DG response.

Proper interpretation of these apparently contradictory findings (enhanced 2-DG-suppressed single-unit activity), however, will require additional work. The fact that mitral/tufted cells are inhibited by a learned odor is not in itself surprising. Mitral/tufted cells are spontaneously active, so that any change in their response pattern to a stimulus provides information about that stimulus. A change from predominant excitation to predominant inhibition is perhaps the most striking change that the system could express, and thus would be a clearly recognizable signal for neurons downstream that a given stimulus was significant. Determination of mechanism will also require additional work, although a working hypothesis has been developed (Leon et al., 1987; Wilson & Leon, 1988). Basically, we suggest that a learning-induced increase in feed-forward inhibition in the olfactory bulb can account for both the enhanced 2-DG uptake and the increased mitral/tufted cells suppression. An increase in the activity or numbers of juxta-glomerular neurons could result in greater metabolic activity in that layer and a subsequent decrease in mitral/tufted cell activity (Leon et al., 1987; Wilson & Leon, 1988).

As these examples demonstrate, combining two or more techniques that provide different types of information in this way can be very useful, if not critical, for extracting information. We have also combined 2-DG autoradiography with neuroanatomical methods to localize changes in olfactory bulb anatomy associated with early learning (Woo, Coopersmith, & Leon, 1987). Furthermore, these techniques are routinely combined with behavioral analyses to correlate changes in the neural response with changes in behavioral response to the learned odor. For example, the degree of change in single-unit responses after learning strongly correlates with the magnitude of the learned behavioral response (Wilson et al., 1987). This is an important correlation to make if the changes we observe in the olfactory bulb are related to the expression of learned behaviors.

PROSPECTS

The work described here begins to outline the changes that occur in the olfactory system after early learning. This work is an example of one approach to understanding the neural correlates of learning and memory. At least two distinct emphases can be distinguished in the work in this field. Our work might most closely be described as a search for neural/synaptic mechanisms of learning and memory—that is, how is information stored? A second approach to understanding early learning, perhaps best displayed by the work of Kucharski and Hall (1987, 1988), can be described as a search for the location or neural pathways of learning and memory—that is, where is information stored? In the extreme, this second approach attempts to describe the path of information flow from the receptor to the effector and where along this pathway learned modifications occur that are critical for the expression of the learned behavior. This approach utilizes techniques such as lesioning, tract tracing, and gross analyses of many CNS structures. At some point, these two approaches need to be combined for a complete understanding of mechanisms of early learning.

Regardless of the theoretical or technical approach that is selected, caution is required in determining the relationship between learned behaviors and associated changes in neural activity. No behavior can be conditioned independently of all of other ongoing activities. For example, aversively conditioning an eyeblink response conditions not only the eyeblink but a variety of other voluntary and autonomic responses as well (Weinberger & Diamond, 1987). The influence of these other behaviors (e.g., increased heart rate or respiration) on the particular neural system being monitored must be determined and controlled for when looking for neural mechanisms of learning. For example, suppose that neural activity is increased in a selected thalamic nucleus after aversive conditioning. This increase may represent storage of learned information regarding the particular behavior being trained or it may simply reflect a sensitivity of that nucleus to heart rate, which has simultaneously been modified by the training procedure. One outcome suggests discovery of an engram, while the other is far less exciting in this regard.

This problem may, in fact, be compounded in neonates that have been demonstrated to have poorly differentiated response systems. For example, a variety of stimuli act as reinforcers in a classical conditioning paradigm in neonates. Many of these stimuli when used with mature animals, however, are aversive (e.g., tail pinch; see above). Thus, at a behavioral level of analysis, the neonate does not appear as effective at distinguishing between aversive and nonaversive stimuli. Similarly, at the neural level, neurons in the noradrenergic nucleus locus ceruleus respond differentially to noxious and nonnoxious stimuli in the mature rat (Foote, Bloom, & Aston-Jones, 1983). Locus ceruleus neurons in immature rats, however, respond equally to noxious and nonnoxious stimuli (Nakamura, Kimura, & Sakaguchi, 1987). Thus, a single manipulation in an immature animal may influence a variety of behavioral and neural systems. These systems may then interact with, or impose upon, the particular structure being studied to create an apparent learned neural change.

Although we have described some learning-associated changes in the neonate brain, to understand how those changes occur, new techniques will have be to adapted to the neonate. Perhaps the greatest innovations need to be made with neurophysiological techniques. All of our single-unit recordings, for example, are made in anesthetized animals. While our anesthetic was chosen for its limited effects on olfactory bulb physiology, the ideal situation would be to record in awake, freely moving pups. Chronic recordings from neonates is currently very difficult for several reasons. First, the electrodes must be firmly attached to the pup, usually by cementing them to the skull. In very young animals, the skull is too thin for the cement headcaps to remain stable for more than a few hours or days. This problem is exaggerated when the electrodes are attached to cables leading from the amplifier or stimulator. These leads need to be very lightweight and flexible in order to avoid added stress to the headcap. Finally, implanting electrodes in a growing brain often results in electrode movement and, thus, loss of optimal recordings. One possible solution to this problem is use of very thin wire "floating" electrodes. These electrodes are attached to the skull, but because of their extremely small size and flexibility, they can keep their position relative to the structure of interest by essentially floating in the brain.

Another technique that could be adapted to the neonate is intracellular recording. While extracellular, single-unit techniques provide information regarding whether neurons increase or decrease their firing rates, intracellular recording can provide information about why firing rates change. For example, if extracellular recording shows a decrease in firing rate of neurons in response to the learned odor, intracellular recording could discriminate between an increase in inhibition or a decrease in excitation impinging on that neuron as the cause of the response change. Discriminating between these two possibilities would greatly extend our knowledge of how early learning modifies the olfactory system.

Finally, understanding how the developing nervous system changes during learning is intimately tied to knowing how the developing nervous system normally functions. There will always be a need for purely descriptive studies of the neuroanatomical, neurochemical, and neurophysiological development of the nervous system.

ACKNOWLEDGMENTS

This work was supported by NSF grant BNS8819189 to D.A.W., BNS8606786 to D. A. W. and M. L., NINCDS grant NS26100 and PEW Foundation funds to R. M. S., and NICHD grant HD24236 and Research Scientist Development Award MH00371 from NIMH to M. L.

REFERENCES

Alberts, J. R. (1976). Olfactory contributions to behavioral development in rodents. In R. L. Doty (Ed.), *Mammalian olfaction, reproductive processes and behavior*. New York: Academic Press.

Astic, L., & Saucier, D. (1982). Metabolic mapping of functional activity of olfactory projections of the rat: Ontogenetic study. *Developmental Brain Research, 2*, 141–156.

Brake, S. C. (1981). Suckling infant rats learn a preference for a novel olfactory stimulus paired with milk delivery. *Science, 211*, 506–508.

Chaput, M., & Lankheet, P. (1987). Influence of stimulus intensity on the categories of single-unit responses recorded from olfactory bulb neurons in awake freely-breathing rabbits. *Physiology and Behavior, 40*, 453–462.

Cohen, D. H. (1984). Identification of vertebrate neurons modified during learning: Analysis of sensory pathways. In D. L. Alkon & J. Farley (Eds.), *Primary neural substrates of learning and behavioral change*. New York: Cambridge University Press.

Coopersmith, R., Henderson, S., & Leon, M. (1986). Odor specificity of the enhanced neural response following odor experience in rats. *Developmental Brain Research, 27*, 191–197.

Coopersmith, R., & Leon, M. (1984). Enhanced neural response to familiar olfactory cues. *Science, 225*, 191–197.

Cotman, C. W., Nieto-Sampedro, M., & Harris, E. W. (1981). Synapse replacement in the nervous system of adult vertebrates. *Physiology Reviews, 61*, 684–784.

Diamond, M. C., Rosenzweig, M. R., Bennett, E. L., Lindner, B., & Lyon, L. (1972). Effects of environmental enrichment and impoverishment on rat cerebral cortex. *Journal of Neurobiology*, *3*, 47–64.

Do, J., Sullivan, R. M., & Leon, M. (1988). Behavioral and neural correlates of postnatal olfactory conditioning: II. Respiration during conditioning. *Developmental Psychobiology*, *21*, 591–600.

Foote, S. L., Bloom, F. E., & Aston-Jones, G. (1983). Nucleus locus ceruleus: New evidence of anatomical and physiological specificity. *Physiology Review*, *63*, 844–914.

Freeman, W. J., & Schneider, W. (1982). Changes in spatial patterns of rabbit olfactory EEG with conditioning to odors. *Psychophysiology*, *19*, 44–56.

Freeman, W. J., Viana Di Prisco, G., Davis, G. W., & Whitney, T. M. (1983). Conditioning of relative frequency of sniffing by rabbits to odors. *Journal of Comparative Physiology*, *97*, 12–23.

Galef, B. G., & Kaner, K. (1980). Establishment and maintenance of preference for natural and artificial olfactory stimuli in juvenile rats. *Journal of Comparative Physiology and Psychology*, *94*, 588–595.

Gallistel, C. R., Piner, C. T., Allen, T. O., Adler, N. T., Yadin, E., & Negin, M. (1983). Computer-assisted analysis of 2-DG autoradiographs. *Neuroscience and Behavior Review*, *6*, 409–420.

Gonzalez-Lima, F., & Scheich, H. (1984). Classical conditioning enhances auditory 2-deoxyglucose pattern in inferior colliculus. *Neuroscience Letters*, *51*, 79–85.

Greenough, W. T., Volkmar, F. R., & Juraska, J. M. (1973). Effects of rearing complexity on dendritic branching in frontolateral and temporal cortex of the rat. *Experimental Neurology*, *41*, 371–378.

Greer, C. A., Stewart, W. B., Kauer, J. S., & Shepherd, G. M. (1981). Topographical and laminar localization of 2-deoxyglucose uptake in rat olfactory bulb induced by electrical stimulation of olfactory nerves. *Brain Research*, *217*, 279–293.

Harrison, T. A., & Scott, J. W. (1986). Olfactory bulb responses to odor stimulation: Analysis of response pattern and intensity relationships. *Journal of Neurophysiology*, *56*, 1571–1589.

Hebb, D. O. (1949). *The organization of behavior. A neuropsychological theory*. New York: Wiley.

Johanson, I. B., & Hall, W. G. (1982). Appetitive conditioning in neonatal rats: Conditioned orientation to an odor. *Developmental Psychobiology*, *15*, 379–397.

Johanson, I. B., & Teicher, M. H., (1980). Classical conditioning of an odor preference in 3-day-old rats. *Behavioral and Neural Biology*, *29*, 132–136.

Kandel, E. R. (1976). *Cellular basis of behavior*. San Francisco: Freeman.

Kehoe, P., & Blass, E. M. (1986). Behaviorally functional opioid systems in infant rats: I. Evidence for olfactory and gustatory classical conditioning. *Behavioral Neuroscience*, *100*, 359–367.

Kucharski, D., & Hall, W. G. (1987). New routes to early memories. *Science*, *238*, 786–788.

Kucharski, D., & Hall, W. G. (1988). Developmental change in the access to olfactory memories. *Behavioral Neuroscience*, *102*, 340–348.

Kucharski, D., Johanson, I. B., & Hall, W. G. (1986). Unilateral olfactory conditioning in 6-day-old rat pups. *Behavioral and Neural Biology*, *46*, 472–490.

Lancet D., Greer, C. A., Kauer, J. S., & Shepherd, G. M. (1982). Mapping of odor-related neuronal activity in the olfactory bulb by high-resolution 2-deoxyglucose autoradiography. *Proceedings of the National Academy of Sciences*, *79*, 670–674.

Leon, M. (1975). Dietary control of maternal pheromone. *Physiology and Behavior*, *14*, 311–319.

Leon, M. (1983). Chemical communication in mother-young interactions. In J. Vandenbergh (Ed.), *Pheromones and mammalian communication*. New York: Academic Press.

Leon, M., Coopersmith, R., Lee, S., Sullivan, R. M., Wilson, D. A., & Woo, C. C. (1987). Neural and behavioral plasticity induced by early olfactory learning. In N. A. Krasnegor, E. M. Blass, M. A. Hofer, & W. P. Smotherman (Eds.), *Perinatal development: A psychobiological perspective*. New York: Academic Press.

Lynch, G. S. (1986). *Synapses, circuits and the beginnings of memory*. Cambridge, MA: MIT Press.

Lynch, G. S., McGaugh, J. L., & Weinberger, N. M. (1984). *Neurobiology of learning and memory*. New York: Guilford Press.

Maier, V., & Scheich, H. (1983). Acoustic imprinting leads to differential 2-deoxy-glucose uptake in chick forebrain. *Proceedings of the National Academy of Sciences*, *80*, 3860–3864.

Mata, M., Fink, D. J., Gainer, H., Smith, C. B., Davidson, L., Savaki, H., Schwartz, W., & Sokoloff, L. (1980). Activity dependent energy metabolism in rat posterior pituitary primarily reflects sodium pump activity. *Journal of Neurochemistry*, *34*, 213–215.

Mirkin, B. L. (1970). Developmental pharmacology. *Annual Review of Pharmacology*, *10*, 255–272.

Moran, T. H., Lew, M. F., & Blass, E. M. (1981). Intracranial self-stimulation in 3-day old rat pups. *Science*, *214*, 1366–1368.

Nakamura, S., Kimura, F., & Sakaguchi, T. (1987). Postnatal development of electrical activity in the locus ceruleus. *Journal of Neurophysiology*, *58*, 510–524.

Pedersen, P. E., Williams, C. L., & Blass, E. M. (1982). Activation and odor conditioning of suckling behavior in 3-day-old albino rats. *Journal of Experimental Psychology: Animal Behavior Processes*, *8*, 829–841.

Ramm, P., & Kulick, J. H. (1985). Principles of computer-assisted imaging in autoradiographic densitometry. In R. T. Mize (Ed.), *The microcomputer in cell and neurobiology research* (pp. 311–334). Amsterdam: Elsevier.

Ramm, P., Kulick, J. H., Stryker, M. P., & Frost, B. J. (1984). Video and scanning micro-densitometer-based imaging systems in autoradiographic densitometry. *Journal of Neuroscience Methods*, *11*, 89–100.

Rosenblatt, J. S. (1976). Stages in the early behavioral development of altricial young of selected species of non-primate mammals. In P. P. G. Bateson & R. A. Hinde (Eds.), *Growing points in ethology*. London: Cambridge University Press.

Schoppman, A., & Stryker, M. P. (1981). Physiological evidence that the 2-deoxyglucose method reveals orientation columns in cat visual cortex. *Nature*, *293*, 574–576.

Sokoloff, L. (1981). Localization of functional ability in the cerebral nervous system by measurement of glucose utilization with radioactive glucose. *Journal of Cerebral Blood Flow and Metabolism*, *1*, 7–36.

Squire, L. R. (1987). *Memory and brain*. New York: Oxford University Press.

Sullivan, R. M., Brake, S. C., Hofer, M. A., & Williams, C. L. (1986). Huddling and independent feeding of neonatal rats can be facilitated by a conditioned change in behavioral state. *Developmental Psychobiology*, *I19*, 625–635.

Sullivan, R. M., & Hall, W. G. (1988). Reinforcement in infancy: Classical conditioning using tactile stroking or intra-oral milk infusions as UCS. *Developmental Psychobiology*, *20*, 215–223.

Sullivan, R. M., Hofer, M. A., & Brake, S. C. (1986). Olfactory-guided orientation in neonatal rats is enhanced by a conditioned change in behavioral state. *Developmental Psychobiology*, *19*, 615–623.

Sullivan, R. M., & Leon, M. (1986). Early olfactory learning induces an enhanced olfactory bulb response in young rats. *Developmental Brain Research*, *27*, 278–282.

Sullivan, R. M., & Leon, M. (1987). One-trial olfactory learning enhances olfactory bulb responses to an appetitive conditioned odor in 7-day-old rats. *Developmental Brain Research*, *35*, 307–311.

Sullivan, R. M., Wilson, D. A., Kim, M. H., & Leon, M. (1988). Behavioral and neural correlates of postnatal olfactory conditioning. I. Effect of respiration on conditioned neural responses. *Physiology and Behavior*, *44*, 85–90.

Sullivan, R. M., Wilson, D. A., & Leon, M. (1989). Norepinephrine and learning-induced plasticity in infant rat olfactory system. *Journal of Neuroscience*, *9*, 3998–4006.

Teyler, T. J., & DiScenna, P. (1984). Long-term potentiation as a candidate mnemonic device. *Brain Research Review*, *7*, 15–28.

Thompson, R. F. (1986). The neurobiology of learning and memory. *Science*, *233*, 941–947.

Weinberger, N. M., & Diamond, D. M. (1987). Physiological plasticity in auditory cortex: Rapid induction by learning. *Progress in Neurobiology*, *29*:1–55.

Wiesel, T. N., & Hubel, D. H. (1965). Comparison of the effects of unilateral and bilateral eye closure on cortical unit responses in kittens. *Journal of Neurophysiology*, *28*, 1029–1040.

Wilson, D. A. (1984). A comparison of the postnatal development of post-activation potentiation in the neocortex and dentate gyrus of the rat. *Developmental Brain Research*, *16*, 61–68.

Wilson, D. A., & Leon, M. (1986). Early apppearance of inhibition in the neonatal rat olfactory bulb. *Developmental Brain Research*, *26*, 289–292.

Wilson, D. A., & Leon, M. (1988). Spatial patterns of olfactory bulb single-unit responses to learned olfactory cues in young rats. *Journal of Neurophysiology*, *59*, 1770–1782.

Wilson, D. A., & Racine, R. J. (1983). The postnatal development of post-activation potentiation in the rat neocortex. *Developmental Brain Research*, *7*, 271–276.

Wilson, D. A., & Racine, R. J. (1985). Barbiturate-enhanced paired-pulse depression in neonatal rats. *Neuroscience Letters, 56*, 101–106.

Wilson, D. A., Sullivan, R. M., & Leon, M. (1985). Odor familiarity alters mitral cell response in the olfactory bulb of neonatal rats. *Developmental Brain Research, 22*, 314–317.

Wilson, D. A. Sullivan, R. M., & Leon, M. (1987). Single-unit analysis of postnatal olfactory learning; Modified olfactory bulb output response patterns to learned attractive odors. *Journal of Neuroscience, 7*, 3154–3162.

Wilson, D. A., Willner, J., Kurz, E. M., & Nadel, L. (1986). Early handling increases hippocampal long-term potentiation in young rats. *Behavioral Brain Research, 21*, 223–227.

Woo, C. C., Coopersmith, R. M., & Leon, M. (1987). Morphological and metabolic changes in the olfactory bulb accompany the enhanced neural response to familiar odors in rat pups. *Journal of Comparative Neurology, 263*, 113–125.

Woody, C. D. (1982). *Memory, learning and higher function: A cellular view.* New York: Springer.

Yarowsky, P. J., & Ingvar, D. H. (1981). Neuronal activity and energy metabolism. *Federation Proceedings, 40*, 2353–2362.

18

Effects of Fetal Methylazoxymethanol Acetate on Neural and Behavioral Development

TIMOTHY H. MORAN AND JOSEPH T. COYLE

The contribution made by a specific neuronal population in either behavioral development or the normal biochemical and neuronal development of the brain is difficult to assess. Traditional lesioning techniques are usually limited to the postnatal period, by which time a large degree of neuronal integration has occurred. Furthermore, other than neurotoxins aimed at neuronal populations using specific transmitters, lesioning techniques are often relatively nonspecific in that a variety of cellular elements are destroyed.

One powerful approach to these questions has been the use of specific autosomal mutations, which result in selective (and often highly specific) failure to develop or early degeneration of certain neuronal populations but a sparing of others that make up a particular brain region. An example of this kind of mutation would be lurcher (gene symbol LC), an autosomal dominant, lethal in the homozygous state but which in the heterozygous state results in degeneration of cerebellar Purkinje cells and the consequent loss of cerebellar granule cells (Caddy & Biscoe, 1979). In this model, the effect of Purkinje cell loss on normal development is clear. The absence of Purkinje cells results in the loss of cerebellar granule cells, and the locomotor behavior of these animals is impaired. While such mutations do exist for a few specific neuronal populations, mutations are not currently available that result in developmental deletions of the majority of the neuronal populations within the brain. To address the role that these other populations play in structural and functional development, the utilization of other research strategies is necessary.

Neuronal development can be viewed as a discontinuous process, with neurons localized to specific nuclei or cortical laminae being generated during discrete developmental "windows." During the latter half of gestation in the rat and mouse, when the bulk of neurogenesis occurs, neuronal populations divide, differentiate, and migrate to form brain structures, usually in a manner in which new regions overlie the previously developed areas. If this process could be temporarily interrupted at the time when a specific cortical layer, for example, is forming, it would be possible to examine the subsequent development of the rest of the brain in the absence of that neuronal population. Such a technique is available and involves the prenatal administration of the antimitotic agent methylazoxymethanol acetate (MAM). As will be demonstrated, MAM administration can result in the partial or complete deletion of specific neuronal populations. While the application of this technique in producing brain lesions is relatively straightforward, the interpretation of the resulting neuronal and behavioral deficits may be quite complex.

In this chapter, we describe the kinds of lesions that MAM can produce, focusing on issues of the specific timing and dosage of MAM administration. We detail the neurochemical and behavioral consequences of MAM administation on gestational Day 15, and we demonstrate MAM 's use as a tool in attempting to assess the contribution of a specific neuronal population in a neonatal behavioral response. Finally, we discuss the potential strengths of this lesion and the interpretational difficulties it presents.

HISTORY AND GENERAL PROPERTIES OF MAM ADMINISTRATION

MAM is a powerful alkylating agent that occurs naturally as a glucoside in the botanical family Cycadacceae (Matsumoto & Strong, 1963). As such, MAM and similar compounds have been studied for their ability to induce tumors. It has been well established that the kidney, liver, and intestinal tract are most sensitive to the carcinogenic actions of MAM (Zedek, Sternberg, McGowan, & Poynter, 1972). In a study correlating the site and size of transplacentally induced neoplasms with the stage of fetal exposure to carcinogens of this type, Spatz and Lacquer (1968) observed that the administration of MAM during the latter part of the gestational period produced microencephaly in rats. This was the original finding that led to the comparative evaluation of the neuroanatomical, neurochemical, and behavior alterations produced by MAM administration at various gestational time points.

Intraperitoneal administration of radiolabeled MAM in pregnant rats results in incorporation in various organs and fluid compartments in the dam and a significant incorporation in the fetus (Nagata & Matsumoto, 1969). Analyses of the fetal incorporation indicated that a significant portion of the recovered radioactivity was present in the DNA of the fetal rat brain. Thus, MAM passes readily through the placental barrier and reacts with nucleic acids and proteins in the fetus. MAM is a powerful methylating agent and has been demonstrated both by in vitro and in vivo methods to kill dividing cells through a process of methylating guanine and other purine and pyrimidine bases in brain nucleic acids (Nagata & Matsumoto, 1969). Since it is depolymerized DNA in dividing cells that is particularly vulnerable to methylation, preventing replication and division, it is cells in this state that are sensitive to MAM administration. The lethal effects of MAM on cells have been demonstrated to be confined to a period extending from 2 to 24 h following the injection, with the maximum effect occurring at 12 h (Matsumoto, Spatz, & Lacquer, 1972).

The effects of MAM on the injected dam at dosages that produce lesions of specific neuronal populations in offspring appear to be minimal. While MAM has maximal carcinogenic effects when administered at dosages of 35 mg/kg and dosages above 50 mg/kg are lethal, dosages of 25 mg/kg and less have no tumor-inducing effects (Zedek et al., 1972).

Timing of MAM Administration

Little information is available on MAM's effects on peripheral tissues or physiological systems. There are no obvious somatic anomalies in rats treated with MAM after 14 days of gestation other than a tendency for decreased body weights (Cannon-Spoor & Freed, 1983; Plonsky, Riley, Lee, & Haddad, 1985). In contrast, the effects of MAM on neural development are readily apparent and have been extensively examined. Within the brain, the sequellae of MAM administration have been demonstrated in the cerebellum (Slevin, Johnston, Biziers, & Coyle, 1982), neocortex (Johnston, Carman, & Coyle, 1981), basal ganglia (Beaulieu & Coyle, 1981), and hippocampus (Singh, 1977, 1978, 1980). The particular cellular population affected by MAM administration depends upon the time of its administration (Johnston & Coyle, 1982) and, as demonstrated in Figure 18.1, is consistent with the timetable of neuronal birth as derived from (³H) thymidine autoradiography (Angevine & Sidman, 1961; Caveness & Sidman, 1973). That is, MAM treatment early in the second gestation week in the rat will predominantly affect the development of brain stem and hindbrain structures such as cerebellum, Purkinje cells, and the locus ceruleus. MAM administration during the latter part of the second week or the beginning of the third week will predominantly affect the development of the cortex. Postnatal administration of MAM will result in relative deletion of cerebellar interneurons, the population that is mitotically active during this period.

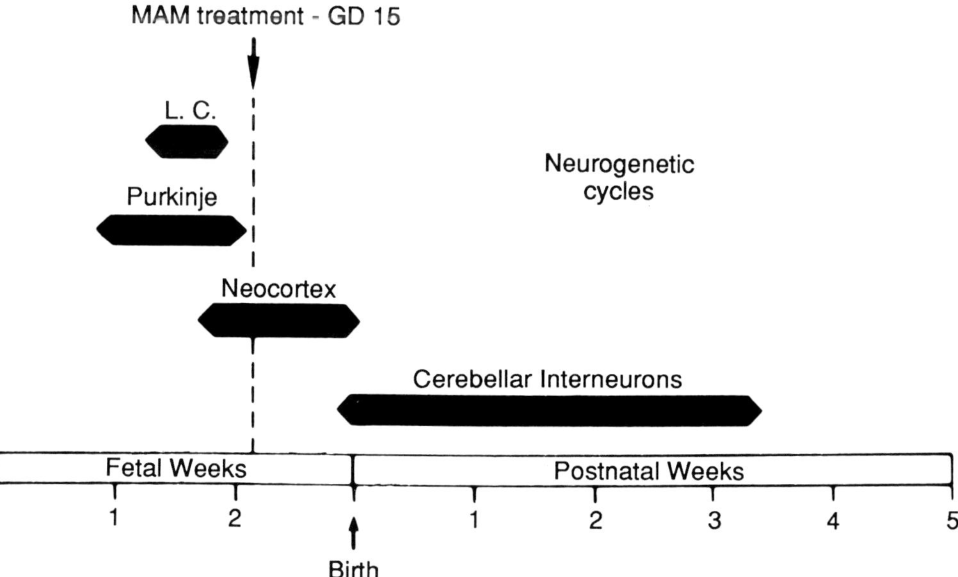

Figure 18.1. Schematic of periods of neurogenesis for selected neuronal populations in the rat. Cortical neurogenesis occurs during the last week of gestation, while many hindbrain neurons, such as those located in the locus ceruleus (LC) and Purkinje cells, divide earlier during gestation. Neurogenesis persists postnatally in the cerebellum and in certain forebrain areas. (From "Cytotoxic Lesions and the Development of Transmitter Systems" by M. V. Johnston and J. T. Coyle, 1982, *Trends in Neuroscience, 5,* p. 155. Copyright Elsevier Biomedical Press. Reprinted by permission.)

MAM Dosage and Administration

In most studies examining the effects of MAM on neural or behavioral development, dosages of 20–25 mg/kg have been administered. MAM (Aldrich Chemical) is soluble in physiological saline. It is generally administered intraperitoneally (i.p.) to the dam, although intravenous administration is effective as well. For i.p. administration, MAM may be diluted to concentration of 10–15 mg/ml. Little work establishing a dose response relationship between the extent of brain lesions and various doses of MAM has been done. The exception is work by Dambska, Haddad, Kozlowski, Lee, and Shek (1982) examining the degree of microencephaly produced by various levels of MAM administration on gestational Day 15. Their results demonstrated that increasing levels of MAM from 0 to 30 mg/kg in the dam resulted in significantly greater loss in brain weight in the progeny. However, the specificity of the lesions was maintained in that the areas of brain affected were similar at all dosages. That is, higher levels of administration produced lesions of the cortex and hippocampus of greater magnitude, but did not produce obvious lesions in other brain regions unaffected by smaller dosages.

ANATOMICAL AND BIOCHEMICAL EFFECTS
OF GESTATIONAL DAY 15 MAM ADMINISTRATION

While the effects of MAM administration at various ages have been examined, the best characterized lesion has been that produced by gestational Day 15 administration. In order to provide a view of the extent of a MAM lesion and the interactions of neuronal systems affected by MAM administration, we will detail the effects of gestational Day 15 MAM administration.

Effects on Cortical Morphology

Early studies demonstrated that while the DNA content in normal brains increases rapidly from gestational Day 15 to birth, MAM treatment at gestational Day 15 retards the rate of brain growth, as measured by this index, for the first three days following administration. Following that period, the rate begins to increase, such that by gestational Day 20, the DNA content of MAM-treated animals is reduced in the whole brain by 25%. The rate of cortical growth remains inhibited in the postnatal period, so that the overall reduction in cortical DNA content is 50% by postnatal Day 7 and remains at this level through postnatal Day 60, the latest point at which it was studied. The DNA content of the rest of the brain is not significantly affected by MAM administration. This pattern of cellular loss expressed as a decrease in DNA content (Matsumoto et al., 1972) is consistent with the high rate of neurogenesis dedicated to formation of the neocortex during the last week of gestation in the rat, thereby rendering precursor neuroblasts in the ventricular germinal zone vulnerable to DNA alkylation by MAM.

The specific morphological and neurochemical sequelae of gestational Day 15 MAM administration have been extensively examined by Johnston and Coyle (1979). The administration of 20 mg/kg MAM on Day 15 of gestation resulted in a 67% reduction in cortical mass in adulthood (Figure 18.2). In response to MAM administration, the weight of cortical slabs was 70% of control levels at birth and decreased

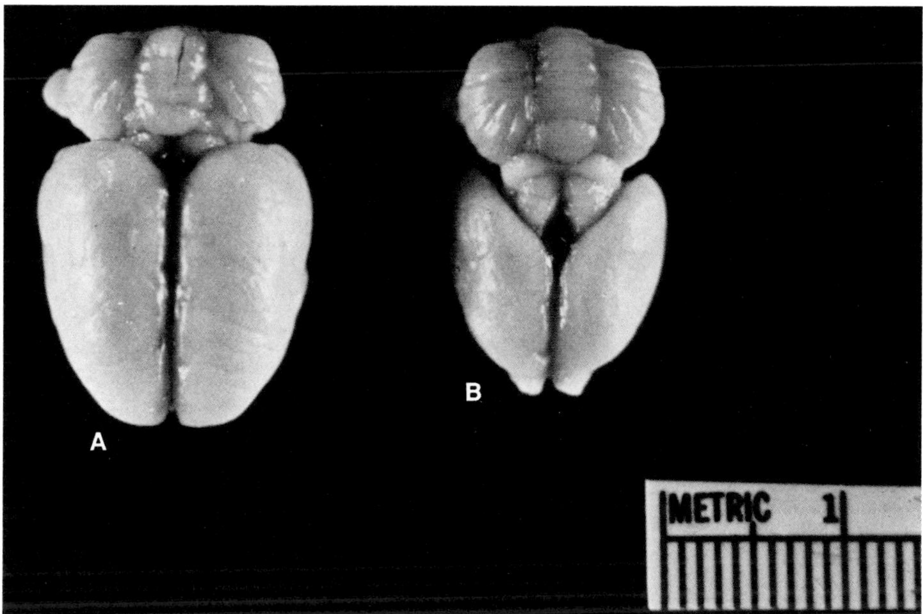

Figure 18.2. Photograph of adult brains from control rat (left) and rat exposed to MAM on gestational Day 15 (right). Size of the forebrain is markedly reduced in the MAM-lesioned brain. Size of the hindbrain is relatively unaffected. (From "Histological and Neurochemical Effects of Fetal Treatment with Methylazoxymethanol Acetate on Rat Neocortex in Adulthood" by M. V. Johnston and J. T. Coyle, 1979, *Brain Research, 170*, p. 140. Copyright Elsevier Biomedical Press. Reprinted by permission.)

to 40% by eight days following birth, owing to the delay in growth rate in the cortex of MAM-treated rats. Even after a growth surge between postnatal days 10 and 25, the weight of cortical slabs in MAM-treated rats was only 35% of that of controls in adulthood. Examination of Nissl-stained sections (Figure 18.3) revealed a marked decrease in cortical thickness that was especially apparent in more dorsal cortex. The cortical thinning was due primarily to an impairment in the formation of cortical layer II to outer layer IV. This pattern of deficit is consistent with the generation of neuronal components of these layers. That is, at Day 15 the cells giving rise to cortical layer II to outer layer IV are undergoing mitosis and are ablated by MAM administration at this time.

Varying the time of MAM administration around gestational Day 15 resulted in different patterns of cortical deficits (Johnston et al., 1981). For example, MAM administration on gestational Day 13 resulted in hypocellularity in cortical layers V and VI, while MAM treatment on Day 17 produced only a subtle neuronal loss restricted to layer II. This changing pattern of deficits from Day 13 to Day 17 is consistent with the "inside out" development of the cortex (Angevine & Sidman, 1961). Gestational Day 13 MAM administration also resulted in significant hindbrain hypoplasia, which was not observed with MAM treatment on subsequent days. Treatment at these early times clearly impinged upon the neurogenesis of more caudal brain structures.

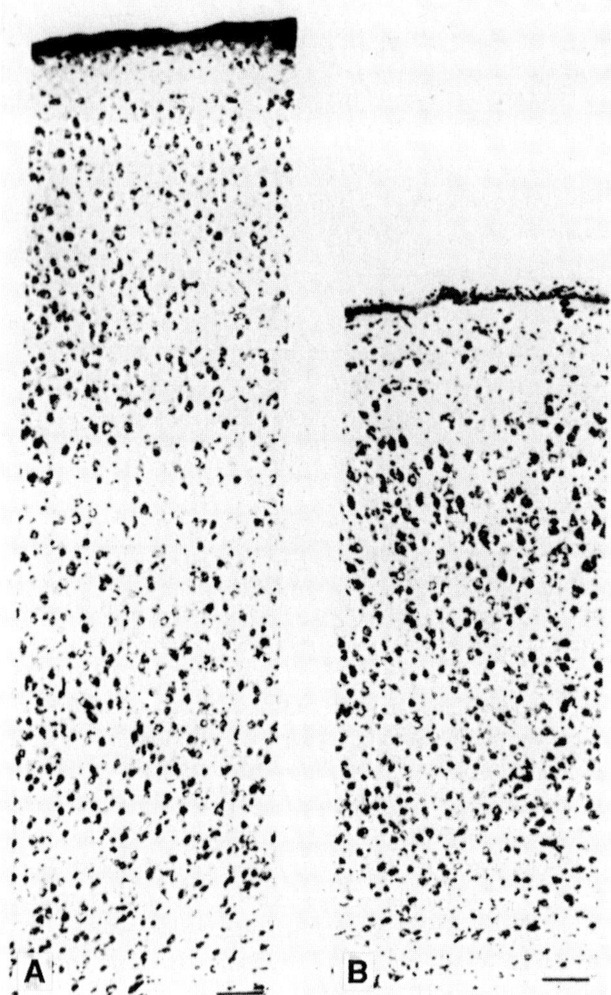

Figure 18.3. Photomicrographs of Nissl-stained sections from control (left) and MAM-lesioned (right) lateral neocortex at the level of the anterior commissure. Rats were 7 weeks old at the time of sacrifice. Cortical layers II–IV are absent in the thinner cortex of the MAM-treated animal. Remaining cortex consists of three layers: a thickened layer II and areas that closely resemble normal layers V and VI. Bar = 100μm. (From "Effects of Fetal Treatment with Methylazoxymethanol Acetate at Various Gestational Dates on the Neurochemistry of the Adult Neocortex of the Rat" by M. V. Johnston, A. B. Carman, and J. T. Coyle, 1981, *Journal of Neurochemistry, 36,* p. 126. Copyright Raven Press. Reprinted by permission.)

Cortical Presynaptic Neurochemical Effects

The cortex receives innervation from a variety of neurotransmitter systems located within the midbrain and brain stem. For example, the cortex is innervated by noradrenergic (NE), serotonergic (5-HT), and cholinergic (ACh) arborizations, while the medial, prefrontal, and cingulate cortex receive a significant dopaminergic (DA) innervation. The cell bodies of these neuronal systems have birth dates prior to gestational Day 15. Thus, MAM administration aimed at affecting cortical development should not affect the neurogenesis of these systems. Gestational Day 15 MAM treatment does, in fact, spare these neuronal systems, as demonstrated by a quantitative analysis of their presynaptic markers within the cortex (Beaulieu & Coyle, 1982, 1983; Johnston & Coyle, 1979; Johnston, Grzanna, & Coyle, 1979; Jonsson & Hallman, 1982; Matsutani, Nagayoshi, Tamura, & Tsukada, 1980).

While these systems are not directly affected by MAM treatment, experiments by Johnston and Coyle (1979) have demonstrated some abnormalities in the levels of several presynaptic markers of these neurotransmitter systems as they innervate the neocortex. Presynaptic markers for the noradrenergic afferents were markedly increased in concentration in neocortex of MAM-lesioned rats at all stages of postnatal development. Conversely, the total content of these markers per cortical slab did not differ from control, suggesting that a quantitatively normal noradrenergic cortical arbor developed in spite of the MAM lesion. In support of this interpretation, immunocytochemical staining for noradrenergic fibers revealed a marked increase in their density in the MAM-lesioned hypoplastic cortex (Johnston, Grzanna, & Coyle, 1979). In contrast to these results for NE, similar analyses for presynaptic cholinergic markers while indicating a relative enrichment in the MAM-lesioned cortex, showed that, the total amount of cholinergic markers was reduced per cortex, suggesting some interaction between the MAM lesion and the ontogeny of this system (Johnston & Coyle, 1980).

Together, these results demonstrated that the atrophic cortex of MAM-treated rats contained a relatively intact noradrenergic innervation compressed within a smaller cortical volume. Thus, normal cortical development is not necessary for this innervation to occur. While the cortical innervation of NE and ACh can be specified in quantitative terms in MAM-treated animals, the nature of that innervation relative to normal is not clear. NE projections may be contacting cells that would not ordinarily receive such projections, or individual cells that receive NE projections could receive multiple inputs. This has not been determined.

In contrast to the pattern of results for noradrenergic markers, the total amount of GABAergic markers was significantly reduced within the cortex while their relative concentration was normal. GABAergic neurons are stellate interneurons with their cell bodies as well as axon terminal fields distributed throughout all the cortical layers (Storm-Mathiesen, 1976; Ribak, 1978). As such, they would be expected to be affected by the alterations in cortical development. The relative lack of effect on the concentration of presynaptic markers for GABAergic neurons in the face of a decrease in their total content suggests that the cortical layers remaining after MAM treatment contained a normal compliment of GABAergic neurons and that their development was not markedly affected by the deletion of the outer cortical layers.

Thus, Day 15 MAM administration results in a significant loss of intrinsic GABAergic cortical neurons and alterations in the innervation of the cortex by ascending neurotransmitter systems. As stated, the innervation of the cortex by the noradrenergic system does not appear to depend upon the integrity of the neurons within the terminal field, so that the total NE content of the cortex is similar to that found in intact brains. Similar results were obtained for dopamine (Hallman & Jonsson, 1984). In contrast, the cellular loss in the cortex may have a partial effect on

Table 18.1 Presynaptic Neurochemical Effects of Gestation Day 15 MAM Administration

	Total Content	Concentration
Cortex		
GABA	↓	—
NE	—	↑
DA	—	↑
5-HT	↓	↑
ACh	↓	↑
Glutamate	↓	—
Striatum		
GABA	↓	—
DA	↓	↑
ACh	↓	—

its innervation by cholinergic (as discussed) and serotonergic fibers (Hallman & Jonsson, 1984). These presynaptic neurochemical results are summarized in Table 18.1.

Cortical Postsynaptic Neurochemical Effects

Alterations in postsynaptic neurotransmitter receptors within the cortex of MAM-treated animals have also been demonstrated (Johnston & Coyle, 1979; Beaulieu & Coyle, 1982). The apparent Bmax (as an expression of the total number of receptor sites) of specific GABA receptor binding was not significantly altered on a per milligram protein basis, but the total number of sites per cortical slab was reduced in proportion to the degree of cortical hypoplasia. The numbers of beta adrenergic and muscarinic receptors were reduced both per milligram protein and per cortical slab. As demonstrated by Beaulieu and Coyle (1982), the decrease in the density of beta adrenergic receptors in the MAM-lesioned cortex may be an adaptive response to the relative noradrenergic hyperinnervation; that is, receptor-down regulation. Thus, the hyperinnervation of ascending monoamine neurons does have postsynaptic consequences.

Striatal and Hippocampal Effects

While these cortical changes represent a number of the major sequelae of MAM administration at this gestational stage, the development of neostriatum and hippocampus were also affected (Beaulieu & Coyle, 1981; Singh, 1977, 1978). MAM treatment at gestational Day 15 results in a 37% decrease in striatal mass. However, histological examination of the hypoplastic striatum revealed no obvious disruption in the cellular organization, although earlier MAM treatment does disrupt cellular organization. The pattern in results produced by MAM in the striatum is somewhat similar to that found in the cortex (Table 18.1). A loss of intrinsic neuronal populations with birth dates corresponding to the time of MAM administration was found, and a relative hyperinnervation by dopaminergic fibers in proportion to the degree of hypoplasia was also observed. However, whereas the total amount of noradrenergic presynaptic markers in the neocortex was unchanged, total amount of presynaptic dopaminergic markers in the striatum was decreased. While neurons with cell bodies within the locus ceruleus cease division well before the time of peak mitotic activity in the cortical terminal field, the development in dopaminergic cell bodies within the substantia nigra exhibits a temporal overlap with the development of intrinsic striatal neurons (Lauder & Bloom, 1974). This overlap may explain the impairment of dopaminergic innervation in the striatum, although other possibilities such as decreased trophic influences cannot be discounted.

In the hippocampus, MAM treatment on gestational Day 15 produces a series of morphological abnormalities (Singh, 1977, 1978, 1980). Among these are the presence of ectopic CA1 pyramidal cells with abnormal dendritic trees and connections. These ectopic cells result in reduced commissural input and a generalized disruption in hippocampal function. The etiology of these deficits is not completely understood, but may depend in part on a reduced afferent input to the hippocampus from the neocortex. MAM treatment also results in the formation of dense mossy fiber terminals in the infrapyramidal region of CA3 (Cheema & Lauder, 1983). This may represent a decreased ratio of pyramidal cells relative to granule cells; however, the etiology of this imbalance is not clear.

Summary

Taken together, these results demonstrate that the administration of MAM on gestational Day 15 produces specific major alterations in brain development. The cytoarchitecture, as well as the pre- and postsynaptic neurochemistry of the telencephalon, are disordered. These changes are most apparent in the neocortex, although significant alterations occur in the striatum and hippocampus. These latter effects of MAM administration may depend in part on interactions with the cortical hypoplasia.

BEHAVIORAL CONSEQUENCES OF GESTATIONAL DAY 15 MAM ADMINISTRATION

Within a year of the original finding that MAM administration could produce microencephaly in animals, Haddad, Rabe, Laquer, Spatz, and Valsamis (1969) reported that adult MAM-lesioned rats exhibited significant learning impairments. Since then, a variety of studies have addressed the nature and extent of the behavioral abnormalities in these animals. The majority of this work has been simply to catalog the deficits produced by this lesion. No extensive integration of the anatomical, neurochemical, and behavioral results has been attempted. In this section, we briefly outline the pattern of behavioral deficits produced by Day 15 MAM administration.

The search for behavioral or cognitive deficits following MAM administration has required careful scrutiny. Rabe and Haddad (1972) reported on a large number of experiments designed to assess the learning and memory abilities of these animals. The behavior of rats that received MAM lesions at gestational Day 15 was not impaired in the acquisition of some commonly used paradigms such as operant conditioning schedules, visual discrimination tasks, or even a maze task. In operant conditioning paradigms, MAM-lesioned animals actually outperformed controls in responding for a food reward, emitting more responses in the test session (Cannon-Spoor & Freed, 1984). MAM-treated rats, however, tended to be resistant to extinction, leading to the suggestion that MAM animals are deficient in response inhibition and tend to perseverate (Weinberg, Haddad, & Dumas, 1983).

Only on more complex learning tests were performance impairments obvious. In a shuttle box active avoidance paradigm, Plonsky et al. (1985) demonstrated facilitated performance in MAM-lesioned rats relative to controls. Furthermore, during the intertrial intervals, the MAM animals crossed between the two chambers significantly more frequently than controls, demonstrating hyperactivity and a lack of response inhibition in their performance. In contrast to their facility of performance in active avoidance paradigms, a deficit in passive avoidance tasks was quite evident in MAM animals. We and others (Sanberg, Moran, & Coyle, 1987; Cannon-Spoor & Freed, 1983; Vorhees, Fernandez, Dumas, & Haddad, 1984) have demonstrated deficits in the acquisition of the passive avoidance paradigm, with MAM-lesioned animals taking significantly longer to acquire the task. No significant differences were found between MAM-lesioned and control animals in retention of the task once the original avoidance response had been acquired. These results suggest that the performance deficits noted in extinction or in passive avoidance may reflect the animals' inability to inhibit responding.

Consistent with this interpretation, hyperactivity in a variety of testing situations has been evident in the performance of MAM rats. In their original report, Rabe and Haddad (1972) demonstrated that MAM rats ambulated and reared more than the controls when tested in photocell chambers. This hyperactivity takes the form of significant increases in activity in brief test situations (Haddad, Lee, Plonsky, & Riley, 1981; Rabe & Haddad, 1972; Cannon-Spoor & Freed, 1984; Vorhees et al., 1984) and significant overall activity increases restricted to the dark cycle (Kiyono, Seo, & Shibagaki, 1980; Sanberg, Johnson, Moran, & Coyle, 1984). When the locomotor behavior of MAM animals is measured only during the light cycle over a 12-h period, no hyperactivity is evident. Utilizing computerized activity chambers, we have demonstrated significantly higher levels of activity on a number of variables, including distance traveled, average speed, and rearing behavior. Some aspects of this hyperactivity are evident as early as Day 10 (Sanberg, Moran, Kubos, Antuono, & Coyle, 1983). Thus, the main behavioral deficit evident in rats having been exposed to MAM at gestational Day 15 is hyperactivity. This hyperactivity may account for the animals' performance in learning and memory paradigms.

The particular neural deficit produced by Day 15 MAM administration that results in hyperactivity has not been identified. The overall pattern of cortical and striatal hypoplasia combined with hyperinnervation of ascending neurotransmitter systems suggests that the interaction of these alterations may produce this behavioral result. The following section investigates this idea more fully.

MAM LESIONS AS A TOOL FOR UNDERSTANDING BEHAVIORAL DEVELOPMENT

Rather than simply identifying the behavioral outcomes of MAM lesions, one can use the lesion as a tool for examining the role of neuronal populations in behavior. Since MAM administration results in a reliable deletion of certain neuronal populations dependent on the time of administration, one can examine how these deletions affect aspects of behavioral development. We have used the MAM lesion for just such an exploration, and in the following section, we outline this strategy, the results, and the potential interpretations of these experiments.

Neonatal pups are activated by a variety of stimuli, exhibiting high levels of behavior at early ages in response to maternal stimulation (Hofer, Shair, & Singh, 1976), and mimicking this stimulation (Pedersen & Blass, 1982; Pedersen et al., 1982), and to oral infusion of milk (Hall, 1979a, 1979b; Johanson & Hall, 1979). With increasing age, the behavioral activation seen in response to these forms of stimulation wanes. Rat pups are also behaviorally activated by electrical stimulation of the medial forebrain bundle (MFB) at the level of the lateral hypothalamus, emitting a complex series of behavioral responses (Moran, Schwartz, & Blass, 1983a). In response to short chains of electrical stimulation, pups reliably emit a series of behaviors ranging from the oral responses of mouthing and licking to full body stretches and responses that resemble the adult sexual behaviors of ear wiggling and lordosis (Moran et al., 1983a). A number of developmental transitions in the pattern of behavioral responses to stimulation are seen. Three-day-old pups demonstrate rapid shifts between various behaviors, while the behavioral responses in Day 10 pups are more temporally segregated to response type, more tied temporally to the electrical stimulation, and can come under the control of environmental influences (Moran, Schwartz, & Blass, 1983b). By Day 15 stimulation of this type no longer produces behavioral activation.

The transitions in the pattern of responses to MFB stimulation outlined above occur during a period in which the pup is undergoing a variety of behavioral, physiological, and neural changes. The complete dependency of the neonate on the dam ebbs, and behavioral independence begins to occur (Moran, 1986). Additional sensory modalities of vision (Rose, 1968) and audition (Crowley & Hepp-Raymond, 1966) are maturing, and the pups are acquiring physiological mechanisms of thermoregulation (Fowler & Kellog, 1975) allowing for this behavioral independence to occur. Rapid cortical development is taking place as the cortex doubles in size from Day 8 to Day 19 (Johnston & Coyle, 1980), and the emergence of cortical inhibitory mechanisms is occurring, as demonstrated by the onset of behavioral changes to spreading cortical depression (Hicks & D'Amato, 1975). Neurotransmitter systems are developing pre- and postsynaptically (Coyle & Henry, 1973) and becoming functional (Cheronis, Erinoff, Heller, & Hoffman, 1979; Horowitz, Heller, & Hoffman, 1982). Any or all of these changes could contribute to the transitions in responding to stimulation.

While the neural bases for these changing patterns of responses to stimulation cannot be specified, there are a number of likely candidates. Electrical stimulation of MFB at the level of the lateral hypothalamus results in activation of a variety of brain areas in adult rats (Roberts, 1980). Among the areas activated by stimulation that may play a role in determining the behavioral responses are the basal ganglia. While the role of the basal ganglia in motivated behaviors remains a topic of investigation, they have been suggested to be sites for integrating motivational and motoric systems. In this view, the basal ganglia play a gating function in the integration of inputs from the cortex and ascending subcortical monoamine pathways (Stricker & Zigmond, 1976; Moran, 1986). This basal ganglia integration results in organized behavioral output, appropriate to environmental demands and the organism's internal state. We had hypothesized that the changing pattern of responses to electrical stimulation in the pup may represent the maturation of cortical inhibitory mechanisms and the functional development of ascending monoamine transmitter systems and their integration at the level of the basal ganglia.

The pattern of neuronal deficits produced by Day 15 MAM administration resulting in cortical hypoplasia and hyperinnervation by ascending monoamines in both the cortex and the basal ganglia provided a tool for assessing the role of basal ganglia integration in the changing pattern of responses to MFB stimulation in development. Thus, we compared the responses of 3-, 10-, 15-, and 20-day-old pups that had been treated with MAM or saline on gestational Day 15 to electrical stimulation of the medial forebrain bundle at the level of the lateral hypothalamus (Moran, Sanberg, Antuono, & Coyle, 1986).

The results of these experiments demonstrated that electrical stimulation of the medial forebrain bundle produced behavioral responses in both MAM- and saline-treated 3- and 10-day-old pups. In response to stimulation, pups exhbited oral and other behaviors that increased with increasing frequency of stimulation. At these ages, there were no differences in the overall number of behavior responses exhibited by MAM or saline pups. However, Day 15 MAM pups continued to be activated by stimulation while saline pups were not. By Day 20, electrical stimulation elicited significantly fewer behavioral responses, although some MAM pups continued to be activated (Figure 18.4).

An examination of responses across individual behaviors demonstrated that at 3 and 10 days of age there were no differences in the frequencies of behavior elicited by electrical stimulation between MAM- and saline-treated pups. However, the temporal organization of response to stimulation remained immature in MAM-treated pups. That is, while in both control and MAM-treated 3-day-old pups rapid shifts in behavioral responses were evident, by Day 10 ingestive or lordosis responses became temporally segregated in control pups, and rapid shifts between these response categories continued to be evident in MAM-treated pups, again demonstrating a developmental lag (Figure 18.5). Only at Day 15 did the separation of stimulation-induced oral and lordosis responses begin to occur in MAM-treated pups.

Thus, there were two major differences in the responses of saline- and MAM-treated pups to electrical stimulation of the medial forebrain bundle. While both saline- and MAM-treated 3- and 10-day-old pups exhibited a reliable of series of behaviors that increased in frequency with increasing frequency of stimulation and demonstrated similiarities in the magnitude and overall pattern of responding, saline pups demonstrated a maturation in response pattern from Day 3 to Day 10 and a lack of responsiveness at Day 15. The responses of MAM-treated pups continued to demonstrate an immature pattern from Day 3 to Day 10, and these pups were still responsive to stimulation at days 15 and 20.

The original goal of these studies was to evaluate MAM as a research tool to identify the role of specific neuronal populations in normal brain and behavioral development. To some extent, the model has been successful—prevention of the

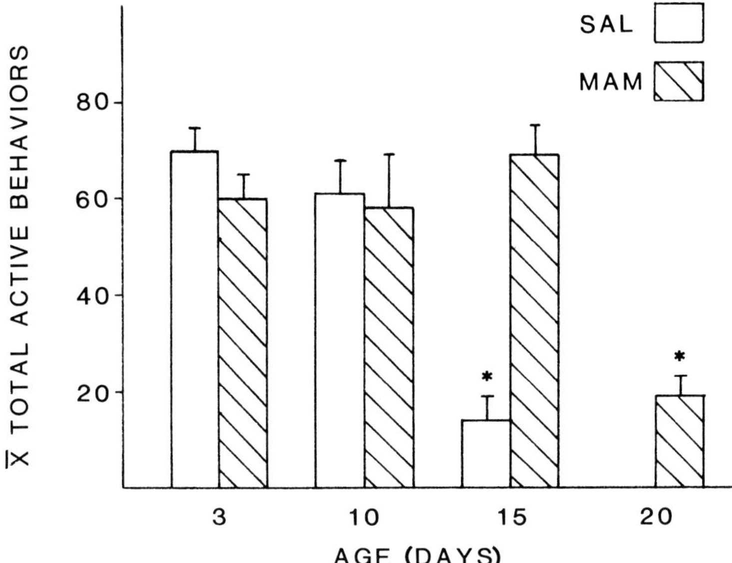

Figure 18.4. Mean number of behavioral responses to MFB stimulation in saline-treated and MAM-lesioned rat pups at different ages. (From "Microencephaly: Cortical Hypoplasia Induced by Methylazoxymethanol" by P. R. Sanberg, T. H. Moran, and J. T. Coyle, 1987, in J. T. Coyle (Ed.), *Animal Models of Dementia* (p. 271). New York: Alan R. Liss. Copyright Alan R. Liss. Reprinted by permission.)

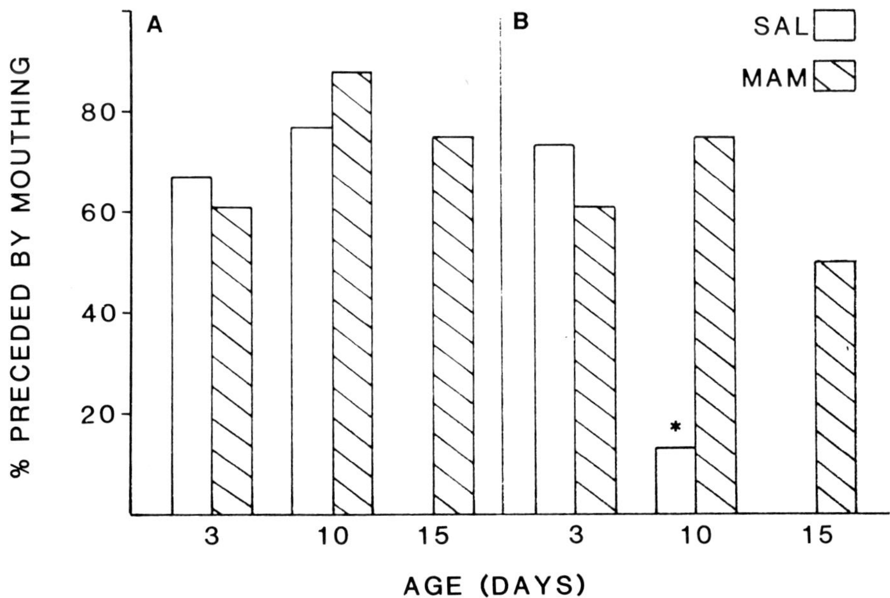

Figure 18.5. Percentage of stretch and lordosis responses preceded by oral behaviors (mouthing or licking) in 3-, 10-, and 15-day-old saline- (open bars) or MAM-treated (hatched bars) pups. (From "Methylazoxymethanol Acetate (MAM) Cortical Hypoplasia Alters the Pattern of Stimulation Induced Behavior in Neonatal Rats" by T. H. Moran, P. R. Sanberg, P. G. Antuono, and J. T. Coyle, 1986, *Developmental Brain Research, 27,* p. 239. Copyright Elsevier Biomedical Press. Reprinted by permission.)

development of specific neuronal populations with MAM administration was achieved, and MAM lesions resulted in quantifiably altered patterns of behavior responding to stimulation. But the overall interpretation of these results with MAM and MFB stimulation is complex. MAM treatment and the lesion it produces clearly altered the developmental transition in the responses to MFB stimulation—the behavioral output that may reflect state of basal ganglia integration was altered. Since a major consequence of MAM administration is an inhibition of cortical development, these results may be viewed as supporting a role for normal cortical development and cortical inhibitory mechanisms in the ontogeny of responses to activating stimuli. However, the remaining cortex and the striatum in MAM rats receives a relative hyperinnervation by monoamine pathways. Thus, the altered response patterns may be a result of hyperstimulation of the cortical tissue that is present. As well, there is both a loss of intrinsic neurons and a dopaminergic hyperinnervation in the striatum, and this may be a critical lesion. Any or all of these neural alterations could be responsible for the altered response patterns of MAM-treated pups. Thus, the particular cellular component underlying these developmental response differences to stimulation cannot be specified.

One way of viewing the behavior of MAM-lesioned pups in response to stimulation is as a result of a lack of normal integration in the basal ganglia. As discussed above, previous behavioral studies have demonstrated that one primary behavioral abnormality in MAM-treated animals is hyperactivity. Although MAM animals have been demonstrated to have deficits in a variety of learning tasks such as passive avoidance and maze learning, it is not clear whether these represent specific impairments or simply an inability to withhold responses (see Sanberg et al., 1987, for a discussion). The latter hypothesis is consistent with a lack of basal ganglia integration arising from dopaminergic hyperstimulation and low levels of cortical inhibitory input. In normal development, activation of the ascending MFB results in behavioral responses that initially are disorganized and encompass a variety of behavioral systems. Impulse activity in ascending dopaminergic pathways is stimulation dependent in the very young rat pup (Cheronis, et al., 1979; Horowitz et al., 1982), and such impulse activity in the absence of both ongoing background activity and cortical inhibitory inputs may precipitate behavioral responses. As development progresses, ongoing impulse activity in the ascending dopamine system is not dependent upon external stimulation, and cortical development has progressed so that integration of these inputs to the basal ganglia occurs. MFB stimulation in older pups results in behavioral responses that are more organized and eventually does not produce overt behavioral responses. In the MAM-treated animal, hyperinnervation of the basal ganglia by ascending dopamine is present, and the development of the cortex is severely limited. These factors may result in alterations of basal ganglia integration in both the pup and the adult, resulting in continued response to stimulation in the pup and hyperactivity in the adult MAM-treated animals.

SUMMARY AND CONCLUSIONS

MAM is a potentially powerful research tool. Administration of MAM at various developmental time points results in alterations of neuroanatomical and neurochemical development, and behavioral abnormalities arising from these alterations can be identified. However, as with other lesioning techniques, interpreting the results of MAM lesions can be problematic. The administration of MAM to pregnant rats produces a developmental lesion. That is, specific neuronal populations are eliminated at their moment of birth. Populations that are not so directly affected develop in this already altered environment such that cellular interactons that may play a role in their normal migration and differentiation do not occur. Thus, seemingly straightforward interpretations of the resulting patterns of development in subtractive terms (total minus A results in normal B) is overly simplistic.

The problem is less serious for the altered patterns of neuroanatomical and neurochemical development produced by MAM administration. The birth dates of neuronal populations can be identified, and the populations affected by the particular timing of MAM administration can be specified. Often, however, optimal timing for preventing the development of a single neuronal population overlaps with the birth date of other populations. Thus, the gestational Day 15 timing of MAM administration to affect cortical neurons also results in lesions of hippocampal and striatal neurons. The resulting neuroanatomical and neurochemical patterns of development within these regions reflect not only a loss of the specific intrinsic neuronal population eliminated by the MAM lesions but also the effects of altered developmental interactions with the other affected brain regions.

The interpretational difficulties become even more complex for the behavioral outcomes. The same intepretational problems are present, but an added layer of difficulty exists as one moves from structural to functional outcomes. The overall integration of the behavioral findings with the pattern of anatomical and neurochemical alterations produced by MAM traeatment is complex. While Day 15 MAM administration produces a relatively selective deletion of neurons destined to the outer layers of the neocortex, resulting in markedly altered patterns of synaptic connections that appear to be physiologically significant, other brain regions are also affected, as we have demonstrated. Thus, the particular neuronal population whose absence is responsible for the behavioral deficits of MAM pups cannot be specified. Also, neuronal development in areas of the brain that do not contain cell bodies with birthdays that would be directly affected by this MAM administration may be affected by differing patterns of interaction with other brain regions. Any or all of these alterations could affect behavioral development. Furthermore, brain regions that would not normally play a primary role in certain behaviors may be recruited in the altered neural environment produced by MAM, so that some behavioral expressions may appear to be relatively intact when the cellular populations usually underlying these behaviors may be quite disordered or completely absent.

Despite these difficulties, the MAM lesion technique has certain strengths and can provide answers to a variety of developmental questions. We have detailed the sequelae of gestational Day 15 MAM administration. While the effects of earlier and later administration of some neural systems have been documented, their full extent has not been investigated, and the behavioral outcomes of these lesions have not been addressed. Thus, the potential of this technique for examining the role of a variety of neural systems in overall brain and behavioral development has only begun to be tapped. One needs to be cautious in interpreting the results of all developmental lesions, but MAM provides a powerful tool, especially for investigating the role of interactions between neuronal populations in brain and behavioral development.

REFERENCES

Angevine, J. B., & Sidman, R. L. (1961). Autoradiographic study of cell migration during histogenesis of cerebral cortex in mouse. *Nature, 192,* 766–768.

Beaulieu, M., & Coyle, J. T. (1981). Effects of fetal methylazoxymethanol acetate lesion on synaptic neurochemistry of the adult striatum. *Journal of Neurochemistry, 37,* 878–887.

Beaulieu, M., & Coyle, J. T. (1982). Fetally-induced noradrenergic hyperinnervation of cerebral cortex results in persistent down-regulation of beta-receptors. *Developmental Brain Research, 43,* 491–494.

Beaulieu, M., & Coyle, J. T. (1983). Postnatal development of aminergic projections to frontal cortex: Effects of cortical lesions. *Journal of Neuroscience Research, 10,* 351–361.

Caddy, K. W., & Biscoe, T. J. (1979). Structural and quantitative studies of the normal C3H and Lurcher mutant mouse. *Philosophical Transactions of the Royal Society of London. Series B: Biological Sciences, 287,* 167–201.

Cannon-Spoor, H. E., & Freed, W. J. (1983). Hyperactivity induced by prenatal administration of methylazoxymethanol: Association with altered performance on conditioning tasks in rats. *Pharmacology, Biochemistry and Behavior, 20,* 189–193.

Caveness, V. S., & Sidman, R. L. (1973). Time of origin of corresponding cell classes in the cerebral cortex of normal and reeler mutant mice: An autoradiographic study. *Journal of Comparative Neurology, 148,* 141–151.

Cheema, S. S., & Lauder, J. M. (1983). Infrapyramidal mossy fibers in the hippocampus of methylazoxymethanol acetate-induced microencephalic rats. *Developmental Brain Research, 9,* 411–415.

Cheronis, J. C., Erinoff, L., Heller, A., & Hoffman, P. C. (1979). Pharmacological analysis of the functional ontogeny of the nigrostriatal dopaminergic neurons. *Brain Research, 169,* 545–560.

Coyle, J. T., & Henry, D. (1973). Catecholamines in fetal and newborn rat brains. *Journal of Neurochemistry, 21,* 61–67.

Crowley, D. E., & Hepp-Raymond, M. C. (1966). Development of cochlear function of the ear in the infant rat. *Journal of Comparative Physiology and Psychology, 62,* 427–430.

Dambska, M., Haddad, R., Kozlowski, P. B., Lee, M. H., & Shek, J. (1982). Telencephalic cytoarchitectonics in the brains of rats with graded degrees of microencephaly. *Acta Neuropathologica (Berlin), 52,* 203–209.

Fowler, S. J., & Kellog, G. (1975). Ontogeny of thermoregulatory mechanisms in the rat. *Journal of Comparative and Physiological Psychology, 89,* 738–746.

Haddad, R. K., Lee, M. H., Plonsky, M., & Riley, E. P. (1981). Activity and exploration following prenatal methylazoxymethanol acetate (MAMac) in rats. *Terotology, 24,* 55A.

Haddad, R. K. Rabe, A., Laqueur, G. L. Spatz, M., & Valsamis, M. P. (1969). Intellectual deficit associated with transplacentally induced microcephaly in the rat. *Science, 163,* 88–90.

Hall, W. G. (1979a). Feeding and behavioral activation in the infant rat. *Science, 205,* 206–209.

Hall, W. G. (1979b). The ontogeny of feeding in rats. I. Ingestive and behavioral responses to oral infusions. *Journal of Comparative and Physiological Psychology, 93,* 977–1000.

Hallman, H., & Jonsson, G. (1984). Monoamine neurotransmitter metabolism in microencephalic rat brain after prenatal methylazoxymethanol treatment. *Brain Research Bulletin, 13,* 383–389.

Hicks, S. P., & D'Amato, C. J. (1975). Motor sensory cortex, cortical spinal systems and developing locomotion and placing in rats. *American Journal of Anatomy, 143,* 1–42.

Hofer, M. A., Shair, H., & Singh, P. (1976). Evidence that maternal ventral skin substances promote suckling in infant rats. *Physiology and Behavior, 17,* 131–136.

Horowitz, J., Heller, A., & Hoffman, P. C. (1982). The effect of development of thermoregulatory function on the biochemical assessment of the ontogeny of neonatal dopaminergic neuronal activity. *Brain Research, 255,* 245–252.

Johanson, I. B., & Hall, W. G. (1979). Appetitive learning in one-day-old rat pups. *Science, 205,* 419–421.

Johnston, M. V., Carman, A. B., & Coyle, J. T. (1981). Effects of fetal treatment of methylazoxymethanol acetate at various gestational dates on the neurochemistry of the adult neocortex of the rat. *Journal of Neurochemistry, 36,* 124–128.

Johnston, M. V., & Coyle, J. T. (1979). Histological and neurochemical effects of fetal treatment with methylazoxymethanol on rat neocortex in adulthood. *Brain Research, 170,* 135–155.

Johnston, M. V., & Coyle, J. T. (1980). Ontogeny of neurochemical markers for noradrenergic GABAergic and cholinergic neurons in neocortex lesioned with MAM. *Journal of Neurochemistry, 34*, 1429–1441.

Johnston, M. V., & Coyle, J. T. (1982). Cytotoxic lesions and the development of transmitter systems. *Trends in Neuroscience, 5*, 153–156.

Johnston, M. V., Grzanna, R., & Coyle, J. T. (1979). Abnormally dense noradrenergic innervation of rat neocortex follows fetal treatment with methylazoxymethanol. *Science, 203*, 369–371.

Jonsson, G., & Hallman, H. (1982). Effects of prenatal methylazoxymethanol treatment on the development of monoamine neurons. *Developmental Brain Research, 2*, 513–530.

Kiyono, S., Seo, M., & Shibagaki M. (1980). Sleep-waking cycle in microencephalic rats induced by prenatal methylazoxymethanol application. *Electroencephalography and Clinical Neurophysiology, 48*, 73–79.

Lauder, J. M., & Bloom, F. E. (1974). Ontogeny of monoamine neurons in the locus coeruleus, raphe nucleus and substantia nigra of the rat. I. Cell differentiation. *Journal of Comparative Neurology, 155*, 469–481.

Matsumoto, H., Spatz, M., & Laquer, G. L. (1972). Quantitative changes with age in the DNA content of methylazoxymethanol-induced microencephalic rat brain. *Journal of Neurochemistry, 19*, 297–306.

Matsumoto, H. & Strong, F. M. (1963). The occurrence of methylazoxymethanol in *Cycas ciccinalis* L. *Archives of Biochemistry and Biophysics, 101*, 299–310.

Matsutani, T. Nagayoshi, M., Tamura, M., & Tsukada, Y. (1980). Elevated monoamine levels in the cerebral hemispheres of microencephalic rats treated prenatally with methylazoxymethanol or cytosine arabinoside. *Journal of Neurochemistry, 34*, 950–956.

Moran, T. H. (1986). Environmental and neural determinants of behavior in development. In E. M. Blass (Ed.), *Handbook of behavioral neurobiology: Developmental psychobiology*, pp. 99–128. New York. Plenum.

Moran, T. H., Sanberg, P. R., Antuono, P. G., & Coyle, J. T. (1986). Methylazoxymethanol acetate (MAM) cortical hypoplasia alters the pattern of stimulation induced behavior in neonatal rats. *Developmental Brain Research, 27*, 235–242.

Moran, T. H., Schwartz, G. J., & Blass, E. M. (1983a). Organized behavioral responses elicited by lateral hypothalamic electrical stimulation in neonatal rats. *Journal of Neuroscience, 3*, 10–19.

Moran, T. H., Schwartz, G. J., & Blass, E. M. (1983b). Stimulation induced ingestion in neonatal rats. *Developmental Brain Research, 7*, 197–204.

Nagata, Y., & Matsumoto H. (1969). Studies on methylazoxymethanol: Methylation of nucleic acids in fetal rat brain, *Proceedings of the Society for Experimental Biology and Medicine, 132*, 383–385.

Pedersen, P. E., & Blass, E. M. (1982). Prenatal and postnatal determinants of the first suckling episode in albino rats. *Developmental Psychobiology, 15*, 349–355.

Pedersen, P. E., Williams, C. L., & Blass, E. M. (1982). Classical conditioning of suckling behavior in three day old rats. *Journal of Experimental Psychology, 8*, 329–341.

Plonsky, M., Riley, E. P., Lee, M. H., & Haddad, R. K. (1985). The effects of prenatal methylazoxymethanol acetate (MAM) on holeboard exploration and shuttle box performance in rats. *Neurobehavior, Toxicology and Teratology, 7*, 221–226.

Rabe, A., & Haddad, R. K. (1972). Methylazoxymethanol-induced microencephaly in rats: Behavioral studies. *Federation Proceedings, 31*, 1536–1539.

Ribak, C. E., (1978). Aspinous and sparsely spinous stellate neurons in the visual cortex of rats contain glutamic acid decarboxylase. *Neurocytology, 7*, 461–478.

Roberts, W. W. (1980). C14 Deoxyglucose mapping of first order projections activated by stimulation of lateral hypothalamic sites eliciting gnawing, eating and drinking in rats. *Journal of Comparative Neurology, 194*, 617–638.

Rose, G. H. (1968). Development of visually evoked electrocortical response in the rat. *Developmental Psychobiology, 1*, 35–40.

Sanberg, P. R., Johnson, D. A., Moran, T. H., & Coyle, J. T. (1984). Investigating locomotion abilities in animal models of extrapyramidal disorders: A commentary. *Physiological Psychology 12*, 48–50.

Sanberg, P. R., Moran, T. H., & Coyle, J. T. (1987). Microencephaly: Cortical hypoplasia induced by methylazoxymethanol. In J. T. Coyle (Ed.), *Animal models of dementia* (p. 253). New York: Alan R. Liss.

Sanberg, P. R., Moran, T. H., Kubos, K. L., Antuono, P. G., & Coyle, J. T. (1983). Prenatal methylazoxymethanol (MAM) treatment produces spontaneous hyperactivity in neonatal and adult rats. *Society of Neuroscience Abstracts, 9*, 938.

Singh, S. G. (1977). Ectopic neurons in the hoppocampus of the postal rat exposed to methylazoxymethanol during fetal development. *Acta Neuropathologica (Berlin), 40*, 111–116.

Singh, S. C. (1978). Redirected perforant and commisural connections of eutopic and ectopic neurons in the hippocampus of methylazoxymethanol acetate-treated rats. *Acta Neuropathologica (Berlin), 44*, 197–202.

Singh, S. C. (1980). Deformed dendrites and reduced spone number on ectopic neurons in the hippocampus of rats exposed to methylazoxymethanol acetate. A Golgi-Cox study. *Acta Neuropathologica (Berlin), 49*, 193–198.

Slevin, J. T., Johnston, M. V., Biziers, K., & Coyle, J. T. (1982). Methylazoxymethanol acetate ablation of mouse cerebellar granule cells: Effects on synaptic neurochemistry. *Developmental Neuroscience, 5*, 3–12.

Spatz, M., & Laquer, G. L. (1968). Transplacental chemical induction of microencephaly in two strains of rats. *Proceedings of the Society for Experimental Biology and Medicine, 129*, 705–710.

Storm-Mathiesen, J. (1976). Distribution of the components of the GABA system in neuronal tissue: Cerebellum and hippocampus effects of axotomy. In T. Roberts, T. Chast, & D. B. Tower (Eds.), *GABA in nervous system function* (pp149–168). New York: Raven Press.

Stricker, E. M., & Zigmond, M. J. (1976). Recovery of functions after damage to central catecholamine containing neurons: A neurochemical model for the lateral hypothalamic syndrome. In J. M. Sprague & A. N. Epstein (Eds.), *Progress in psychobiology and physiological psychology*, pp. 121–188. New York: Academic Press.

Vorhees, C. V., Fernandez, K., Dumas, R. M., & Haddad, R. K. (1984). Pervasive hyperactivity and long-term learning impairments in rats with induced microencephaly from prenatal exposure to methylazoxymethanol. *Developmental Brain Research, 15*, 1–10.

Weinberg, J., Haddad, R., & Dumas, R. (1983). Acquisition and extinction of a conditioned taste aversion in microencephalic rats. *Society of Neuroscience Abstracts, 9*, 637.

Zedek, M. S., Sternberg, S. S., McGowan, J., & Poynter, R. W. (1972). Methylazoxymethanol acetate: Induction of tumors and early effects on RNA synthesis. *Federation Proceedings, 31*, 1485–1492.

19

Neuropharmaco-ontogeny: Concepts and Methods of Study

GORDON A. BARR

The types of experiments that are conducted by any individual reflect personal scientific biases. Thus, I will begin this chapter by describing briefly my biases about brain and behavior relationships. My work in developmental neuropsychobiology has been guided by the fact that the complexity of the brain is due to its anatomical and neurochemical diversity and the interrelatedness of its structure. In the past 30 or so years, largely because of the advances in neuroanatomical methods used to map neurotransmitter systems, it has become apparent that neural tracts which contain the same neurotransmitter may serve very different functions. This is perhaps best worked out for the monoamine neurotransmitter, dopamine. Anatomically, dopamine pathways can be divided into five separate systems (Cooper, Bloom, & Roth, 1986). Each has a distinct function. For example, the ascending nigrostriatal pathway is involved in motor control (Ungerstedt, 1974), the arcuate-hypophysis system in regulation of anterior pituitary control (Lindvall & Björklund, 1974), and the mesencephalic limbic dopamine system in locomotor and motivational processes (Moore & Kelly, 1978). Conversely, single tracts and even single neurons contain multiple transmitters (Hökfelt, Johansson, & Goldstein, 1984; Hökfelt et al., 1986). To understand accurately the behavioral function of distinct neurochemical systems, we require methods that assess the functional specificity of neuroanatomically and neurochemically diverse systems.

The techniques I describe in this chapter are designed to tap into specific neurochemical pathways to determine the function of these systems. These methods include chemical and electrical stimulation of specific brain targets and a technique for intrathecal administration of drugs. There are technical and conceptual limits to these procedures, as there are with any methods. But the methods described offer distinct advantages over techniques that do not take into account the differential function of anatomically distinct neurotransmitter systems. Peripheral or intraventricular administration of drugs does have its place. The types of information that can be gained, however, about the chemical coding of behavior by use of those techniques alone is limited. These ideas are certainly not original with me. The investigation of electrical and chemical stimulation of specific brain loci is almost a century old (Hashimoto, 1915; Myers, 1974; Spencer, 1894). Its use in the study of brain behavior relationships in adults is fully appreciated. But the use of these techniques in developmental work is far more recent and certainly less common. It is hoped that this chapter will encourage more of these types of studies.

In this chapter I discuss, first, the reasons I believe the study of development is critical in understanding central nervous system behavior relationships. Then I describe general steretaxic procedures for infant rats, including implantation of electrodes and cannulas, chemical and electrical stimulation of the brain, and histological verification of the implant. Next I describe methods for the intraspinal administration of drugs in developing rats. Then I detail methods to implant and perfuse the brain and spinal cord for analysis of the release of neurotransmitters. Each method was developed out of a need to answer a specific question about the neuropharmacological function of the central nervous system during development. The last section is a discussion of the methodological and conceptual limitations of these procedures and of new questions these methods have raised.

WHY DEVELOPMENTAL STUDIES?

Developmental History of Behavior

One strategy used to understand the neural bases of any behavior is to study the developmental history of the behavior. Determining the sequential organization during maturation of physiological and behavioral processes provides clues about the neural organization of those functions. The appearance of neural circuits may correspond to specified behavioral functions, or neurobehavioral processes that are intertwined in adults may have separate developmental histories. Of course, the behavior of the infant may be different from that of the adult or may not be fully developed. This developmental strategy is illustrated in several chapters in this book. The reader is directed to the work of Phifer and Williams. The work of Hall and his collaborators (reviewed recently in Hall, 1985; see also Phifer, chap. 11 in this volume) provides an exceptionally rich source for this approach.

One example of this method has been the recent studies in our laboratory by Capuano, who examined the development of noradrenergic and peptidergic regulation of independent feeding by the paraventricular and perifornical nuclei of the hypothalamus (Capuano, Barr, & Leibowitz, 1986a, 1986b, 1986c; Capuano, 1987). The question we asked was, What is the role that the paraventricular nucleus of the hypothalamus (PVN) plays in independent feeding during development, and how does that changing function relate to what is known about the physiological control of feeding during maturation and to its function in the adult? We have shown that some systems, such as the noradrenergic and neuropeptide Y (NPY) systems, were mature by 3 days of age (the earliest tested), while others, such as cholecystokinin, did not function until 14 days of age. Thus, the discrete subsystems regulating feeding develop differentially. This points to multiple and possibly separate regulation of ingestion by these processes.

Studies of Behaviors Unique to Development

A second and equally important reason for studying the neural bases of behavior in development is to understand behaviors and processes that are unique to young animals. The developing animal has a specific ecological and physiological niche and provides important examples of the adaptive strategies the young use to thrive in this unique setting. It also allows us to study how learning and the environment modify these early behaviors. Perhaps the best studied behavior is sucking for milk, which apparently is under different physiological control than is adult or infant feeding (Blake & Henning, 1986; Capuano, 1987; Ellis, Axt, & Epstein, 1983; Hall, 1985; Rodriguez-Aendejas, Chambert, Lora-Vilchis, Epstein, & Russek, 1986; see also Brake, chap. 3, this volume). The study of developmentally distinctive behaviors is discussed in this volume by Shair, (chap. 4), Wilson, Sullivan and Leon (chap. 17), and Stehouwer (chap. 8), but to my knowledge no studies have focused on the specific central neurotransmitters of any developmentally specific behaviors.

Issue of Neural Plasticity

A large literature suggests that the central nervous system of the developing animal responds differently to damage than does that of the adult animal (for examples, see Barr, Eckenrode, & Murray, 1987; Bregman & Goldberger, 1982, 1983a, 1983b, 1983c; D'Amato & Hicks, 1978; Goldman, 1974; Hebb, 1949; Weber & Stelzner, 1977. Depending on such factors as the type and extent of damage, age of the animal, and other variables (see Goldberger & Murray, 1985, for a detailed discussion), the immature nervous system can be more vulnerable to damage than that of the adult animal (Bleier, 1969; Risling, Culhein, & Hieldebrand, 1983). This increased vulnerability has been attributed to a variety of developmental differences, including lack of sustaining collaterals, less available trophic factors, and lower metabolic capabilities (see Jacobson, 1978). On the other hand, the central nervous system of the young animal can demonstrate greater neuroplasticity in response to injury. The young animal demonstrates an increased ability to recover functionally from physiological insult. This has been defined as the ''infant lesion effect'' and is seen in the young of a variety of species (Bregman & Goldberger, 1983b; Goldman, 1974; Schneider, 1970; Stelzner, Ershler, & Weber, 1975; Weber & Stelzner, 1977), including humans (Witelson, 1987). The possible reasons for this ability to recover include the ability to project axons around the damage to innervate appropriate targets and the ability to project to normally inappropriate targets, to decreased retraction of excessive (exuberant) projections, or to less glial scarring (for reviews, see Bregman, 1986; Goldberger, 1986; Goldberger & Murray, 1985; Stelzner, Weber, & BrzGornia, 1986). Many neurotransmitter systems demonstrate this plasticity, but the opioid system of the preweanling rat is an example of a neural system that is uniquely responsive to insult at early ages. Treatment with opiates during development results in permanent changes in morphological, behavioral, and biochemical measures (reviewed in McDowell & Kitchen, 1987). Because much of this plasticity is greater and different in the infant, it is important to understand the differences in the regulatory rules and processes between the mature and immature nervous systems that permit the increased plasticity in infancy.

A recent series of experiments focused on developmental differences in up and down regulation of opioid receptors following pre- and postnatal treatment with opiates and opiate antagonists. The question asked was whether the opioid receptors were more sensitive early in life, when that increased sensitivity declined to adult levels, and what possible mechanisms could account for it. We found that prenatal treatment with naloxone produced a 115% increase, and postnatal treatment produced a 78% increase in μ receptor ([³H] D-Ala², N-Me-Phe⁴, Gly-ol⁵-enkephalin, DAMGO) density in preweanling rats. This finding is comparable to the findings in adult animals. Unlike adult animals, however, pre- or early postnatal morphine resulted in a 30–35% decrease in μ receptor density. There were no effects on receptor affinities nor on ∂ or k opioid receptors (Tempel, Habis, Paredes, & Barr, 1988). The ability of morphine to down regulate μ opioid receptors was limited to the first two week of life, and after that time morphine did not result in receptor down regulation. Down regulation of opioid receptors in infants contrasts to results in adults that fail to demonstrate any change in opioid receptor binding following chronic agonist treatment (Brunello, Volterra, DiGiulio, Cuomo, & Racagni, 1984; Hitzemann, Hitzemann, & Loh, 1974; Klee & Streaty, 1974; Perry, Rosenbaum, & Sadee, 1982). The important question is how the infant central nervous system is more easily regulated by changes in the internal and external environment of the organism. Comparisons of plasticity in the adult and the infant may provide information on the processes that enhance or inhibit recovery of function following injury.

Development as a Model System

Another reason for studying developing animals is that the immature organism provides a model system for analyzing the relationships between neurotransmitters and behavior. The use of maturing animals can provide information that cannot be easily obtained in adult animals. The developing animal is not simple; very likely it is not even simpler than the adult. But as the animal matures in its behavior or in the response to drugs, those developmental changes can be related to neural development. For instance, an important technical advance in neuropharmacology has been the development of neurotoxins that can be used to deplete specific neurotransmitters (e.g., 5,7-DHT; 6-OHDA, DSP-4). These neurotransmitters are often lacking or not fully mature in the developing animal and therefore provide a naturally occurring "depletion." Thus, this natural depletion can be taken advantage of rather than using more artificial systems. Furthermore, development of even a single neurotransmitter system is not homogeneous. Different tracts may develop with different time courses (e.g., nigrostriatal vs. mesocortical dopamine systems; Coyle & Campochiaro, 1976; Loren, Björklund, & Lindvall, 1976; Schmidt, Björklund & Lindvall, 1982); different targets may be innnervated at different times (e.g., 5-HT projections to the dorsal cervical vs. lumbar spinal cord; Bregman, 1987); and different subunits of a single system may mature at different ages (e.g., κ, μ and δ opioid receptors, or pre- vs. postsynaptic processes; Barr, Paredes, Erickson, & Zukin, 1986; Loughlin, Massamirï, Kornblum, & Leslie, 1985; McDowell & Kitchen, 1987; Pasternak, Zhang, & Tecott, 1980; Petrillo, Tavani, Verolla, Robson, & Kosterlitz, 1985; 1987; Spain, Roth, & Coscia, 1985; Tavani, Robson, Zhang & Pasternak, 1981).

Sometimes differences in the function of a neural system that are not apparent in adults appear unexpectedly in infants. Such differences, which may be attributable to the differential maturity of some neural systems compared with others, can provide opportunities to study the organization of neural circuits in a manner not possible in the adult. For instance, recent work on the changing role of the periaqueductal gray (PAG) in development of opiate-induced analgesia may provide insights on the function of the nucleus in analgesia because it appears to work quite differently in infants than in older animals. In the adult, stimulation of the ventral aspects of this nucleus electrically or with glutamate or opiates produces profound analgesia. The analgesia produced by electrical or glutamate stimulation is naloxone reversible and cross tolerant with morphine and is clearly an opiate-mediated effect. But in the 3–10-day-old pup, stimulation of the PAG by morphine produces profound analgesia while glutamate is ineffective (Tive & Barr, 1988). It is likely, therefore, that analgesia produced by morphine and glutamate is mediated by different populations of cells with different development courses.

THE METHODS

In the late 1970s, we became interested in ways to measure changes in forebrain dopamine function in developing animals to assess whether behaviors thought to be mediated by these circuits in adults were similarly coded early in development. We specifically wanted to be able to (1) assess damage to the mesolimbic and mesostriatal dopamine systems in animals that had been exposed to various pharmacotherapeutic drugs in utero and (2) study the development of these neural systems in relation to the natural reinforcement processes. Because we wanted to assess specific projection systems, we needed to be able to stimulate discrete tracts. We decided to use intracranial self-stimulation, which in adults is quite sensitive to manipulations of dopamine systems (see Wise & Bozarth, 1984). While self-stimulation had been demonstrated in both infant dogs (Bacon & Wong, 1972) and chickens (Andrews, 1967), the methods for implanting electrodes in rat pups had not been described. We developed crude but simple methods to implant in-dwelling electrodes or cannulas into very young pups (Lithgow & Barr, 1982). The following sections describe the procedures we are currently using, which have been modified from our earlier techniques and from the methods of a number of other investigators. The general procedures are applicable to a variety of stereotaxic surgical methods in the infant.

Stereotaxic Surgery

This section describes the procedures for implanting any type of small electrode or cannula into the brain of infant rats. These techniques are the minimum necessary to begin to be able to localize the function of specific neural pathways during development. Beginning on the day of implant, pups are weighed and marked with indelible ink. A simple but permanent tattooing method is described by Geller and Geller (1966). Surgical implantation of cannulas and electrodes is standard in our laboratory and has been described previously (Barr & Lithgow, 1986; Capuano, 1987; Lithgow & Barr, 1982, 1984). Recent modifications from the methods described in our earlier work (Barr & Lithgow, 1986; Lithgow & Barr, 1982, 1984), particularly the use of the infant stereotaxic (Heller, Hutchens, Kirby, Karapas, & Fernandez; 1979) rather than our original alginate mold, enable us to implant electrodes accurately and reproducibly in small neural structures (e.g., the dorsal periaqueductal gray and paraventricular nucleus of the hypothalamus). Briefly, pups are anesthetized with methoxyfluorane (Pitman-Moore, Inc.) and a midsagittal incision made in the scalp. Because the skull is not ossified, care must be taken to use gentle pressure to avoid cutting through the skull. The underlying fascia are scraped away, and visualization of the junctions of the midsagittal suture with bregma and lambda are enhanced by application of a small drop of green food dye. Two or three small holes that do not completely penetrate the skull are drilled into the cartilaginous skull, and stainless steel screws are inserted and anchored with cyanoacrylate cement (for a discussion of the vagaries of this type of glue, see Hand, 1989). (Despite my firm instructions to the contrary, at least three of my graduate students—Chris Capuano, Nina Goodless, and Leslie Tive—have insisted on using shortcuts and not implanting skull screws. I do not recommend this, although I must admit that their implants appear to be as well anchored to the skull as mine. Likely, a small, lightweight device that is not subjected to repeated torque will not require the skull screws.) The pup is then mounted in a stereotaxic instrument with lambda and bregma in the same horizontal plane. Using bregma as a reference, a small hole is drilled through the skull with care taken not to penetrate the dura. The dorsal-ventral reference is taken from the dura, which is then punctured with a sharp point. The guide cannula or bipolar electrode constructed of Teflon-coated wire (75 μm, A-M Systems, Inc.) connected to Amphenol connectors is lowered to the implant site (or to 1 mm above the implant site for the cannula). The implant is anchored to the skull and to the embedded screws by a thin layer of Caulk-Grip resin cement followed by a thicker layer of acrylic cement. The use of resin is crucial. Acrylic cement invariably separates from the skull, presumably because of its inflexibility (Lithgow & Barr, 1982). The surgical procedure takes less than 30 min, and we lose fewer than 1% of our subjects to all causes (death, loss of implant, etc.). The implants remain patent for 2–3 days even with growth of the skull, although testing is done sooner. (It may not be possible to test very young animals much later than 18–24 h because the rapid growth of the brain might result in movement of the implant relative to the site of stimulation.) Accuracy in placement for an experienced surgeon for animals of any age with which we have had experience is greater than 95% and is nearly comparable to results with adults. Pups are kept in a warm, moist (34 °C) incubator until testing or until they can be returned to the mother. If pups are kept in the incubator, they can be fed milk, using any of the independent feeding paradigms (see Phifer, chap. 13, this volume). Pups do not gain weight as rapidly as they normally would, but they gain more weight than nonfed pups. We have had no difficulty with the dam accepting pups if she is used to being handled and if the pups are fully mobile when returned to her. Typically, she finds the implant a convenient handle when carrying pups.

Determining Coordinates

At present, there are only three atlases for the infant rat brain. Two are published as journal articles and thus are limited in scope (Heller et al., 1979; Valenstein, Case, & Valenstein, 1969). The Heller et al. atlas is included with the description of the stereotaxic device that has been helpful in our work. It is the only detailed atlas for 3-day-old pups, but suffers from a limited number of sections and a lack of anatomic detail within sections. The Valenstein et al. paper is limited to the hypothalamus, but does provide coordinates for 1-, 7-, and 14-day-old pups. It contains no photomicrographs. However, it gives a fairly accurate correction scheme for calculating coordinates for in-between ages and regions. The only full atlas is that of Sherwood and Timiras (1970), which contains a fairly complete series of transverse sections rostral to the midbrain for 10-, 21-, and 33-day-old rats. While it is the most complete atlas, it lacks more caudal structures and plates of horizontal or sagittal sections. Furthermore, for our work the failure to include ages younger than 10 days is an important limitation. In 1970, such a need was probably less urgent than it is today. To determine coordinates, we typically start with the atlas that most closely matches the site that we wish to "hit," do several implants, evaluate the histology, and then readjust the coordinates. Usually within two to three iterations we can accurately determine the coordinates.

Perfusions

Perfusions are performed identically to those for the adult animal except for the necessary modifications required due to the size of the pup. After the pup is administered an overdose of an anesthetic, the thoracic cavity is opened with rounded scissors, the surgeon taking care not to puncture the heart. To help prevent heart puncture, a small wooden dowel is placed under the pup to arch the chest upward. For standard Nissl stains we perfuse with saline followed by buffered formalin; for other histological procedures the method may differ. A 27-gauge butterfly needle is used to puncture the left ventricle. Care must be taken not to insert the needle into the right ventricle, especially in very young pups. Flow of the perfusate is controlled by a peristaltic pump (Cole Palmer Instrument Co.). For 3-day-old pups we perfuse at a flow rate of 5–8 ml/min, although this can vary for certain procedures. Removal of the brain or spinal cord, embedding the tissue, and sectioning are done as for the adult.

Histological Verification

Histological verification of the site of implant (or lesion) is essential, but provides unique difficulties compared with the adult. Frozen sectioning is technically more difficult in the infant than in the adult because the tissue is more fragile. Also, because of a relative lack of myelin the tissue adheres less well to slides during staining, especially if the tissue is unfixed. Mounting tissue on well-subbed slides and adequate drying of the sections on the sides are essential. The major difficulty we have faced is that our implants are in young pups and are relatively shortterm. Both factors conspire to make verification of the sites of implant difficult because of the lack of gliosis around the site. Following behavioral tests, we verify the cannula placements by injecting the same volume of india ink as drug through the cannula. The verification of the electrode placements provides a more difficult problem. By careful histological examination of Nissl-stained frozen sections (cut every 18–30 μm) accurate identification is possible. A more precise method is to section paraffin-embedded tissue (5–10 μm). Methods for paraffin embedding and sectioning in infants are similar to those used in adults. Small restricted lesions are the most difficult, but again, careful sectioning, staining each section, and through serial examination under the microscope permit exact localization of lesions.

We present our histology on the drawings from the atlas of Sherwood and Timiras (1970). A difficulty arises if the ages of the subjects do not match those in the atlas. There are discrepancies in the relative sizes and locations of certain structures when the ages do not match. These inaccuracies for the most part are relatively minor, and the advantages provided by the completeness of the atlas outweighs its limitations. However, if large discrepancies are found, the atlas sections should be modified or new sections drawn to reflect the histology.

Stimulation Methods—Brain

Electrical Stimulation

We fashion bipolar electrodes from 75-μm insulated tungsten wire connected to Amphenol connectors (Medwire Corporation). We loop the wire and clamp the cut ends in a vise. The loop is held by round nose pliers and twisted. Both the vise and pliers are taped to minimize the possibility of damage to the insulation. The looped end is then cut. Each twisted wire is inspected visually under a low-power microscope for breaks in the insulation and tested with an ohmmeter before being inserted into gold Amphenol pins. The pins, in turn, are snapped into a female Amphenol connector. Excess plastic from the plug is trimmed with a knife to minimize the weight of the electrode. Alternatively, commercially available electrodes of 125-μm wire may be used (Plastic One, Inc.). Sample electrodes with connectors are shown in Figure 19.1.

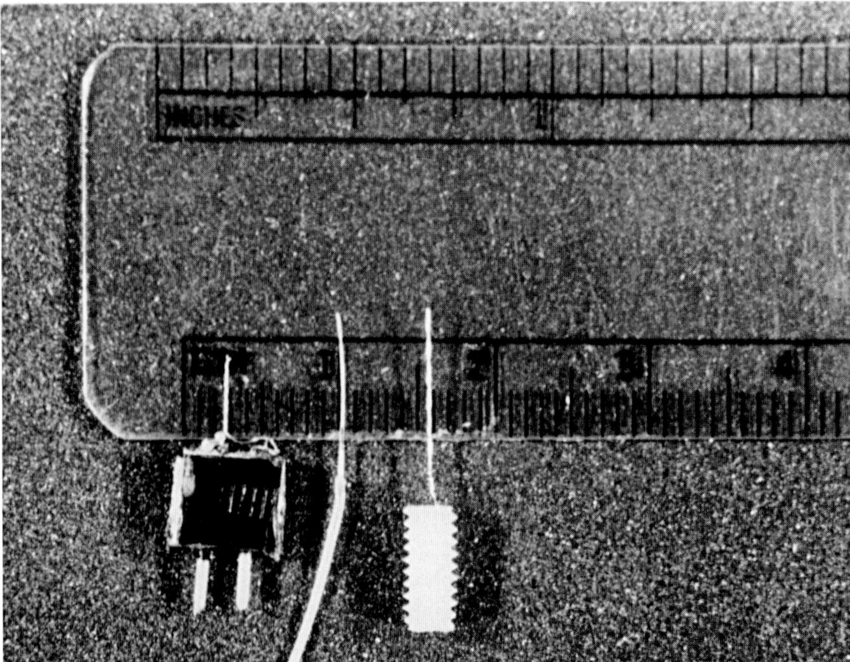

Figure 19.1. The electrodes used for electrical stimulation of the infant rat brain and the catheter for the intrathecal injections. On the left is the electrode that we construct and on the right is the commercially available electrode (Plastics One). We use both, depending on the application. In the center is the intrathecal catheter constructed of dialysis tubing (at the top) and either silastic or PE-10 tubing (at the bottom).

The implanted electrode is connected to a constant current stimulator. The choice of parameters depends on many factors, but in behavioral stimulation studies the parameters appear to be similar to those in the adult (Barr & Lithgow, 1986; Goodless, 1988; Lithgow & Barr, 1984; Moran, Lew, & Blass, 1981; Viscardi, 1988). Ranck (1975) has provided a thorough and critical evaluation of electrical stimulation of the brain in adults. Issues specifically related to infants have been recently reviewed (Moshe, 1987; Prince & Kriegstein, 1989; Schwartzkroin, 1984). Clearly, the electrophysiological properties of the immature brain, including membrane properties, synaptic properties, and response to electrical stimulation, are quite different from those in the adult brain. Compared with that of the adult, the infant brain is characterized by high membrane input resistance, longer action potentials, and depolarizing potentials with absent or late hyperpolarizing potentials. These properties are characteristic of increased excitability of the immature nervous system. Therefore, negative results must be viewed with caution unless careful parametric studies are done. Failure to obtain effects may be due to the wrong choice of stimulation parameters.

Chemical Stimulation

One question that has been extensively studied is how the physiological controls of adultlike independent ingestion develop (see Brake, chap. 3, and Phifer, chap. 11, this volume, and Hall, 1985). Because of the substantial data on the role of the hypothalamus in feeding, we chose to study the paraventricular nucleus and the perifornical area of the hypothalamus, and we have developed for the infant modifications of existing chemical stimulation methods used in adults.

The injector cannula (Plastics One) is connected to PE-10 tubing (Clay Adams), which in turn is connected to a 1.0-μl syringe. Drugs are dissolved in sterile saline, and the injector cannula, tubing, and syringe are filled. A small air bubble is left in the tubing so that during the injection its course can be used to calibrate the volume of fluid injected (Myers, 1974). The stylet is gently removed from the implanted guide cannula and the injector cannula inserted. The drug is injected by hand or syringe pump over a 60-s period and the injector cannula left in for an additional 2 min. The injector is removed and the stylet replaced. It is not possible to determine exactly the spread of the injected fluid in the brain. Variables such as the volume injected, site of injection, characteristics of the drug (e.g., lipid solubility, more extracellular space) are important in the adult (Myers, 1974). In the infant, however, different characteristics of the neural microenvironment (e.g., fewer lipids, differences in neural density) are additional factors. We attempt to minimize spread as much as possible by injecting very small volumes of fluid (0.05–0.1 μl) through small cannulas (see Figure 19.2, which shows a sample section with ink localized to the interpeduncular nucleus in a 3-day-old rat pup). In addition, placements in the adjacent ventricles or surrounding tissue are required controls. In our experience, young pups require an order of magnitude higher dosages of drugs for any effect than do adults, possibly because they have fewer receptor sites (Capuano, 1987; Tive, 1988). We choose doses that range from the lower limits effective in adults to approximately 10 times the maximum effective dose used in adults. However, in all experiments the use of a wide dose range is required to ensure a full dose response curve that includes both ineffective doses and doses that produce maximum effects.

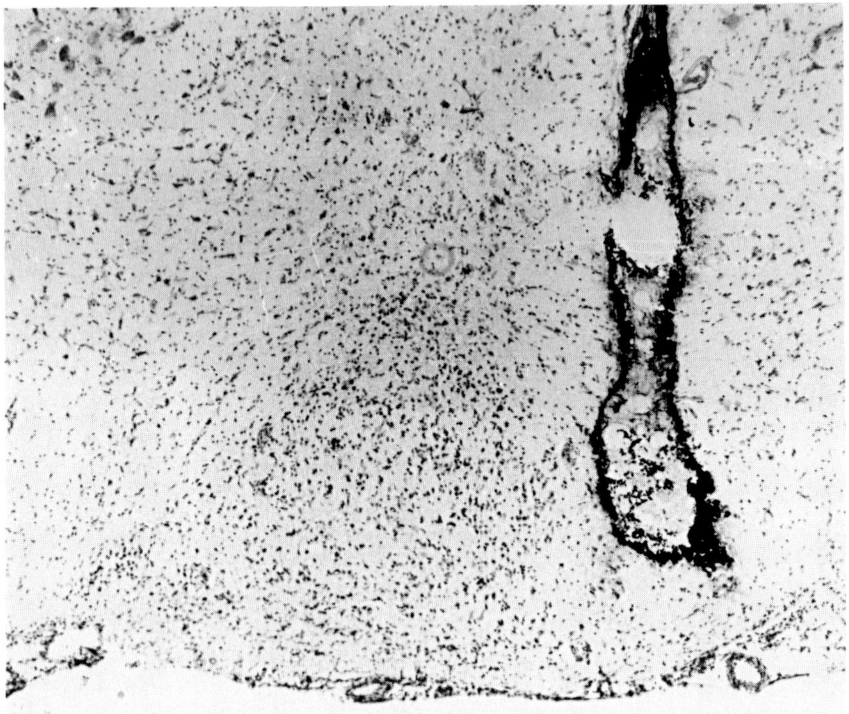

Figure 19.2. This figure depicts 0.10μL of India ink injected into the interpeduncular nucleus of the 3-day-old rat pup. Note the limited spread of the ink laterally or ventrally, but the relatively large spread dorsally along the cannula shaft.

Chemical Stimulation—Spinal Cord

Pain mechanisms are much less well understood in the infant than in the adult, although clearly they are different (e.g., Bronstein, Mitteldorf, Sadeqhi, Kirby, & Lytle, 1986; Hamm & Knisely, 1984, 1987a, 1987b). The development of a technique to inject drugs into the intrathecal space of adult animals was a major advance in understanding the contribution of spinal cord pharmacology for pain, gastrointestinal transit, and other physiological processes (Porreca, Mosberg, Hurst, Hruby, & Burk, 1984; Yaksh, 1986). The administration of drugs directly into the spinal cord was first described by Yaksh and Rudy (1976). Modifications have been published for the mouse (Fu & Dewey, 1981). In our hands both methods produced significant damage to the spinal cord when used on young rats. Thus, we developed methods that allow us to inject small quantities of drug into the spinal cerebrospinal fluid (CSF) with minimal trauma to the spinal cord (Hughes & Barr, 1988). We have applied this method to the study of spinal opioid and monoamine receptors that produce analgesia (Barr, Miya, & Paredes, 1987; Hughes, 1988; Hughes & Barr, 1988). Potentially, there are a broad number of applications of this method for studying the development of other spinal reflexes as well as the spinal autonomic system function.

Catheter Construction and Implantation

The catheter is a flushed 1-cm length of dialysis tubing (5,000 molecular weight cutoff; MWCO, Spectrum Medical) inserted into a 4-cm length of Dow Corning Silastic medical-grade tubing or PE-10 tubing (Figure 19.1). The dialysis tubing is fastened to the Silastic by cyanoacrylate applied at the interface of the two tubes. Prior to implantation, the catheter is filled with saline to check for possible leaks at the interface and to provide the dialysis tubing with a measure of firmness. The entire catheter holds 3 μl of fluid. The dialysis end of the catheter is dipped into blue methylene dye to enhance visibility prior to implant. Subjects are anesthetized with methoxyfluorane, and the skin is incised just above the thoracic and lumbar vertebrae. The underlying muscle and fascia are gently scraped away, exposing vertebrae T8 to T11. A laminectomy is performed, with care taken not to injure the underlying dura. Removal of the dorsal aspects of three vertebrae exposes about 0.5 cm of spinal cord in the 2–3 day-old pup. The dura is punctured slightly off midline by moving the tip of a curved 30-gauge hypodermic needle caudally parallel to the spinal axis until the tip pierces the dura. Approximately 0.25 cm of the needle tip is inserted into the subarachnoid space with a gentle upward pressure to prevent damage to neural tissue. Clear spinal fluid leaks immediately after removal of the needle. The catheter is held with forceps at the dialysis/Silastic interface, and the dialysis tubing is completely inserted caudally into the subdural space through the hole. The dura apparently seals around the tubing because the CSF stops leaking. Slow and steady advancement of the catheter prevents crimping of the dialysis tubing. The angle of the catheter is parallel to the dorsolateral surface of the cord, and the inserted catheter tip lies visibly above the cord. The Silastic portion of the catheter is anchored by one drop of cyanoacrylate to the first whole vertebra rostral to the laminectomized area. The glue is permitted to dry before the incision is closed to prevent the catheter from adhering to the overlying tissue. The free end of the catheter remains outside the animal to permit microinjections of drug directly to the surface of the spinal cord. The entire implantation procedure requires about 15 min per animal. After surgery, subjects are incubated at 32–34 °C and housed individually to prevent dislodgement of the catheter by sibling contact. While we have not replaced the pup with the dam, it is likely possible with minor modifications. For example, a longer length of Silastic tubing could be run subcutaneously and connected to a Plastics One pedestal anchored to the skull. A diagram schematicizing the implant is shown in Figure 19.3.

There is remarkably little damage to the spinal cord with this procedure. A coronal section of the implant in a 3-day-old is shown in Figure 19.4. Little if any deficit or change in motor or sensory function is detectable. Some animals do show decreased tactile sensitivity ipsilateral to the implant, due to disruption of the dorsal roots. We have never detected any change contralateral to the catheter.

Injection

Twenty-four hours later, pups can be injected with soluble compounds in 2–4 μl of saline. Animals rarely but occasionally display hyporeflexia, clonic flexion of the hindlimbs, tremors, or rigidity, and those that do are excluded from the experiments. To inject, a 10-μl microsyringe is filled with 5–7 μl of drug solution and inserted 0.25 cm into the exposed end of the catheter. Three μl of solution are injected into the

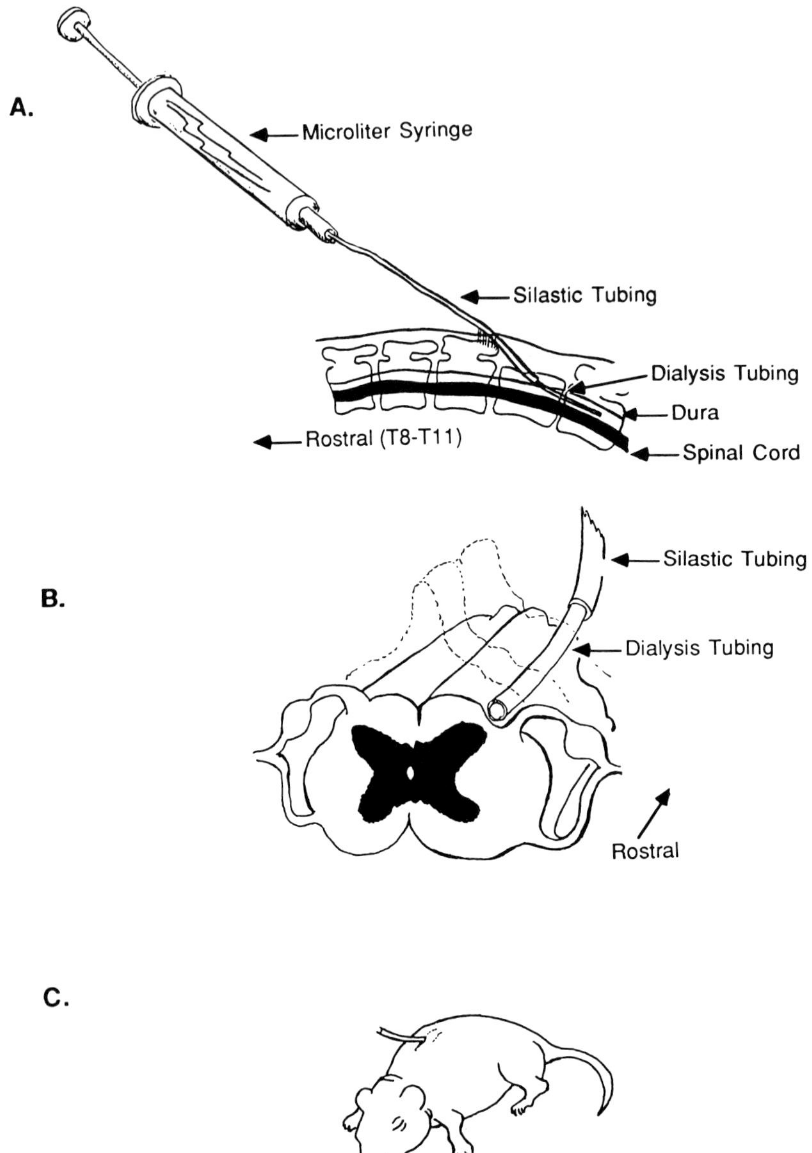

Figute 19.3. A schematized depiction of the intrathecal catheter implanting procedure. (A) is a parasagittal section which shows the tubing inserted under the dura and exiting from the dorsal aspect of the spinal column. Note the laminectomy where the dialysis tubing enters the dura and the attachment of the PE-10 tubing to the next most rostral vertebra. Only the dialysis tubing enters the dura. (B) is a transverse section of the spinal cord demonstrating the location of the dialysis tubing on the lateral dorsal surface of the spinal cord. (C) is a cartoon demonstrating the rat pup with the tube exiting the skin above the implant. There are no noticeable motor or sensory deficits after surgery and recovery from anesthesia and only occasional changes in latency to respond to noxious or non-noxious stimuli on formal testing. These sensory deficits, if they occur, are always restricted to the side of implant and are likely due to disruption of the dorsal roots and dorsal root ganglia.

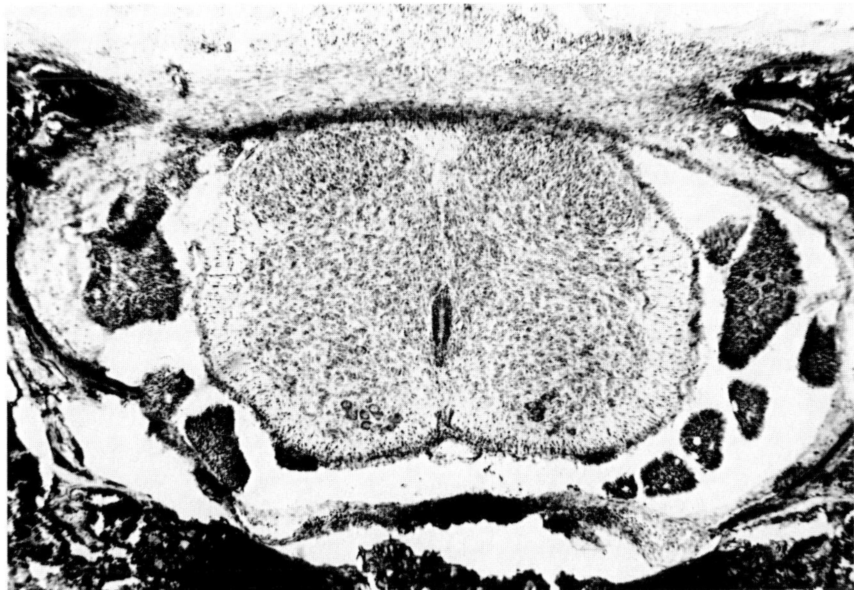

Figure 19.4. A Nissl-stained coronal section of spinal cord still within the vertebral column of a 3-day-old pup 24 hours after the intrathecal catheter implant. The location of the catheter was at the upper right of the spinal cord, dorso-lateral to the dorsal horn. The spinal cord appears intact with only a slight flattening at the site of the catheter. Note the disruption of the dorsal roots on the side of the implant. There is no damage on the side opposite the implant.

catheter to fill the dead space followed by a steady injection of 2–4 μl solution at the rate of one 1μl per 15 s. A speck of cyanoacrylate is used to seal the exposed end of the Silastic to prevent the remaining solution in the catheter from seeping into the cord. Following testing, each animal is intrathecally injected with the same volume of methylene blue dye using the same procedure implemented for drug injection. Pups are killed by pentobarbital overdose after a period of time equal to the time elapsed between drug injection and testing. The dorsal aspects of all vertebrae are removed, exposing the spinal cord, brain stem, and cerebellum. Litters containing any animal that does not demonstrate a profuse spread of dye from the caudal cervical segments of the cord to the lumbar/sacral region are excluded from the data analysis. Animals that exhibit any supraspinal staining are also excluded.

Controls

Several controls are required for the intraspinal injection procedures. First, although assessment for the spread of dye is necessary, it is not sufficient to localize the spread of drug. Thus, injection of equal volumes and doses of the drug into the lower brain stem by intracisternal injection is needed to control for spread to those loci. The assumption is that the potency of the drug will be greater at the primary site of action even if there is spread of drug either to the brain stem following intraspinal injection or to the spinal cord following intracisternal injection. A second control is intravenous or intraarterial injection of the agent. Because the spinal catheter disrupts the local vasculature and because the blood-brain barrier is incomplete in the young animal, the effects observed could be due to diffusion from the spinal cord to the periphery or even to higher brain loci. Intravenous or intraarterial injections of equal volumes and doses of the drug control for this. We have been unable to inject intravenously rat pups younger than 10 days of age because of the small size of the vasculature; thus, intraspinal injection in younger pups should be viewed more cautiously because of the limitation. In the 10-day-old pups, however, we have never found intravenous injections to have an effect, probably due to the small volumes and very low doses administered intrathecally. This finding argues against the effect being due to spread from the spinal cord to the peripheral circulation. I suspect that with the dialysis catheter, younger pups could be tested but we have not attempted that yet.

We have begun to use this method to study spinal receptors that are involved in pain regulation. Some of this work has been published (Hughes & Barr, 1988). In other studies, we compared directly the analgesic actions of opiates when administered to the brain or to the spinal cord (Barr, Miya, & Paredes, 1987). In these latter studies, spinal and brain opiate administration produced very different patterns of analgesia. Intraspinally administered morphine and ketocyclazocine were both effective in pups as young as 4 days of age for both noxious stimuli and showed a uniform topographic effect. Morphine administered to the lateral ventricle by acute injection produced analgesia in 3-day-old through 14-day-old pups. The pattern of analgesia changed with age. In the youngest animals, the effect was limited to a thermal stimulus applied to the forepaw; there were no effects on either a thermal stimulus to the hindpaw or tail or to a mechanical stimulus to any appendage. As the pup matured, morphine became effective against both stimuli in all body parts. Ketocyclazocine was ineffective when given intracerebraventricularly. Thus, for both morphine and ketocyclazocine, the analgesic effect depended on the site of administration.

METHODOLOGICAL CONSIDERATIONS

Systems Analysis

Behavior is the result of the integrated function of a number of neural circuits. These include input (sensory function), central processing, and output (motor function). The functional integrity of any neural system is dependent on the maturation of each component of that system (see Hyson & Rubel, chap. 14, this volume). In the extreme, the function is dependent on the last-developing neuron in the circuit. To draw from one of my favorite analogies, an interstate highway system can be almost 100% complete yet not be able to carry traffic because one critical bridge is not completed. Only when the highway system is totally built (or the neural system fully mature in all components) does it function. At a different developmental stage, the function of the system may differ depending on the completeness of the maturation of the system. To again draw a parallel to a highway, only three of four lanes may be complete, or there might be a detour around the construction site. The system functions but less well, or at least differently, than the fully mature pathways. But it does function. Thus, if a specific neural system is not functional, or functions differently, when tested, it is not known whether that specific system is immature or whether some outflow component, several neurons away, has yet to develop. Further, subcomponents of a particular system may differ in the maturation. The development of functional postsynaptic receptors may precede presynaptic release of the ligand. Receptor agonists may be effective while drugs that act at the presynaptic terminal may not be, and the neural system is not physiologically functional, although some components are functional and some drugs are active.

This is apparently the case for the opioid-feeding system. Naloxone, an opioid receptor antagonist, fails to inhibit feeding until 14 days of age in rats. This led to the erroneous conclusion that the responsible opioid receptor did not function until that age (Aroyewun & Barr, 1982, 1983). However, morphine, a receptor agonist, enhanced feeding as early as 3 days of age (Capuano, Giordano, Bowling, & Barr, 1985). Thus, the relevant opioid receptors are functional early. The failure of naloxone to work is likely not due to the late development of receptors but to the slow maturation of the presynaptic release of the opioid ligand. Without the release, naloxone is ineffective because there is no activity at the receptor to antagonize. We could have avoided our initial error of interpretation if we had assessed completely the functional maturity of this neural system, minimally by examining the pre- and postsynaptic components of that system as well as its inputs and outputs. The use of the methods described above, along with specific phramacological tools and biochemical and anatomical methods, are required to map fully the function of even the simplest system.

Neural Plasticity

As described above, the infant offers unique opportunities to study the rules that govern neuroplasticity. However, plasticity also provides possible problems in interpreting experiments. For example, any study that alters physiological functioning either phramacologically or physiologically could result in a permanent change in a neural response. While this is more important in more chronic studies, there is evidence in the adult of very short-term regulatory changes following insult (e.g., acute tolerance or tachyphylaxis; see Levine, 1983, for a discussion of drug tolerance). Less is known of these processes in the infant, but given the generally greater plasticity of the infant it is possible, if not likely, that these responses exist in the immature nervous system. Therefore, cross-sectional tests rather than repeated testing may be required.

Drug Specificity

While drugs may target specific receptor types or binding sites in adults, are they as specific in infants? It is possible that the confirmation of the receptor or binding site is different, so that the drug-receptor interaction is more or less specified. Studies on the development of neurotransmitter receptors have repeatedly found that the affinity (K_d) of the ligand for its receptor remains constant while the density (β_{max}) of binding sites changes. However, the number of drugs for which this has been demonstrated is limited, and the failure to demonstrate that specificity and selectivity of the drug in immature animals is conspicuously absent from almost all the development pharmacological literature.

On the other hand, there are clearly differences in drug actions due to the immaturity of the neonate. I will consider just two here. First, because drugs typically act at multiple receptor sites, if receptor populations develop differentially, the drugs may act differently, and any results will require different interpretations. While this is an advantage of the development approach, it is also a problem. Ideally, one should know the specificity of the ligand, the population of binding sites in the locus of stimulation, and the developmental time course of the receptor population at those loci. In practice, these data are rarely available. For example, stimulation of the ventral PAG with morphine produces analgesia (Tive & Barr, 1988). In the adult, morphine is a ligand for μ δ and κ receptors. But in the infant these receptor populations are incomplete. Therefore, the receptor mechanisms through which morphine acts cannot be known without further studies to assess biochemically and anatomically the maturation of these three receptor types in the PAG and to use more specific ligands to assess the role of each opioid receptor in analgesia in this region.

A second problem is that the physiology of the infant differs from that of the adult. Thus, the pharmacokinetics of drugs administered centrally or peripherally will differ. Care must be taken not to accept dose and time course effects from adults uncritically. A particularly important factor is the immaturity of the blood-brain barrier. Drugs that do not cross from the periphery into the brain (or vice versa) in adults may readily do so in infants. Some investigators have claimed only a peripheral action for drugs that do not cross the blood-brain barrier in adults. For example, Hamm and Knisely (1987a), who have done substantial work on the development of analgesia, reported that the muscarinic cholinergic agonist oxotremorine when injected simultaneously with the cholinergic blocker methylscopolamine failed to produced analgesia in 10- and 28-day-old rats but did so in 90-day-old rats. In the adult, methylscopolamine does not cross the blood-brain barrier and is often used to control for peripheral effects of cholinergic agonists such as oxotremorine. However, because the blood-brain barrier is not fully developed in the 10-day-old (Saunders & Møllgård, 1984), an alternative and simpler explanation for their data is that methylscopolamine entered the brain and blocked the effects of the agonist, resulting in the observed failure to obtain analgesia. Without pharmacokinetic data to establish that methylscopolamine did not enter the brain, the conclusions that the authors drew, that the cholinergic system was not involved in analgeia in young pups, is not justified. Clearly, these drugs could be acting centrally in the absence of a formed barrier. A more subtle problem is the assessment of dose-dependent effects during development following peripheral drug administration. Because the immaturity of the blood-brain barrier, gastric function, liver enzymes, and so forth, drugs will reach the central nervous system of the infant in different concentrations than in that of the adult. Thus, any conclusions about pharmacodynamic changes requires a full consideration of pharmacokinetic actions.

NEW DIRECTIONS

The methods and experiments described above are aimed at understanding the function of specific populations of neurons in behavior. We have for the most part studied the effects of specific neurotransmitters and their direct agonists and antagonists at the site of stimulation but not examined the maturation of the entire system. We have yet to describe the afferents, synaptology, and efferents for any brain site for any behavior. In part, the work is too new; in part, the brain is too complex. But the questions that our experiments and methods have raised are always the same: How are the neural systems at each level of the neural axis organized to function at different ages? For the most part, postsynaptic receptors are present and functional at the earliest age (3 days) that we test. For example, this is true for norepinephrine and NPY in the PVN and for dopamine and epinephrine in the perifornical hypothalamus (Capuano, 1987). Therefore, the exception, CCK, becomes important because it points to differences in neural organization. Is it simply that CCK receptors are not present, or does the action of CCK receptors require the function of some other, perhaps presynaptic, neuron that is not mature? If the latter, what does that tell us about the synaptic organization of CCK relative to other transmitters, and what does that tell us about the function of CCK in the PVN? Likewise, from data using the intrathecal administration of drugs to 3-day-olds, we know that the receptors in the spinal cord for norepinephrine and serotonin are present and functional in blocking afferent pain projections. But the presynaptic axons have not yet descended. This again raises a number of questions: What is the function of these receptors in the absence of normal input? How does this function change upon innervation? How does the late innervation explain the different actions of analgesic drugs during maturation?

The ability to administer drugs directly to neural tissue allows for the study of specific receptor populations and even presynaptic function within that structure. But other methods are required to answer some of the above issues. In vivo dialysis techniques will provide information about the normal release of transmitters at a site. Anatomical methods are required to map both the afferents and efferents. It is important not only to know the path of the tract but also its neurotransmitter(s). Knowledge of the functional maturation of that system is important, measured by the presence of important biochemical markers (e.g., enzymes), functional markers (e.g., ^{3}[H]-2-deoxyglucose autoradiography, cytochrome oxidase), and electrophysiological events. Most, if not all, of this information will be required before the changing function of any neural system during development is clarified.

ACKNOWLEDGMENTS

The work described here is largely the effort of a number of talented graduate students and collaborators. In particular, the original electrical stimulation work was conceived and worked out by Ted Lithgow and carried on by Nina Goodless and Eunjee Lee-Viscardi. The chemical stimulation work was done by Chris Capuano and Leslie Tive. Harry Hughes and Bill Paredes were largely responsible for the data on intraspinal administration of drugs. Others who contributed in many ways include Ola Aroyewun, Jim Giordano, Myra Joyce, and Dorene Miya. Finally, none of this work would have been possible without the extraordinary technical talent of Bill Paredes, who developed each of these methods himself or who had a major role in their perfection. The work described here has been supported by NIDA grant DA-15213, grants from the Research Foundation of the City University of New York, and funds from the Biopsychology Doctoral Program, CUNY and Hunter College. During preparation of this chapter, the author was supported in part by Grant DA-06600. This chapter is dedicated to the memory of Dr. Nansie S. Sharpless, whose recent death has left her collaborators and friends personally saddened and scientifically poorer.

REFERENCES

Andrews, R. J. (1967). Intracranial self-stimulation in the chick. *Nature, 213*, 847–848.

Aroyewun, O., & Barr, G. A. (1982). The effects of opiate antagonists on milk intake of preweanling rats. *Neuropharmacology, 21*, 757–762.

Aroyewun, O., & Barr, G. A. (1983). Effects of chronic antenatal and postnatal narcotics on naloxone-induced anorexia in preweanling rats. *Neuropharmacology, 22*, 329–336.

Bacon, W. E., & Wong, I. G. (1972). Reinforcement value in electrical brain stimulation in neonatal dogs. *Developmental Psychobiology, 5*, 195–200.

Barr, G. A., Eckenrode, T. C., & Murray, M. (1987a). Normal development and effects of early deafferentation on choline acetyltransferase, substance P, and serotonin-like immunoreactivity in the interpeduncular nucleus. *Brain Research, 418*, 301–313.

Barr, G. A., & Lithgow, T. (1986a). Pharmaco-ontogeny of reward. Enhancement of self-stimulation by d-amphetamine and cocaine in 3- and 10-day old rats. *Developmental Brain Research, 24*, 193–202.

Barr, G. A., Miya, D., & Paredes, W. (1987). Changing patterns of analgesia induced by lateral ventricle or spinal injections of morphine or ketocyclazocine in developing rats. *Neuroscience Abstracts, 13*, 1001.

Barr, G. A., Paredes, W., Erickson, K. L., & Zukin, R. S. (1986b). K-opioid receptor-mediated analgesia in the developing rat. *Developmental Brain Research, 29*, 145–152.

Blake, H. H., & Henning, S. J. (1986). Control of protein intake in the young rat. *Physiology and Behavior, 38*, 607–611.

Bleier, R. (1969). Retrograde transynaptic cellular degeneration in mammillary and ventral tegmental nuclei following limbic decortication in rabbits of various ages. *Brain Research, 15*, 365–393.

Bregman, B. S. (1986). Neural tissue transplants modify central neurons' responses to damage. In A. Gorio, M. E. Goldberger & M. Murray, (Eds.), *Development and plasticity of the mammalian spinal cord* (Fidia Research Series, Vol. 3). New York: Springer.

Bregman, B. S. (1987). Development of serotonin immunoreactivity in the rat spinal cord and its plasticity after neonatal spinal cord lesions. *Developmental Brain Research, 34*, 245–263.

Bregman, B. S., & Goldberger, M. E. (1982). Anatomical plasticity and sparing of function after spinal cord damage in neonatal cats. *Science, 217*, 553–555.

Bregman, B. S., & Goldberger, M. E. (1983a). Infant lesion effect. I. Development of motor behavior following neonatal spinal cord damage in cats, *Developmental Brain Research, 9*, 103–117.

Bregman, B. S., & Goldberger, M. E. (1983b). Infant lesion effect. II. Sparing and recovery of function after spinal cord damage in newborn and adult cats. *Developmental Brain Research, 9*, 119–135.

Bregman, B. S., & Goldberger, M. E. (1983c). Infant lesion effect. III. Anatomical correlates of sparing of recovery of function after spinal cord damage in newborn and adult cats. *Developmental Brain Research, 9*, 137–154.

Bronstein, D. M., Mittledorf, P., Sadeghi, M. M., Kirby, K., & Lytle, L. D. (1986). Visceral nociception in developing rats. *Developmental Psychobiology, 19*, 473–487.

Brunello, N., Volterra, A., DiGiulio, A. M., Cuomo, V., & Racagni, G. (1984). Modulation of opioid system in C57 mice after repeated treatment with morphine and naloxone: biochemical and behavioral correlates. *Life Sciences, 34*, 1669–1678.

Capuano, C. A. (1987). *The pharmaco-ontogeny of hypothalamic receptor systems mediating independent feeding in the rat*. Unpublished doctoral dissertation, City University of New York.

Capuano, C. A., Barr, G. A., & Leibowitz, S. F. (1986a). Early development of independent feeding via α_2-receptors in the paraventricular nucleus of the hypothalamus. *Neuroscience Abstracts, 12*, 593.

Capuano, C. A., Barr, G. A., & Leibowitz, S. F. (1986b). *Inhibition of independent feeding by catecholaminergic stimulation of the perifornical hypothalamus.* Paper presented at the meeting of the International Society for Developmental Psychobiology, Annapolis, MD.

Capuano, C. A., Barr, G. A., & Leibowitz, S. F. (1986c). *Neuropeptide Y and cholecystokinin: Effects on independent feeding in preweanling rats via hypothalamic stimulation.* Paper presented at the meeting of Eastern Psychological Association, New York.

Capuano, C. A., Giordano, J., Bowling, D., & Barr, G. A. (1985). Morphine enhances independent feeding in infant rats. *Neuroscience Abstracts, 11,* 1128.

Cooper, J. R., Bloom, F. E., & Roth, R. H. (1986). *The biochemical basis of neuropharmacology.* New York: Oxford University Press.

Coyle, J. T., & Campochiaro, P. (1976). Ontogenesis of dopaminergic-cholinergic interactions in the rat striatum: A neurochemical study. *Journal of Neurochemistry, 27,* 673–678.

D'Amato, C. J., & Hicks, S. P. (1968). Normal development and postraumatic plasticity of corticospinal neurons in rats, *Experimental Neurology, 60,* 557–569.

Ellis, S., Axt, K., & Epstein, A. N. (1983). The arousal of ingestive behaviors by chemical injection into the brain of the suckling rat. *Journal of Neuroscience, 4,* 945–955.

Fu, T., & Dewey, W. L. (1981). A new technique for the in vivo bioassay of opiates and other drugs on mouse spinal cord. *European Journal of Pharmacology, 74,* 239–242.

Geller, L. M., & Geller, E. S. (1966). A simple technique for permanent marking of newborn albino rats. *Phsycological Reports, 18,* 221–222.

Giordano, J., & Barr, G. A. (1987). Morphine- and ketocyclazocine-induced analgesia in the developing rat: Differences due to type of noxious stimulus and body topography. *Developmental Brain Research, 32,* 247–253.

Goldberger, M. E., & Murray, M. (1985). Recovery of function and anatomical plasticity after damage to the adult and neonatal spinal cord. In C. W. Cotman (Ed.), *Synaptic plasticity* (pp. 77–110), New York: Guilford.

Goldberger, M. E. (1986). Autonomous spinal motor function and the infant lesion effect. In M. E. Goldberger, A. Gorio, & M. Murray (Eds.), *Development and plasticity of the mammalian spinal cord* (pp. 363–380). New York: Springer.

Goldman, P. S. (1974). An alternative to developmental plasticity: Heterology of CNS structure in infants and adults. In D. G. Stein, J. J. Rosen, & N. Butters (Eds.), *Plasticity and recovery of function in the C.N.S.* (pp. 149–174). New York: Academic Press.

Goodless, N. L. (1988). *Neuroanatomical and chemical changes within the nucleus accumbens paralleling patterns of behavioral activation in the developing rat.* Unpublished doctoral dissertation, City University of New York.

Hall, W. G. (1985). What we know and don't know about the development of independent ingestion in rats. *Appetite, 6,* 333–356.

Hamm, R. J., & Knisely, J. S. (1984). Developmental changes in environmentally-induced analgesia. *Developmental Brain Research, 14,* 93–99.

Hamm, R. J., & Knisely, J. S. (1987a). Development differences in the analgesia produced by the central cholinergic system. *Developmental Psychobiology, 20,* 345–354.

Hamm, R. J., & Knisely, J. S. (1987b). Ontogeny of an endogenous, nonopioid and hormonally mediated analgesic system. *Developmental Psychobiology, 20,* 539–548.

Hand, A. J. (1989). Secrets of the superglues. *Popular Science, 234*(2), 80–104.

Hashimoto, M. (1915). Fieberstudien, I. Mitteilung: Über die spezifische überempfindlichkeit des wärmzentrans an sensibilisierten. *Archiv fuer Experimentelle Pathologie und Pharmakologie, 70,* 370–393.

Hebb, D. O. (1949). *The Organization of behavior: A neuropsychological theory.* New York: Wiley.

Heller, A., Hutchens, J. O., Kirby, M. L., Karapas, F., & Fernandez, C. (1979). Stereotaxic electrode placement in the neonatal rat. *Journal of Neuroscience Methods, 1,* 41–76.

Hitzemann, R. J., Hitzemann, B. A., & Loh, H. H. (1974). Binding of ^{3}H-naloxone in the mouse brain: Effect of ions and tolerance development. *Life Sciences, 14,* 2393–2404.

Hökfelt, T., Holets, V. R., Staines, W., Meister, B., Melander, T., Schalling, M., Schultzberg, M., Freedman, J., Björkland, H., Olson, L., Lindh, B., Elfvin, L. G., Lundberg, J., Lindgren, J. A., Samuelsson, B., Terenius, L., Post, C., Everitt, B., & Goldstein, M. (1986). Coexistence of neuronal messengers—An overview. *Progress in Brain Research, 68,* 33–70.

Hökfelt, T., Johansson, O., & Goldstein, M. (1984). Chemical anatomy of the brain. *Science, 225*, 1326–1334.

Hughes, H. E. (1988). *The pharmaco-ontogeny of spinal noradrenergic receptor systems mediating behavioral analgesia in the rat*. Unpublished doctoral dissertation, City University of New York.

Hughes, H. E., & Barr, G. A. (1988). Analgesic effects of intrathecally applied noradrenergic compounds in the developing rat: Differences due to thermal vs. mechanical nociception. *Developmental Brain Research, 41*, 109–120.

Jacobson, M. (1978). *Developmental neurobiology*, New York: Plenum.

Klee, W. A., & Streaty, R. A. (1974). Narcotic receptor sites in morphine-dependent rats. *Nature, 248*, 61–63.

Levine, R. R. (1983). *Pharamacology: Drug actions and reactions*. Boston: Little, Brown.

Lindvall, O., & Björklund, A. (1974). The organization of the ascending catecholamine neuron systems in the rat brain as revealed by the glyoxylic acid fluroescence method. *Acta Physiologica Scandinavica. Supplement, 412*, 1–48.

Lithgow, T., & Barr, G. A. (1982). A method for stereotaxic implanation in neonatal rats. *Developmental Brain Research, 2*, 315–320.

Lithgow, T., & Barr, G. A. (1984). Self-stimulation in 7- and 10-day-old rats. *Behavior and Neuroscience, 98*, 479–486.

Lorén, I., Björklund, A., & Lindvall, O. (1976). The catecholaminergic systems in the developing rat brain: Improved visualization by a modified glyoxylic acid-formaldehyde method. *Brain Research, 117*, 313–318.

Loughlin, S. E., Massamiri, T. R., Kornblum, H. I., & Leslie, F. M. (1985). Postnatal development of opioid systems in rat brain. *Neuropeptides, 5*, 469–472.

McDowell, J., & Kitchen, I. (1987). Development of opioid systems: Peptides, receptors and pharmacology. *Brain Research Reviews, 12*, 397–421.

Moore, K. E., & Kelly, P. H. (1978). Biochemical pharmacology of mesolimbic and mesocortical dopaminergic neurons. In M. A. Lipton, A. DiMascio, & K F. Killam (Eds.), *Psychopharmacology: A generation of progress* (pp. 41–98). New York: Plenum.

Moran, T. H., Lew, M. F., & Blass, E. M. (1981). Intracranial self-stimulation in 3-day-old rat pups. *Science, 214*, 1366–1368.

Moshe, S. L. (1987). Epileptogenesis and the immature brain. *Epilepsia, 28 (Supplement 1)*, S3–S15.

Myers, R. D. (1974). *Handbook of drug and chemical stimulation of the brain*. New York: Van Nostrand Reinhold.

Pasternak, G. W., Zhang, A. Z., & Tecott, L. (1980). Developmental differences between high and low affinity opiate binding sites: Their relationship to analgesia and respiratory depression, *Life Sciences, 27*, 1185–1190.

Perry, D. C., Rosenbaum, J. S., & Sadee, W. (1982). *In vitro* binding of [³H]etorphine in morphine-dependent rats. *Life Sciences, 31*, 1405–1408.

Petrillo, P., Travani, A., Verolla, D., Robson, L. E., & Kosterlitz, H. W. (1987). Postnatal development of μ-, δ-, and κ-opioid binding sites in rat brain. *Developmental Brain Research, 428*, 53–58.

Porreca, F., Mosberg, H. I., Hurst, R., Hruby, V. J., & Burks, T. F. (1984). Roles of μ, δ, and κ opioid receptors in spinal and supraspinal mediation of gastrointestinal transit effects and hot-plate analgesia. *Journal of Pharmacology and Experimental Therapy. 230*, 341–348.

Prince, D. A., & Kriegstein, A. R. (1989). Electrophysiological studies of immature neocortical neurons. In P. Kellaway, & J. L. Noebels, (Eds.), *Problems and concepts in developmental neurophysiology*. Baltimore, MD: John Hopkins University Press.

Ranck, J. B., Jr. (1975). Which elements are excited in electrical stimulation of mammalian central nervous system: A review. *Brain Research, 98*, 417–440.

Risling M., Culhein, S., & Hiedelbrand, C. (1983). Reinnervation of the ventral root L7 from ventral horn neurons following intramedullary axotomy in adult cats. *Brain Research, 280*, 15–23.

Rodriguez-Aendejas, A. M., Chambert, G., Lora-Vilchis, M. C., Epstein, A. N., & Russek, M. (1986). Ontogeny of epinephrine-induced anorexia in rats. *American Journal of Physiology, 250*, R313–R317.

Saunders, N. R., & Møllgård, K. (1984). Development of the blood-brain barrier. *Journal of Developmental Physiology, 6*, 45–57.

Schmidt, R. H., Björklund, A., Lindvall, O., & Lorén, I. (1982). Prefrontal cortex: Dense dopaminergic input in the newborn rat. *Developmental Brain Research*, 5, 222–228.

Schneider, G. E. (1970). Mechanisms of functional recovery following lesions of visual cortex or superior colliculus in neonate and adult hamsters. *Brain Behavior and Evolution*, 3, 295–323.

Schwartzkroin, P. A. (1984). Epileptogenesis in the immature central nervous system. In P. A. Schwartzkroin & H. V. Wheal (Eds.), *Electrophysiology of epilepsy* (pp. 389–412). London: Academic Press.

Sherwood, N. M., Timiras, P. (1979). *A stereotaxic atlas of the developing rat brain*. Berkeley: University of California Press.

Spain, J., Roth, B., & Coscia, C. (1985). Differential ontogeny of multiple opioid receptors (μ, δ, and κ). *Journal of Neuroscience*, 5, 584–588.

Spencer, W. G. (1984). The effect produced upon respiration by faradic excitation of the cerebrum in monkey, dog and rabbit. *Phiolosophical Proceedings of the Royal Society of London*, 185, 609–657.

Stelzner, D. J., Ershler, W. B., & Weber, E. D. (1975). Effects of spinal transection in neonatal and weaning rats: Survival of function. *Experimental Neurology*, 46, 156–177.

Stelzner, D. J., Weber, E. D., & BryzGornia, W. F. (1986). Sparing of function in developing spinal cord: Anatomical substrate. In M. E. Goldberger, A. Gorio, & M. Murray (Eds.), *Development and plasticity of the mammalian spinal cord* (pp. 81–99). New York: Springer.

Tavani, A., Robson, L. E., & Kosterlitz, H. W. (1985). Differential postnatal development of μ-, δ-, and κ-opioid receptors in mouse brain. *Developmental Brain Research*, 23, 306–309.

Tempel, A., Habis, J., Paredes, W., & Barr, G. A. (1988). Morphine-induced down regulation of μ opioid receptors in neonatal rat brain. *Developmental Brain Research, 41*, 129–133.

Tive, L. A., & Barr, G. A. (1988). Differential development of the effects of focal morphine and glutamate administration to the periaqueductal grey of the rat. *Neuroscience Abstracts*, 14, 716.

Ungerstedt, U. (1974). Brain dopamine neurons and behavior. In F. O. Schmitt & F. G. Worden (Eds.), *The neurosciences, third study program*, (pp. 695 703). Cambridge, MA: MIT Press.

Valenstein, T., Case, B., & Valenstein, E. S. (1969). Stereotaxic atlas of the infant rat hypothalamus. *Developmental Psychobiology*, 2, 75–80.

Viscardi, E-J. (1988). *The role of the dopaminergic system in behavioral activation of infant rats*. Unpublished doctoral dissertation, City University of New York.

Weber, E. D., & Stelzner, D. J. (1977). Behavioral effects of spinal cord transection in the developing rat. *Brain Research*, 125, 241–255.

Wise, R. A., & Bozarth, M. A. (1984). Brain reward circuitry: Four circuit elements "wired" in apparent series. *Brain Research Bulletin*, 12, 203–208.

Witelson, S. F. (1987). Neurobiological aspects of language in children. *Child Development*, 58, 653–688.

Yaksh, T. L., & Rudy, T. A. (1976). Chronic catheterization of the spinal subarachnoid space. *Physiology and Behavior*, 17, 1031–1036.

Yaksh, T. L. (1986). The effects of intrathecally administered opioid and adrenergic agents on spinal function. In T. L. Yaksh (Ed.), *Spinal afferent processing* (pp. 165–195). New York: Plenum.

Zhang, A-J., & Pasternak, G. W. (1981). Ontogeny of opioid pharmacology and receptors: High and low affinity site differences. *European Journal of Pharmacology*, 73, 29–40.

LIST OF VENDORS

VENDORS	FROM CHAPTERS NUMBERED	VENDORS	FROM CHAPTERS NUMBERED
Abbott Laboratories Abbott Park North Chicago, IL 60064 (800) 323-9100	17	Braintree Scientific, Inc. P.O. Box 361 Braintree, MA 02184 (617) 843-2202	4, 11
Air Shields, Inc. 330 Jacksonville Road Hatboro, PA 19040 (215) 674-4600	11	Carl Zeiss, Inc. 1 Zeiss Drive Thornwood, NY 10594 (914) 747-1800	2, 5, 15
Aldrich Chemicals 1001 West St. Paul Avenue Milwaukee, WI 53233 (800) 558-9160	18	Chrontrol Timer Lindberg Enterprises 4888 Ronson Court San Diego, CA 92111 (619) 566-5656	11
Allied Electronics, Inc. Many local distributors For information call: (800) 433-5700	3	Circon Microtechnology Corporation 749 Ward Drive Santa Barbara, CA 93111 (805) 967-0404	2
A-M Systems, Inc. 11627-A Airport Road Everett, WA 98204 (800) 426-1306	8, 17, 19	Clay Adams Division of Becton-Dickenson Parsippany, NJ 07054 (201) 848-6800	2, 3, 4, 11, 16, 19
A. R. Vetter Company P.O. Box 143 Rebersburg, PA 16872 (814) 349-5461	5	Cole-Palmer Instrument Company 7425 North Oak Park Avenue Chicago, IL 60648 (800) 323-4340	2, 19
Aus Jena: U.S. Distributor Seiler Instrument & Manufacturing Co. 170 East Kirkham Avenue St. Louis, MO 63119 (314) 968-2282	2	Columbus Instruments 950 North Hague Avenue Columbus, OH 43204 (614) 488-6176	17
Baxter Hospital Supply 202 Great Southwest Parkway Grand Prairie, TX 75050 (800) 444-1166	17	Cooner Wire Company 9186 Independence Avenue Chatsworth, CA 91311 (213) 882-8311	4
Bioquant. *See* R & M Biometrics			

VENDORS	FROM CHAPTERS NUMBERED
Dow Corning Corporation Medical Products Division Midland, MI 48640 (517) 496-4000	2, 4, 10, 11, 12, 14, 16
Ealing Scientific Ltd. 6010 Vanden Abeele Street St. Laurent, Quebec, H4S IR9, CANADA (800) 361-1905	11
Elkins-Sinn, Inc. Subsidiary of A. H. Robbins, Company Cherry Hill, NJ 08034 (609) 424-3700	4
Ernst Leitz, Inc. 24 Link River Rockleigh, NJ 07647 (201) 767-1100	2
Ethicon, Inc. (A division of Johnson & Johnson)	2
Eutectic Electronics, Inc. 8608 Jersey Court Raleigh, NC 27613	15
Fine Science Tools 1512 Industrial Way, Unit 10 Belmont, CA 94002 (800) 521-2109	2, 17
Fisher Scientific (Main Office) 50 Fadem Road Springfield, NJ 07081 (201) 467-6400	2, 4, 12, 16, 17
General Valve Corporation 19 Gloria Lane Fairfield, NJ 07006 (201) 575-4844	5
Goodfellow Technology The Science Park Cambridge, CB4 4DJ, ENGLAND	2

VENDORS	FROM CHAPTERS NUMBERED
Gould, Inc. (Main Office) 8333 Rockside Road Valley View, OH 44125 (216) 328-7000	5
Gould-Statham. *See* Spectramed	
Grass Instrument Company P.O. Box 516 101 Old Colony Avenue Quincy, MA 02169 (617) 773-0002	3, 4, 5, 17
Harvard Apparatus 22 Pleasant Street South Natick, MA 01760 (800) 272-2775	2, 11, 16
Imaging Research, Inc. Brock University St. Catherine's Ontario, CANADA L2S 3A1 (416) 688-2040	17
Interfan, Inc. Airport Boulevard Burlingame, CA 94010 (415) 347-1203	11
Jandel Scientific 65 Koch Road Corte Madera, CA 94925 (415) 924-8640	3
Johnson & Johnson P.O. Box 4000 New Brunswick, NJ 08903 (908) 562-3000	2
Karl Storz Endoscopy-America, Inc. 10111 West Jefferson Culver City, CA 90232 (213) 558-1500	9
Kopf Instruments 7324 Elmo Street P.O. Box 636 Tujunga, CA 91042–0636 (818) 352-3274	12, 17

VENDORS	FROM CHAPTERS NUMBERED	VENDORS	FROM CHAPTERS NUMBERED
Leica, Inc. P.O. Box 123 Buffalo, NY 14240 (716) 891-3000	2, 17	Richmond Dental Cotton Distributor: *See* Schein	
Leitz. *See* Ernst Leitz		Roboz Surgical Instrument Company, Inc. 1000 Connecticut Avenue, N.W. Post Office Box 19148 Washington, DC 20036 (202) 393-1234	2, 4
Medical Systems, Inc. One Plaza Road Greenvale, NY 11548 (800) 654-5406	17		
Medwire Corp. 121 S. Columbus Ave. Mt. Vernon, NY 10553 (914) 664-5300	2, 3	Schein, Inc. 5 Harbor Park Drive Port Washington, NY 11050 (516) 621-4300	2, 4
Merry X-Ray 4100 Will Rogers Parkway Suite 800 Oklahoma City, OK 73108 (405) 947-7209	17	Sentel Associates, Inc. 1125 C Stewart Drive Sunnyvale, CA 94086 (408) 980-2500	10
Mini-Mitter, Inc. P.O. Box 3386 Sunriver, OR 97707 (503) 593-8639	10	Sigma Chemical Company P.O. Box 14508 St. Louis, MO 63178 (800) 325-3010	14, 16, 17
MPCS Video 514 West 57th Street New York, NY 10019 (212) 586-3690	12	Sigma Scan. *See* Jandel Scientific	
Pamotor. *See* Interfan, Inc.		Small Parts, Inc. 6901 N.E. Third Avenue Miami, FL 33238 (305) 751-0856	11, 12
Pitman-Moore, Inc. 421 East Hawley Street Mundelein, IL 60060 (800) 525-9480	2, 4, 11, 19	Spectramed 1900 Williams Drive Oxnard, CA 93030 (805) 983-1300	3, 4
Plastics One, Inc. P.O. Box 12004 Roanoke, VA 24022 (703) 772-7950	4, 19	Spectrum Medical Industries, Inc. 1100 Rankin Road Houston, TX 77073 (800) 634-3300	19
R & M Biometrics, Inc. 5611 Ohio Avenue Nashville, TN 37209 (615) 350-7866	5	Stoelting Company 620 Wheat Lane Wood Dale, IL 60191 (708) 860-9700	9
Reichert Scientific. *See* Leica		Storz. *See* Karl Storz	